AF352267

Viral Hepatitis:
Molecular Biology, Diagnosis, Epidemiology and Control

PERSPECTIVES IN MEDICAL VIROLOGY

Volume 10

Series Editors

A.J. Zuckerman

Royal Free and University College Medical School
University College London
London, UK

I.K. Mushahwar

Abbott Laboratories
Viral Discovery Group
Abbott Park, IL, USA

Viral Hepatitis:
Molecular Biology, Diagnosis, Epidemiology and Control

Editor

I.K. Mushahwar

Abbott Laboratories
Viral Discovery Group
Abbott Park, IL, USA

2004

ELSEVIER

Amsterdam – Boston – Heidelberg – London – New York – Oxford – Paris
San Diego – San Francisco – Singapore – Sydney – Tokyo

ELSEVIER B.V.
Sara Burgerhartstraat 25
P.O. Box 211, 1000 AE
Amsterdam, The Netherlands

ELSEVIER Inc.
525 B Street, Suite 1900
San Diego, CA 92101-4495
USA

ELSEVIER Ltd
The Boulevard, Langford Lane
Kidlington, Oxford OX5 1GB
UK

ELSEVIER Ltd
84 Theobalds Road
London WC1X 8RR
UK

First edition 2004

Library of Congress Cataloging in Publication Data
A catalog record is available from the Library of Congress.

British Library Cataloguing in Publication Data
A catalogue record is available from the British Library.

ISBN: 0-444-51487-2
ISSN: 0168-7069

♾ The paper used in this publication meets the requirements of ANSI/NISO Z39.48-1992 (Permanence of Paper).
Printed in The Netherlands.

Contents

Preface

The undisputed medical impact of human viral hepatitis on public health resources and national economy of both developed and developing nations is frightening. Globally, it is estimated that at least 300 million people are chronically infected with hepatitis B virus (HBV) and 170 million people are chronically infected with hepatitis C virus (HCV). Chronic hepatitis due to either HBV or HCV eventually leads to liver cirrhosis and hepatocellular carcinoma. Also, exposures to enterically transmitted hepatitis viruses, namely, hepatitis A virus (HAV) and hepatitis E virus (HEV) especially in developing countries account for thousands of hepatitis cases every year. Both HAV and HEV are major causes of epidemic hepatitis and of acute, sporadic and sometimes fulminant hepatitis in these countries.

Over the past 40 years, increase in our knowledge of human viral hepatitis has been explosive in nature. It started with the discovery of Australia antigen in the mid 1960s, followed by the identification and characterization of HAV and HBV. The development of sensitive, specific, rapid and precise immunological assay techniques for the detection of hepatitis B surface antigen (HBsAg) and antibodies to HAV in the early 1970s, led to the identification of a third hepatitis virus by exclusion that was tentatively named post-transfusion non-A, non-B hepatitis virus (PT-NANBH). This was followed by the discovery of hepatitis delta virus (HDV) in the late 1970s. With the advent of molecular biology and the introduction of the polymerase chain reaction (PCR) the viral genome of PT-NANBH virus was molecularly cloned in 1989 and was designated HCV. This important accomplishment led in turn to the introduction of serodiagnostic and PCR assays for HCV infection. A variety of new molecular cloning techniques such as sequence-independent, single-primer amplification (SISPA) and representational difference analysis (RDA) led also to the discovery of HEV and the GB viruses A, B and C and to the TT virus and the development of immunological and molecular probe assays for the detection and study of these viruses.

Significant progress has also been made in the last 20 years in the introduction of efficacious vaccines for the prevention of both HAV and HBV, and also in therapy of HBV and HCV by antiviral drugs such as interferon, lamivudine, ribavirin, serine protease inhibitors and combination therapies with interferon.

In this book, leading researchers in the field from Great Britain, Italy, Japan, Switzerland and the United States of America report on the up-to-date advances in the molecular biology, immunology, biochemistry, pathology, diagnosis and treatment of the human hepatitis viruses including two related blood-borne viruses, namely, GB virus C and TT virus. It is hoped that the contents of this book will encourage new and innovative approaches to future viral hepatitis research, and offer directions for a better understanding of these viruses.

Isa K. Mushahwar
Emeritus Distinguished
Research Fellow
Virus Discovery
Abbott Laboratories
Abbott Park, IL, USA

Viral Hepatitis
I.K. Mushahwar (editor)
© 2004 World Health Organisation. All rights reserved.

Hepatitis A

Nicoletta Previsani and Daniel Lavanchy
World Health Organization, Geneva, Switzerland

Günter Siegl
Institut fur Klinische Mikrobiologie und Immunologie, St. Gallon, Switzerland

Hepatitis A—an introduction

Hepatitis is a general term meaning inflammation of the liver and can be caused by a variety of different viruses such as hepatitis A, B, C, D and E. Since the development of jaundice is a characteristic feature of liver disease, a correct diagnosis can only be made by testing patients' sera for the presence of specific antiviral antibodies [22,40].

Hepatitis A, one of the oldest diseases known to humankind, is a self-limited disease which results in fulminant hepatitis and death in only a small proportion of patients. However, it is a significant cause of morbidity and socio-economic losses in many parts of the world [18,40].

Transmission of hepatitis A virus (HAV) is typically by the faecal–oral route [18,23, 39,40]. Infections occur early in life in areas where sanitation is poor and living conditions are crowded. With improved sanitation and hygiene, infections are delayed and consequently the number of persons susceptible to the disease increases. Under these conditions explosive epidemics can arise from faecal contamination of a single source.

Hepatitis A was formerly called Infectious hepatitis, Epidemic hepatitis, Epidemic jaundice, Catarrhal jaundice, Type A hepatitis, HA [18,40].

What causes the disease?

Hepatitis A is caused by infection with the HAV, a nonenveloped, positive-stranded RNA virus, first identified by electron microscopy (EM) in 1973, classified within the genus hepatovirus of the picornavirus family [18,23].

The virus interferes with the liver's functions while replicating in hepatocytes. The individual's immune system is then activated to produce a specific reaction to combat and possibly eradicate the infectious agent. As a consequence of pathological damage, the liver becomes inflamed.

How is HAV spread?

HAV is transmitted from person-to-person via the faecal–oral route [18,23]. As HAV is abundantly excreted in faeces, and can survive in the environment for prolonged periods of time, it is typically acquired by ingestion of faeces-contaminated food or water. Direct person-to-person spread is common under poor hygienic conditions [22].

Occasionally, HAV is also acquired through sexual contact (anal–oral) and blood transfusions [22].

Who is susceptible to infection?

People who have never contracted HAV and who are not vaccinated against hepatitis A, are at risk of infection.

The risk factors for HAV infection are related to resistance of HAV to the environment, poor sanitation in large areas of the world, and abundant HAV shedding in faeces [18].

In areas where HAV is highly endemic, most HAV infections occur during early childhood.

Where is HAV a problem?

The virus is present worldwide, and the risk of infection is inversely proportional to levels of sanitation and personal hygiene [23].

In developing countries with poor environmental hygienic conditions, nearly all children are infected with HAV before the age of 9. There is substantial underestimation of hepatitis A cases in these areas because HAV infections for young children are mostly asymptomatic and therefore unrecognised. As sanitation conditions improve, transmission shifts to older age groups and the incidence of symptomatic disease increases. In most developed countries, endemic HAV transmission is unlikely.

World distribution map

The illustration given below has been taken from Ref. [10].

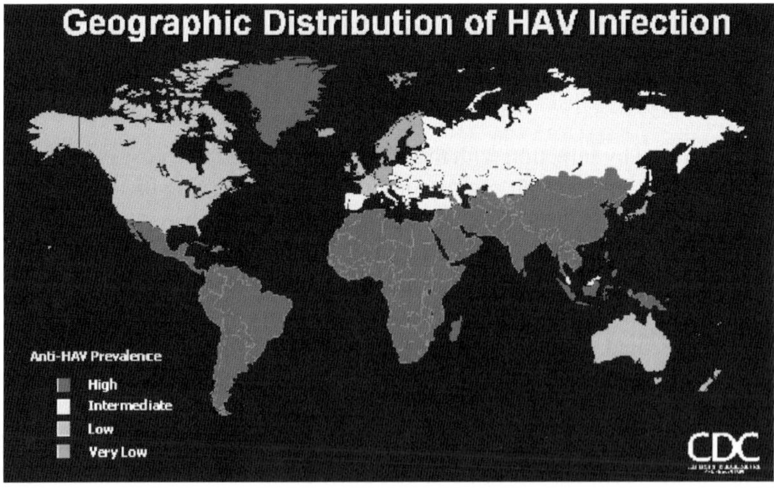

From Ref. [10], http://www.cdc.gov/ncidod/diseases/hepatitis/slideset/hep16.gif

Endemicity patterns (low, intermediate and high) of HAV infection worldwide. (Note: this map generalises available data and patterns may vary within countries) [9].

When is hepatitis A contagious?

In persons who develop clinically apparent hepatitis A, secretion of virus in stool at high titres begins 1–3 weeks prior to onset of illness, and may continue for several weeks at lower titres after jaundice occurs [21].

Although the level of virus shedding does not correlate with the severity of liver disease, faeces are highly infectious and therefore extremely contagious during all of this period.

Why is there no treatment for the acute disease?

Hepatitis A is a viral disease, and as such, antibiotics are of no value in the treatment of the infection. Antiviral agents, as well as corticosteroids, have no effect in the management of the acute disease [18].

The administration of immune globulins (IG) may help preventing or improving the clinical manifestations of the disease if given within two weeks of infection, but it is of no help in the acute phase of hepatitis A [18,39]. Therapy can only be supportive and aimed at maintaining comfort and adequate nutritional balance [18]. Complete recovery without therapy is generally the rule [18].

The hepatitis A virus (HAV)

HAV, first identified in 1973, is a nonenveloped, spherical, positive-stranded RNA virus, classified within the genus hepatovirus of the picornavirus family [18,21–23].

HAV infection does not lead to chronic or persistent hepatitis [18,23,40].

HAV strains recovered from widely separated regions of the world are antigenically similar. In humans, a single serotype of HAV exists [18,21,23,39,40].

HAV is known to produce disease in humans and non-human primates. *In vitro*, the wild-type virus is generally difficult to grow and no cytopathic effect is observed. Attenuated HAV strains adapted to cell culture have been used to develop vaccines [18,21,23,40].

HAV infection induces lifelong protection against reinfection [18].

4

Electron microscopy picture

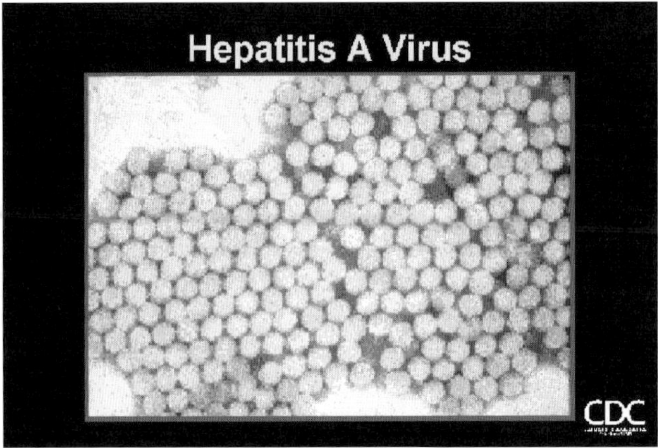

From Ref. [10], http://www.cdc.gov/ncidod/diseases/hepatitis/slideset/hep06.gif

EM picture of human HAV.

Morphology and physicochemical properties

HAV is among the smallest and structurally simplest of the RNA animal viruses. The virion is nonenveloped and, with a diameter of 27–32 nm, it is composed entirely of viral protein and RNA. EM analyses show particles with icosahedral symmetry although no structural details could be discerned. Morphologically, HAV particles are indistinguishable from other picornaviruses [18,22,40]. Full virions have a buoyant density of 1.32–1.34 g/cm^3 in CsCl and a sedimentation coefficient of 156–160 S in neutral sucrose solutions [18,21].

Empty capsids, abundant in faeces collected during early infection, band at 1.20 and 1.29–1.31 g/cm^3, with sedimentation coefficients ranging from 50 to 90 S, predominantly 70 S [18,21].

Genome and proteins

The hepatitis A genome consists of a linear, single-stranded, positive-sense RNA of approximately 7.5 kb containing a 5′-nontranslated region with complex secondary and tertiary structure [18,21,22,40].

The 5′-end represents a noncoding region (NCR) extending over 10% of the genome, it is uncapped and covalently linked to the viral protein VPg (2.5 kDa) [18,21,22,40]. A single large polyprotein is expressed from a large open reading frame extending through most of the genomic RNA. This polyprotein is subsequently cleaved by a viral protease (3C^{pro}) to form three (possibly four) capsid proteins and several nonstructural proteins [18,21–23,40]. The 3′-end terminates with a poly(A)tail of 40–80 nucleotides [18,22].

Hepatitis A capsids contain 60 copies of VP1 (30–33 kDa), VP2 (24–30 kDa) and VP3 (21–28 kDa). Exposed parts of VP1 (residues Ser102 and Ser114) and of VP3 (residue Asp70) on the capsid surface define the conformational immunodominant antigenic site of HAV [18,22,23,40].

Sequences for known human HAV isolates are highly similar even when geographic and temporal origins are widely separated, yet seven distinct genotypes have been identified to date [18].

HAV genomic replication occurs exclusively in the cytoplasm of the infected hepatocyte by a mechanism involving an RNA-dependent RNA polymerase [21].

Genetic structure of HAV

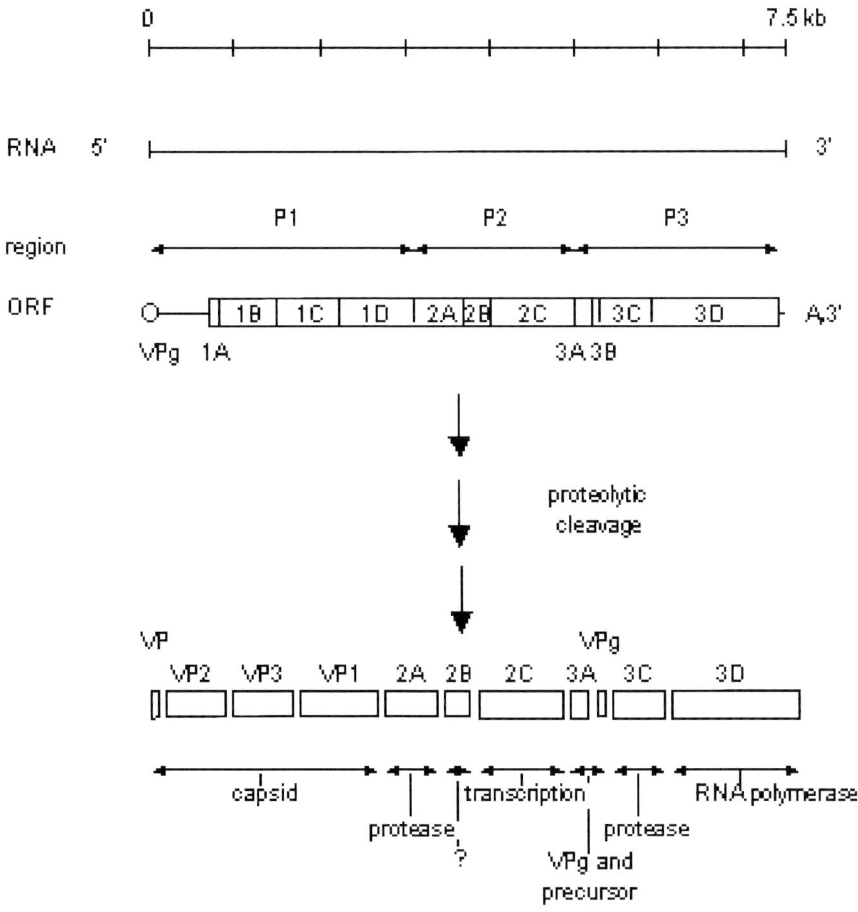

Delineation, to scale, of the genome of HAV with its 5′-linked VPg protein, 5′-nontranslated region, single long open reading frame, 3′-nontranslated region and polyadenylated 3′-end. The RNA is translated into a precursor polyprotein that is cleaved to generate mature proteins [6].

Proposed secondary structure of the 5′ NCR of HAV RNA

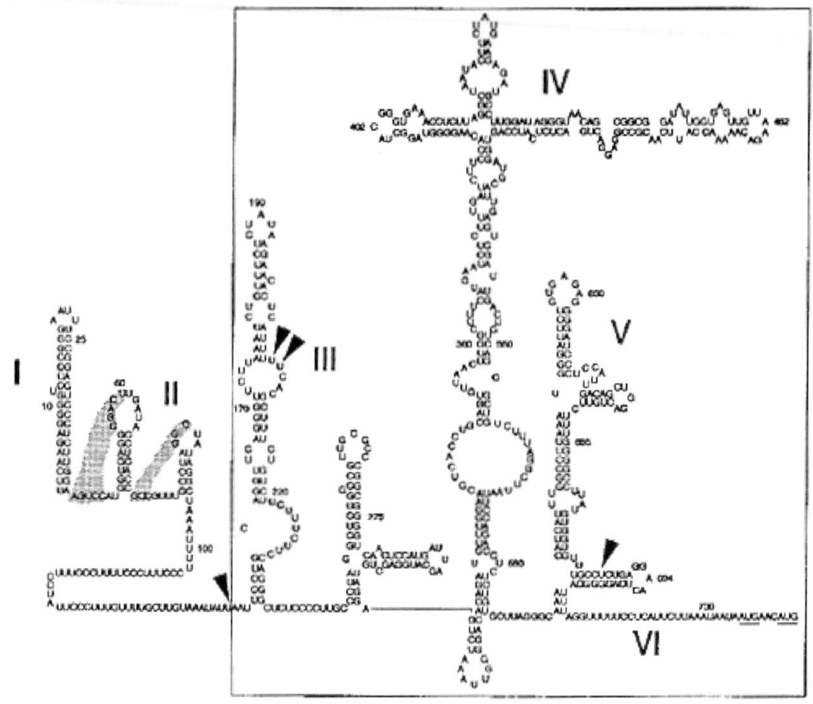

From Ref. [25], with permission.

Proposed secondary structure of the 5′ NCR of HAV RNA (strain HM175/wild-type) (see also Ref. [25]). This model is based on a combination of phylogenetic comparisons, thermodynamic predictions, and nuclease digestions of synthetic RNA between nucleotides 300 and 735. Major structural domains are indicated by Roman numerals beginning at the 5′ terminus. Domains contributing to the HAV internal ribosome entry site are included in the box, although the boundaries are not precisely determined. Sites of mutations that appear to enhance HAV replication in cell culture are indicated (arrows). Two possible pseudoknots are indicated by shaded interactions near the 5′ terminus. The two initiation codons for the open reading frame (nucleotides 735–737 and 741–743) are underlined [18].

Antigenicity

HAV has only one known serotype, and one neutralisation site is immunodominant. Different viral strains show similar reactivity to monoclonal anti-HAV antibodies [18,22,39,40].

Antigens of the intact virion are conformational and different from those of isolated proteins. Antibodies to purified capsid proteins or to synthetic peptides have weak or no detectable neutralising activity [18,23].

HAV is neutralised by both anti-HAV IgG and anti-HAV IgM. No serologic or hybridising cross-reactivity between HAV and other viral hepatitis agents, including hepatitis E virus (HEV), have been observed [18,22]. The nonstructural proteins of HAV are also immunogenic during natural and experimental infections [18].

Stability

HAV has no lipid envelope and is stable when excreted from the infected liver to the bile to enter the gastrointestinal tract. It has been found to survive in experimentally contaminated fresh water, seawater, wastewater, soils, marine sediment, live oysters, and creme-filled cookies.

HAV is extremely resistant to degradation by environmental conditions, a property that allows its maintenance and spread within populations [18,22,39,40].

HAV is resistant to:

- Thermal denaturation (survives at 70°C for up to 10 min).
- Acid treatment (pH 1 for 2 h at room temperature), 20% ether, chloroform, dichlorodifluoromethane, and trichlorotrifluoroethane.
- Perchloracetic acid (300 mg/l for 15 min at 20°C).
- Detergent inactivation (survives at 37°C for 30 min in 1% SDS).
- Storage at -20°C for years.

HAV is inactivated by:

- Heating to 85°C for 1 min.
- Autoclaving (121°C for 20 min).
- Ultraviolet radiation (1.1 W at a depth of 0.9 cm for 1 min).
- Formalin (8% for 1 min at 25°C).
- β-Propriolactone (0.03% for 72 h at 4°C).
- Potassium permanganate (30 mg/l for 5 min).
- Iodine (3 mg/l for 5 min).

8

- Chlorine (free residual chlorine concentration of 2.0–2.5 mg/l for 15 min).

- Chlorine-containing compounds (3–10 mg/l sodium hypochlorite at 20°C for 5–15 min).

- Shellfish from contaminated areas should be heated to 90°C for 4 min or steamed for 90 s.

The disease

The course of hepatitis A may be extremely variable [18]. Patients with inapparent or subclinical hepatitis have neither symptoms nor jaundice. Children generally belong to this group. These asymptomatic cases can only be recognised by detecting biochemical or serologic alterations in the blood [18,22,40]. Patients may develop anicteric or icteric hepatitis and have symptoms ranging from mild and transient to severe and prolonged, from which they recover completely or develop fulminant hepatitis and die (see below). The severity of the disease increases with age at time of infection [18,23].

The course of acute hepatitis A can be divided into four clinical phases [18, 21–23,40]:

- An incubation or preclinical period, ranging from 10 to 50 days, during which the patient remains asymptomatic despite active replication of the virus. In this phase, transmissibility is of greatest concern.

- A prodromal or preicteric phase ranging from several days to more than a week, characterised by the appearance of symptoms like loss of appetite, fatigue, abdominal pain, nausea and vomiting, fever, diarrhoea, dark urine and pale stools, followed by:

  - An icteric phase, during which jaundice develops at total bilirubin levels exceeding 20–40 mg/l. Patients often seek medical help at this stage of their illness. The icteric phase generally begins within 10 days of the initial symptoms. Fever usually improves after the first few days of jaundice. Viremia terminates shortly after hepatitis develops, although faeces remain infectious for another 1–2 weeks. Extrahepatic manifestations of hepatitis A are unusual. Physical examination of the patient by percussion can help to determine the size of the liver and possibly reveal massive necrosis. The mortality rate is low (0.2% of icteric cases) and the disease ultimately resolves. Occasionally, extensive necrosis of the liver occurs during the first 6–8 weeks of illness. In this case, high fever, marked abdominal pain, vomiting, jaundice and the development of hepatic encephalopathy associated with coma and seizures, are the signs of fulminant hepatitis, leading to death in 70–90% of the patients. In these cases mortality is highly correlated with increasing age, and survival is uncommon over 50 years of age. Among patients with chronic hepatitis B or C or underlying liver disease, who are superinfected with HAV, the mortality rate increases considerably.

- A convalescent period, where resolution of the disease is slow, but patient recovery uneventful and complete. Relapsing hepatitis occurs in 3–20% of patients 4–15 weeks after the initial symptoms have resolved. Cholestatic hepatitis with high bilirubin levels persisting for months is also occasionally observed. Chronic sequelae with persistence of HAV infection for more than 12 months are not observed.

Predicted outcome of HAV infection

Parameter	Predicted outcome		
	Children (<5 years)		Adults
Inapparent infection	80–95%		10–25%
Anicteric or icteric disease	5–20%		75–90%
Complete recovery	99 + %		98 + %
Chronic disease		None	
Mortality rate (years):			
#14		0.1%	
15–39		0.3%	
∃40		2.1%	

From Ref. [18], with permission.

Diagnosis

Since both clinically and biochemically, acute hepatitis due to HAV cannot be distinguished from that due to the other hepatitis viruses, serologic tests are necessary for a virus-specific diagnosis [18,21].

Diagnosis of hepatitis is made by biochemical assessment of liver function (laboratory evaluation of: urine bilirubin and urobilinogen, total and direct serum bilirubin, ALT and/or AST, alkaline phosphatase, prothrombin time, total protein, serum albumin, IgG, IgA, IgM, complete blood count) [18,21–23,40].

The specific routine diagnosis of acute hepatitis A is made by finding anti-HAV IgM in the serum of patients. A second option is the detection of virus and/or antigen in the faeces [21,23].

Virus and antibody can be detected by commercially available RIA, EIA or ELISA kits. These commercially available assays for anti-HAV IgM and total anti-HAV (IgM and IgG) for assessment of immunity to HAV are not influenced by the passive administration of IG because the prophylactic doses are below detection level [23].

At the onset of disease, the presence of IgG anti-HAV is always accompanied by the presence of IgM anti-HAV. As IgG anti-HAV persists lifelong after acute infection, detection of IgG anti-HAV alone indicates past infection [18,21,40].

Virus may still be present in the absence of detectable HAV antigen, as demonstrated by the use of more sensitive methods [18].

If laboratory tests are not available, epidemiologic evidence can help in establishing a diagnosis.

Host immune response

In acute hepatitis A, the presence of anti-HAV IgM is detectable about 3 weeks after exposure, its titre increases over 4–6 weeks, then declines to nondetectable levels generally within 6 months of infection [18,21–23,40].

Anti-HAV IgA and IgG are detectable within a few days of the onset of symptoms. IgG antibodies persist for years after infection and provide lifelong immunity [18,21–23, 39,40].

The development of antibody to HAV coincides with a decrease in quantity of viremia and faecal shedding of virus.

Saliva and faeces generally do not contain neutralising antibodies [22,40].

Typical serologic course

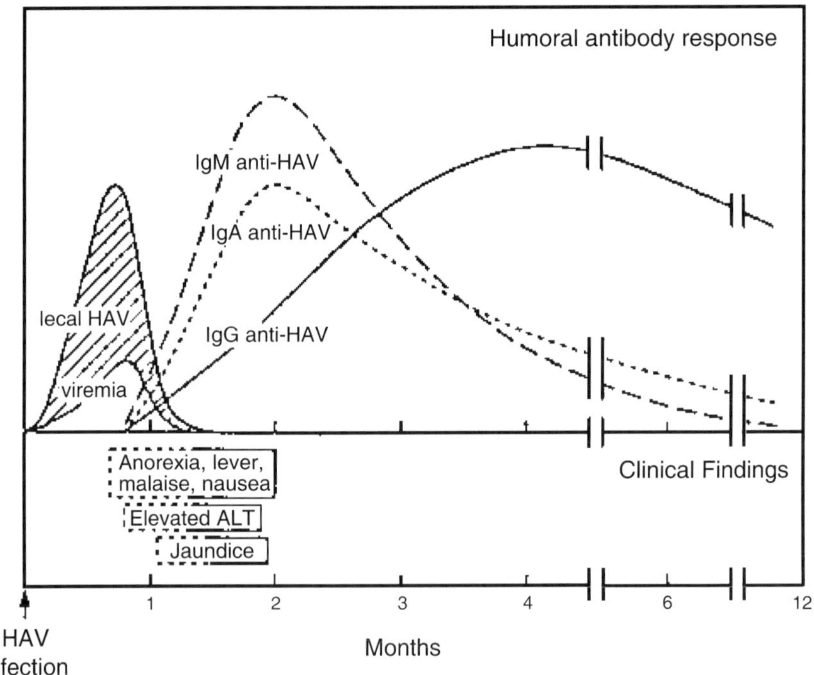

From Ref. [40] with permission

Summary of clinical, virologic, and serologic findings in uncomplicated acute hepatitis A.

Prevalence

The highest prevalence of faecal–oral infection occurs in regions where low standards of sanitation promote the transmission of the virus [22].

In most industrialised nations, where hepatitis A is no longer considered a childhood disease, infections with HAV are increasingly contracted by adults [31,40].

Despite the high prevalence of antibody in highly endemic populations, the virus perpetuates in the region due to its high physical stability.

Pathogenesis

Virus-induced cytopathology may not be responsible for the pathologic changes seen in HAV infection as liver disease may result primarily from immune mechanisms. Antigen-specific T-lymphocytes are responsible for the destruction of infected hepatocytes [18, 21–23,39].

Increased levels of interferon have been detected in the serum of HAV-infected patients and are presumably responsible for the reduction in virus burden seen in patients following the onset of clinical disease and in their symptoms [18]. Rarely, patients with acute viral hepatitis A develop features of cholestasis [18]. Confluent hepatic necrosis may lead to fulminant hepatitis and death in 30–60% of cases. Death appears to be inevitable when necrosis involves more than 65–80% of the total hepatocyte fraction. In patients who survive an episode of acute fulminant hepatic failure, neither functional nor pathologic sequelae are common, despite the widespread necrosis [18]. During the recovery stage, cell regeneration is prominent. The damaged hepatic tissue is usually restored within 8–12 weeks [18].

Transmission

HAV is generally acquired by the faecal–oral route by either person-to-person contact or ingestion of contaminated food or water. Hepatitis A is an enteric infection spread by contaminated excreta [9,18,23,40].

High concentrations of virus are shed in the stools of patients during 3–10 days prior to the onset of illness till 1–2 weeks after the onset of jaundice. Faecal excretion of HAV persists longer in children and in immunocompromised persons (up to 4–5 months after infection) than in otherwise healthy adults. Communicability is highest during this interval [18].

Hepatitis A may be acquired from faecally contaminated food or water and from wastewater-contaminated drills or water supplies [18,22]. HAV present in sewage-contaminated fresh or salt water can be concentrated by mollusc-like oysters and clams, which can be an important source of infection if eaten raw or inadequately cooked. Cooked food may become recontaminated after cooking during inappropriate handling [18].

12

Transmission by blood transfusion is rare: the donor must be in the viremic prodromal phase of infection at the time of blood donation. Current blood practices do not include screening of donors for evidence of active HAV infection [18,22,23,40].

Substantial viremia persisting for several weeks suggests the possible role of needle-borne transmission of virus among intravenous drug users, although HAV concentrations in blood are manifold lower than in faeces.

Outbreaks (1992) have occurred among haemophiliacs receiving factor VIII concentrates prepared by a solvent–detergent inactivation process which did not reduce the infectivity of nonenveloped viruses [18,23,38].

HAV is not transmitted from infected mothers to newborn infants, as anti-HAV IgG antibodies present during initial stages of HAV infection cross the placenta and provide protection to the infant after delivery.

Transmission by exposure to urine, nasopharyngeal secretions or aerosol of infected persons is improbable, regardless of the stage of infection. Transmission of HAV by biting insects is conceivable [18].

Role of non-human primates in the transmission of HAV

Various monkey species such as chimpanzees, owl monkeys, cynomolgus monkeys, rhesus monkeys, stump-tailed monkeys, African green monkeys, tamarins, marmosets and squirrel monkeys are susceptible to HAV [8,15,18,20,22,24,29,36]. HAV-induced disease in non-human primates resembles human disease, but is usually milder, or subclinical, followed by complete recovery [8,18,20,29]. HAV can be transmitted experimentally to these animals, but the presence of anti-HAV antibody in the sera of newly captured monkeys shows that infection may also spread in the natural habitat of non-human primates. HAV isolates from several naturally infected monkeys were shown to represent strict simian HAV strains, closely related antigenically to human HAV strains [8,14,20,22,24].

A molecular comparison of the human HM175 and the simian PA21 and PA33 strains of HAV has shown that, despite major divergence at the nucleotide level (>10%), the viruses share immunodominant neutralisation epitopes. Infection of owl monkeys with either virus provides high, although incomplete (mild symptoms, relapsing hepatitis) protection against later intravenous challenge with the other virus [14,24].

Well documented is the natural transmission of human HAV from experimentally infected animals to humans. Still unknown is the susceptibility of humans to true simian HAV strains. If it could be shown that simian strains do not induce disease in humans despite virus replication and subsequent seroconversion, simian strains might be used as live, attenuated vaccines. If, on the other hand, HAV-immune people challenged with simian HAV strains developed signs of hepatitis, even a global immunisation programme could never achieve the eradication of HAV, because monkeys would constantly represent a natural reservoir of virus.

Risk groups

Certain groups can be defined as high risk for contracting HAV [18,21,23,41,45]:

- People in household/sexual contact with infected persons.
- Medical and paramedical personnel in hospitals
- International travellers from developed countries to regions of the world where HAV is endemic (3/1000 to 20/1000 people per month's stay abroad).
- Persons living in regions with endemic hepatitis A.
- Persons residing in areas where extended community outbreaks exist.
- Preschool children attending day-care centres, their parents and siblings.
- Day-care centre employees.
- Residents and staff of closed communities (institutions).
- Refugees residing in temporary camps following catastrophes.
- Homosexually active men.
- Injecting drug users using unsterilised injection needles.
- Persons with clotting factor disorders.
- Persons with chronic liver disease.
- Food-service establishments/food handlers.
- Persons working with non-human primates.

Risk factors remain unidentified in as much as 50% of hepatitis A cases [21,45].

Hepatitis A is contracted at least 100 times more frequently than typhoid fever or cholera.

Persons falling into any of the above-mentioned categories should consider being vaccinated as a preventive measure.

Surveillance and control

Surveillance and control procedures should include:

- Providing safe drinking water and proper disposal of sanitary waste.
- Monitoring water beds where shellfish are harvested.
- Monitoring disease incidence.
- Determining sources of infection.

- Identifying contacts of case-patients for post-exposure prophylaxis.
- Detecting outbreaks.
- Containing spread.

Endemicity

Geographic areas can be characterised by high, intermediate or low levels of endemicity patterns of HAV infection. The levels of endemicity correlate with hygienic and sanitary conditions of each geographic area [18,21,31,41,45].

- *High*: In developing countries with very poor sanitary and hygienic conditions (parts of Africa, Asia and Central and South America), infection is usually acquired during early childhood as an asymptomatic or mild infection. Reported disease rates in these areas are therefore low and outbreaks of disease are rare. Reported disease incidence may reach 150 per 100,000 per year.

- *Intermediate*: In developing countries, countries with transitional economies and some regions of industrialised countries where sanitary conditions are variable (Southern and Eastern Europe, some regions in the Middle East), children escape infection in early childhood. Paradoxically, these improved economic and sanitary conditions may lead to a higher disease incidence, as infections occur in older age groups, and reported rates of clinically evident hepatitis A are higher.

- *Low*: In developed countries (Northern and Western Europe, Japan, Australia, New Zealand, USA, Canada) with good sanitary and hygienic conditions, infection rates are generally low. In countries with very low HAV infection rates, disease may occur among specific risk groups such as travellers.

Worldwide endemicity of HAV infection

HAV endemicity	Regions by epidemiological pattern	Average age of patients (years)	Most likely mode of transmission
Very high	Africa, parts of South America, the Middle East and of South-East Asia	Under 5	Person-to-person; contaminated food and water
High	Brazil's Amazon basin, China and Latin America	#5–14	Person-to-person; outbreaks/contaminated food or water

(continued)

(continuation)

HAV endemicity	Regions by epidemiological pattern	Average age of patients (years)	Most likely mode of transmission
Intermediate	Southern and Eastern Europe, some regions of the Middle East	5–24	Person-to-person; outbreaks/contaminated food or water
Low	Australia, USA, Western Europe	5–40	Common source outbreaks
Very low	Northern Europe and Japan	Over 20	Exposure during travel to high endemicity areas, uncommon source

Worldwide endemicity of HAV infection [5,11,42,44,45].

Incidence/epidemiology

Hepatitis A occurs sporadically and epidemically worldwide, with a tendency to cyclic recurrences [22].

Epidemics are uncommon in developing countries where adults are generally immune. Improved sanitation and hygiene conditions in different parts of the world leave large segments of the population susceptible to infection, and outbreaks may result whenever the virus is introduced [22,31,37].

Common-source epidemics, related to contaminated food or water, may evolve explosively, as did the largest mollusc-linked epidemic in Shanghai, in 1988, involving about 300,000 people [31].

Worldwide, HAV infections account for 1.4 million cases annually [45].

Estimated number of cases per continental region

Region	1990 Population (in millions)	Incidence (per 100,000 per year)	Cases (per year)
North America	275	10	28,000
Central and South America	453	20–40	162,000
Europe	791	5–60	278,000
Africa and Middle East	827	20–60	251,000
Asia	2893	10–30	676,000
Oceania	28	15–30	5,000
Total			1,399,000

From: Hadler SC. Global impact of hepatitis A virus infection changing patterns. In: Hollinger FB, Lemon SM, and Margolis HS, Eds. Viral Hepatitis and Liver Disease. Baltimore: Williams & Wilkins, 1991: 14–20, with permission.

Trends

As nations develop public sanitation, the age at which individuals will become infected is delayed until adulthood, at which time the likelihood of developing symptomatic illness is considerably higher [16,18,31].

In the United States, nationwide outbreak cycles appear every decade, as observed in 1961, 1971 and 1989.

Hepatitis A appears to follow a minor cyclic phase, with a peak occurring during fall and winter, possibly as a result of exposure during summer holidays spent in endemic countries. Smaller epidemics are present in different parts of the world, with cases increasing slightly during the past several years. Decreasing are cases in Greece and Italy [28,31].

The rate in males is about 20% higher than in females [18].

Costs

Although most infected persons recover completely and a significant proportion remain asymptomatic, HAV infection causes considerable morbidity and mortality and imposes a large economic burden throughout the world [7,43].

On average, adults miss 30 days of work. Young children tend to suffer only flu-like symptoms, if any, but infection of children can initiate and perpetuate community-wide outbreaks.

Both medical treatment and work loss account in the United States for an estimated annual US$ 500 million (1997) costs for 63,500 cases of acute hepatitis A. For each hospitalised case, medical care costs sum up to about US$ 6900 [7].

Worldwide, an estimated 1.4 million cases of acute hepatitis A annually cost US$ 1.5–3 billion [4,18,40].

An HAV antibody screening test and its subsequent evaluation are estimated to cost US$ 43 per case [43].

The costs of vaccination are estimated to be US$ 40 for one 720 EL.U. dose plus US$ 15 for the administration. Two doses are required for a complete vaccination [43].

For passive immunisation, the purchase and administration of one IG dose is estimated at US$ 41 [43].

Cost-effective analyses performed in Ireland showed that where HAV immunity is 45% or less, vaccination is the strategy of choice, and when immunity is greater than 45%, then screening followed by vaccination should be used [34].

Immune prophylaxis

Until recently, passive immunisation with pooled IG was the only option available for preventing hepatitis A. HAV research has led to the development of inactivated vaccines. Their use is being encouraged and preferred to the administration of IG for pre-exposure prophylaxis when repeated exposure is anticipated [21].

Passive immunisation

The administration of IG can reduce the incidence of hepatitis A up to 90%, and it is most effective if given before exposure. Its use is declining now that HAV vaccines are being used more widely [21]. IG is still used for post-exposure prophylaxis. If administered within two weeks of exposure it will either prevent development or reduce the severity of the disease [21,22,39].

Passive immunisation is safe for adults and children, pregnant or lactating women and immunosuppressed persons, but it only provides a limited duration of protection after a single IG dose of 100 IU (6 months), leaving susceptibles available for infection following another exposure.

Because of its short duration of action, IG must be readministered on a regular basis to maintain its effectiveness and ensure continuous protection. It is therefore expensive, logistically complicated and unreliable to supply long-term protection, and considered obsolete, except for situations in which immediate protection is required. Moreover, IG can interfere with immune response to live, attenuated vaccines [measles, mumps, rubella (MMR) and varicella]. The administration of MMR should be delayed at least 3 and 5 months for varicella after administration of IG. On the other hand, IG should not be administered within 2 weeks after administration of live, attenuated vaccines.

Active immunisation

At least four inactivated vaccines (Havrix$^®$, Vaqta$^®$, Epaxal$^®$, and Avaxim$^®$) are presently commercially available in some parts of the world.

Inactivated HA vaccines are safe, highly immunogenic, and provide long-term protection from HAV infection (20 years). They can be administered simultaneously with a number of other vaccines (diphtheria, polio, tetanus, oral typhoid, cholera, Japanese encephalitis, rabies, yellow fever and hepatitis B) without affecting the rates of seroconversion [12].

Hepatitis A is the most common immunisation-preventable infection in travellers [41,47].

Recommendations for use of hepatitis A vaccine and IG

Pre-exposure prophylaxis

Hepatitis A vaccination provides pre-exposure protection from HAV infection. It is recommended for persons who are at increased risk for infection and for any person wishing to obtain immunity.

Persons who seek immunological protection, but are allergic to vaccine components should receive IG. The administration must be repeated if protection is required for periods exceeding 5 months. For persons who require repeated IG, screening of their immune status is useful to avoid unnecessary doses of IG.

Post-exposure prophylaxis

Persons who have been exposed to HAV and who have not previously been vaccinated should be administered a dose of IG (0.02 ml/kg) within two weeks of exposure [39]. Persons who have received a dose of hepatitis A vaccine at least 2 weeks before exposure to HAV do not need IG.

Mass post-exposure vaccination to contain the spread of HAV in established outbreaks has been well documented and proven efficacious in ceasing emerging epidemics [21].

Serologic screening of contacts of infected individuals for anti-HAV before they are given IG is not recommended because screening is more costly than IG and would delay its administration.

The prospective duration of antibody persistence can be estimated to last at least 20 years. However, a booster vaccination after 10 years is recommended for protection, as long as no long-term follow-up data are available [19,21,27].

Recommended doses of IG for hepatitis A pre- and post-exposure prophylaxis

Setting	Duration of coverage	IG Dose (ml/kg)
Pre-exposure	Short term (1–2 months)	0.02
	Long term (3–5 months)	0.06[a]
Post-exposure		0.02

Doses should be given as intramuscular injections into deltoid or gluteal muscle. For children <2 years of age, injections should be given into anterolateral thigh muscle.
Whenever possible, HAV vaccination should be the procedure of choice.
[a]Repeat every 5 months if continued exposure to HAV occurs [9].

Vaccines

The use of gamma globulin has provided passive, short-term protection. Vaccines give active and long-lasting protection against hepatitis A [16,19,23,27].

Live, attenuated HAV vaccine

Inexpensive, live, attenuated vaccines have been produced in China, and millions of Chinese may have been vaccinated, although little information about these preparations is currently available. The H2-strain vaccine does not induce seroconversion if given orally, but nearly all of the individuals that were given the vaccine subcutaneously developed antibodies. This attenuated HAV is not transmitted orally although it is shed in stools in little amounts. The vaccine gave 100% protection against HAV infection during a 4-year period at 11 primary schools [4,18,21,28,40].

Inactivated HAV vaccines

The first inactivated HAV vaccine (Havrix®, SmithKline Beecham) became available for i.m. injection in Europe in 1991 and was approved in the United States in 1995. The second inactivated vaccine came in 1995 (Vaqta®, Merck). Both are whole-virus preparations, produced by growth of attenuated HAV strains in cell culture, inactivated with formalin, adsorbed to aluminium as adjuvant. Havrix® is preserved in 2-phenoxyethanol. Both vaccines are highly effective and provide seroconversion rates of more than 99.4% when given as a single primary immunisation, followed by a booster dose 6–12 months later [18,19,21,23,27,45].

A third vaccine (Epaxal®, Berna) developed in Switzerland and currently marketed in Switzerland and Argentina, incorporates immunogenic formalin-inactivated HAV particles within immunopotentiating reconstituted influenza virosomes that facilitate antigen delivery to immunocompetent cells [3,18,21,26,33,45].

Another inactivated vaccine (Avaxim®, Pasteur Merieux) has given excellent results since its introduction in France, the Netherlands, Sweden and the United Kingdom in 1997 [45].

A combined hepatitis A and B vaccine (Twinrix®, SmithKline Beecham) has been introduced in Australia, Canada and some countries in Europe in 1997. In its adult formulation it contains 720 ELISA Units (EL.U.) of hepatitis A antigen (Havrix®) and 20 μg of hepatitis B surface antigen (Engerix®-B) adsorbed onto aluminium salts [45]. Twinrix® is licensed for use in children aged 1 year or older in several countries and is given as a 3-dose series, using a 0-, 1-, 6-month schedule. The immunogenicity of the combined vaccine has been compared to the immunogenicity of simultaneously or separately applied single vaccines. The results of the study recommend the use of the combined vaccine for subjects at risk for both hepatitis A and hepatitis B [13].

In the United States, IG is still recommended for protection from HAV infection in children less than 2 years of age because residual anti-HAV passively acquired from the mother may interfere with vaccine immunogenicity [21].

If not specified otherwise, manufacturers suggest that paediatric formulations contain half the antigenic mass of the adult formulation and are given to children and adolescents between 1 and 18 years of age. Immunisation priorities differ between countries, as do vaccination schedules, local production capabilities, and demands for specific vaccine combinations.

The systematic use of Havrix® in Alaska in 1996, covering at least 80% of eligible persons, indicates that the vaccine can efficiently stop established outbreaks and prevent epidemics of hepatitis A in communities where cases of hepatitis A are documented [21,30,45].

Havrix® is the only vaccine specifically licensed for active immunisation against the HAV in chronic liver disease patients.

Recommended dosages of available vaccines

Havrix® (SmithKline Beecham Biologicals)					
Group	Age (years)	Dose	Volume (ml)	No. doses	Schedule (months)
Children and adolescents	2–19	720 EL.U.	0.5	2	0,6–12
Children and adolescents	1–18	360 EL.U. (US$ 19.50)[a]	0.5	3	0,1,6–12
Adults	>18	1440 EL.U. (US$ 56.90)[a]	1.0	2	0,6–12

Vaqta® (Merck & Co., Inc.)					
Group	Age (years)	Dose	Volume (ml)	No. doses	Schedule (months)
Children and adolescents	2–17	25 U	0.5	2	0,6–18
Adults	>17	50 U	1.0	2	0,6

0 months represents timing of the initial dose. Subsequent numbers indicate months after the initial dose.
[a]Prices are the average wholesale costs per single-unit dose in the United States [9].

For travellers who seek medical advice less than 2 weeks before travelling, a double dose is recommended (two injections, single dose; or one injection, double dose). This induces antibodies in over 90% of individuals within 2 weeks, and will most probably protect against infection. Alternatively, a dose of IG (0.02 ml/kg body weight) may be given with the first dose of vaccine.

When hepatitis A vaccine is administered concomitantly with pooled IG for immediate protection, separate syringes and different sites must be used.

Persons allergic to vaccine components should follow the recommendations for the use of IG.

Persons who have anti-HAV from prior infection do not need to be vaccinated, but do not react adversely to immunisation.

Studies for the development of safe and efficacious live-attenuated vaccines are currently in progress.

HAV vaccines are stable and can be stored for at least two years at 4°C without loss of immunogenicity [21].

Vaccine safety

Havrix®, Vaqta®, Avaxim®, Epaxal® and Twinrix® have excellent safety profiles and are highly immunogenic in humans [3,18,19,32]. Nearly 100% of vaccinees will develop protective levels of antibody within 1 month of the first dose of vaccine.

Side effects are local, of low intensity and short duration, involving a generally clinically insignificant soreness at the injection site.

Safety of the vaccines has not been determined during pregnancy.

Because the available vaccines are inactivated, no special precautions need to be taken in vaccination of HIV-1 positive or otherwise immunocompromised persons, although they may be less responsive [21].

Vaccination strategies

The best vaccination strategy for a region depends on the epidemiology of HAV, the risk groups involved, the duration of protection, the possibility of post-exposure protection, and the cost of the intervention [45].

Groups at high risk of HAV infection as a result of behaviour, lifestyle or occupation should be the primary target of a hepatitis A vaccination programme [45].

It is important to note that for many cases of hepatitis A risk factors and sources of infection cannot be identified. Immunisation programmes directed only to high risk groups would miss those with unidentified risks, and would therefore not reduce the impact of the disease or eliminate hepatitis A [17].

In most developing countries hepatitis A is not a real public health priority, since it is acquired in early childhood when infections are usually asymptomatic. These countries do not at present need to consider universal hepatitis A immunisation programmes [17].

Prevention and treatment

Since antivirals have never been as successful for the treatment of viral infections as antibiotics have been for the treatment of bacterial infections, prevention of viral diseases remains the most important weapon for their control.

Prevention

As almost all HAV infections are spread by the faecal–oral route, good personal hygiene, high quality standards for public water supplies and proper disposal of sanitary waste have resulted in a low prevalence of HAV infections in many well-developed societies [18,22].

Within households, good personal hygiene, including frequent and proper hand washing after bowel movement and before food preparation, are important measures to reduce the risk of transmission from infected individuals before and after their clinical disease becomes apparent [18].

For pre-exposure protection, the use of hepatitis A vaccines instead of IG is now highly recommended. Immunisation should be a priority for persons at increased risk of acquiring hepatitis A.

For post-exposure prophylaxis of non-vaccinated people, the passive administration of IG can modify the symptoms of infection, provided it is given within 2 weeks of exposure [21,40].

No special precautions are demanded for vaccinated persons.

Universal immunisation would successfully control hepatitis A, although at present, high costs and limited availability of vaccines preclude such a recommendation [17,21,23].

Eradication, however, can only be achieved through universal vaccination policies as long as HAV is not endemic in primates.

Treatment

As no specific treatment exists for hepatitis A, prevention is the most effective approach against the disease [4,40].

Therapy should be supportive and aimed at maintaining adequate nutritional balance (1 g/kg protein, 30–35 cal/kg). There is no good evidence that restriction of fats has any beneficial effect on the course of the disease. Eggs, milk and butter may actually help provide a correct caloric intake. Alcoholic beverages should not be consumed during acute hepatitis because of the direct hepatotoxic effect of alcohol. On the other hand, a modest consumption of alcohol during convalescence does not seem to be harmful. Hospitalisation is usually not required [18,40].

Adrenocortical steroids (corticosteroids) and IG are of no value in acute, uncomplicated hepatitis A, since they have no effect on the resolution of the underlying disease [18].

Antiviral agents have no beneficial clinical effect because a specific antiviral agent is not available and hepatic injury appears to be immunopathologically mediated [40].

Patients who are taking oral contraceptives do not need to discontinue their use during the course of the disease.

Referral to a liver transplant centre is appropriate for patients with fulminant hepatitis A, although the identification of patients requiring liver transplantation is difficult. A good proportion of patients (60%) with grade 4 encephalopathy will still survive without transplantation. Temporary auxiliary liver transplantation for subacute liver failure may be a way to promote native liver regeneration [6,18,40].

Guidelines for epidemic measures

1. Determination of mode of transmission, whether person-to-person or by common vector (vehicle).

2. Identification of the population exposed to increased risk of infection. Elimination of common sources of infection.

3. Improvement of sanitary and hygienic practices to eliminate faecal contamination of food and water.

4. Hepatitis A vaccination has been shown to be effective in controlling outbreaks of infection in communities that have high or intermediate rates of infection, provided a sufficient percent of the target population is reached [30,45].

5. Passive immunisation provides temporary protection, but it is not effective in controlling HAV on a community level.

Future considerations

Appropriate vaccine doses and schedules in the first two years of life need to be determined to overcome the reduced immune response observed among infants who have passively acquired maternal anti-HAV.

The duration of protection following a single dose of vaccine should be investigated.

Combination vaccines that integrate hepatitis A vaccine into existing childhood vaccination schedules need to be determined.

Most effective vaccination strategies for interrupting and preventing community wide outbreaks need to be defined.

Countries are encouraged to carry out studies addressing the cost-effectiveness of HAV prevention strategies to determine the feasibility of vaccination programmes [7,34].

The development of attenuated HAV vaccines capable of offering cost, production and administration advantages should be considered.

An assay to detect antibody to nonstructural proteins may serve, in future, to distinguish between natural infection and vaccine-induced antibody formation [23,35,39].

Glossary

Adjuvant any substance which, when mixed with an antigen, increases the immune response to that antigen [1].

Albumin a water soluble protein. Serum albumin is found in blood plasma and is important for maintaining plasma volume and osmotic pressure of circulating blood. Albumin is synthesised in the liver. The inability to synthesise albumin is a predominant feature of chronic liver disease.

Alkaline phosphatase any of group of phosphatases showing activity at alkaline pH, which are normally measured collectively in blood serum. Serum levels are elevated in various conditions, among which hepatobiliary disorders.

ALT alanine aminotransferase an enzyme that interconverts L-alanine and D-alanine. It is a highly sensitive indicator of hepatocellular damage. When such damage occurs, ALT is released from the liver cells into the bloodstream, resulting in abnormally high serum levels. Normal ALT levels range from 10 to 32 U/l; in women, from 9 to 24 U/l. The normal range for infants is twice that of adults.

Antibody a protein molecule formed by the immune system which reacts specifically with the antigen that induced its synthesis. All antibodies are immune globulins [1].

Antigen any substance which can elicit in a vertebrate host the formation of specific antibodies or the generation of a specific population of lymphocytes reactive with the substance. Antigens may be protein or carbohydrate, lipid or nucleic acid, or contain elements of all or any of these as well as organic or inorganic chemical groups attached to protein or other macromolecule. Whether a material is an antigen in a particular host depends on whether the material is foreign to the host and also on the genetic makeup of the host, as well as on the dose and physical state of the antigen [1].

AST aspartate aminotransferase the enzyme that catalyzes the reaction of aspartate with 2-oxoglutarate to give glutamate and oxaloacetate. Its concentration in blood may be raised in liver and heart diseases that are associated with damage to those tissues. Normal AST levels range from 8 to 20 U/l. AST levels fluctuate in response to the extent of cellular necrosis [1].

Bilirubin is the chief pigment of bile, formed mainly from the breakdown of haemoglobin. After formation it is transported in the plasma to the liver to be then excreted in the bile. Elevation of bile in the blood causes jaundice [46].

Capsid the protein coat of a virion, composed of large multimeric proteins, which closely surrounds the nucleic acid [1].

Cholestasis impairment of bile flow at any level from the canaliculus to the duodenum. The clinical condition resulting therefrom, characterised by icterus and pruritus, is due to the accumulation in blood and tissues of substances normally secreted in bile, particularly bilirubin, bile salts, and cholesterol [1].

Complete blood count chemical analysis of various substances in the blood performed with the aim of (i) assessing the patient's status by establishing normal levels for each individual patient, (ii) preventing disease by alerting to potentially dangerous levels of blood constituents that could lead to more serious conditions, (iii) establishing a diagnosis for already present pathologic conditions, and (iv) assessing a patient's progress when a disturbance in blood chemistry already exists.

Cytopathic effects include morphological changes in the cell appearance (rounding up of cells), agglutination of red blood cells (haemagglutination assay with influenza-virus), zones of cell lysis on monolayers of tissue culture or finally immortalization of animal cell lines (foci formation).

Encephalopathy an acute reaction of the brain to a variety of toxic or infective agents, without any actual inflammation such as occurs in encephalitis [1].

Endemic continuously prevalent in some degree in a community or region [46].

Enzyme any protein catalyst, i.e. substance which accelerates chemical reactions without itself being used up in the process. Many enzymes are specific to the substance on which they can act, called substrate. Enzymes are present in all living matter and are involved in all the metabolic processes upon which life depends [46].

Epidemic an outbreak of disease such that for a limited period a significantly greater number of persons in a community or region suffer from it than is normally the case. Thus an epidemic is a temporary increase in prevalence. Its extent and duration are determined by the interaction of such variables as the nature and infectivity of the casual agent, its mode of transmission and the degree of pre-existing and newly acquired immunity [46].

Epitope or antigenic determinant. The small portion of an antigen that combines with a specific antibody. A single antigen molecule may carry several different epitopes [1].

Fulminant describes pathological conditions that develop suddenly and are of great severity [46].

Genotype the genetic constitution of an individual.

Hepatocytes are liver cells [1].

Humoral pertaining to the humors, or certain fluids, of the body [1].

Icterus jaundice [1].

IgA antibodies IgA has antiviral properties. Its production is stimulated by aerosol immunisations and oral vaccines.

IgG antibodies IgG is the most abundant of the circulating antibodies. It readily crosses the walls of blood vessels and enters tissue fluids. IgG also crosses the placenta and confers passive immunity from the mother to the fetus. IgG protects against bacteria, viruses, and toxins circulating in the blood and lymph.

IgM antibodies IgMs are the first circulating antibodies to appear in response to an antigen. However, their concentration in the blood declines rapidly. This is diagnostically useful because the presence of IgM usually indicates a current infection by the pathogen causing its formation. IgM consists of five Y-shaped monomers arranged in a pentamer structure. The numerous antigen-binding sites make it very effective in agglutinating antigens. IgM is too large to cross the placenta and hence does not confer maternal immunity.

Immune globulin (IG) is a sterile preparation of concentrated antibodies (IG) recovered from pooled human plasma processed by cold ethanol fractionation. Only plasma that has tested negative for (i) hepatitis B surface antigen (HBsAg), (ii) antibody to human immunodeficiency virus (HIV), and (iii) antibody to hepatitis C virus (HCV) is used to manufacture IG. IG is administered to protect against certain diseases through passive transfer of antibody. The IGs are broadly classified into five types on the basis of physical, antigenic and functional variations, and labelled respectively IgM, IgG, IgA, IgE and IgD.

Immunocompetent capable of responding immunologically to an antigen, as by producing antibodies or by developing cell-mediated immunity [1].

Immunodominant describing those epitopes in a molecule, or those components in an antigenic mixture, to which an immune response is produced preferentially [1].

Immunogenic capable of eliciting an immune response.

Incidence the number of cases of a disease, abnormality, accident, etc., arising in a defined population during a stated period, expressed as a proportion, such as x cases per 1000 persons per year [1].

Interferon a class of proteins processing antiviral and antitumour activity produced by lymphocytes, fibroblasts and other tissues. They are released by cells invaded by virus and are able to inhibit virus multiplication in noninfected cells. Interferon preparations have been shown to have some clinical effect as antiviral agents. The preparations so far available have produced side effects, such as fever, lassitude, and prostration, not dissimilar from those accompanying acute virus infection itself [46].

Jaundice a yellow discoloration of the skin and mucous membranes due to excess of bilirubin in the blood, also known as icterus [46].

Lymphocyte a leukocyte of blood, bone marrow and lymphatic tissue. Lymphocytes play a major role in both cellular and humoral immunity, and thus several different functional and morphologic types must be recognised, i.e. the small, large, B-, and T-lymphocytes, with further morphologic distinction being made among the B-lymphocytes and functional distinction among T-lymphocytes [1].

Monoclonal antibodies preparations containing only one kind of antibody molecules, with specificity for a single antigen. Monoclonal antibodies are of great therapeutic and diagnostic potential [46].

Necrosis death of tissue [46].

Peptide a compound of two or more amino acids linked together by peptide bonds [46].

Pleomorphic distinguished by having more than one form during a life cycle [1].

Prevalence the number of instances of infections or of persons ill, or of any other event such as accidents, in a specified population, without any distinction between new and old cases [46].

Prodrome an early symptom suggestive of the onset of an attack or disease; a premonitory event [1].

Prophylaxis the prevention of disease, or the preventive treatment of a recurrent disorder [46].

Protein large molecule made up of many amino acids chemically linked together by amide linkages. Biologically important as enzymes, structural protein and connective tissue.

Prothrombin time a test used to measure the activity of clotting factors I, II, V, VII, and X. Deficiency of any of these factors leads to a prolongation of the prothrombin time. The test is basic to any study of the coagulation process, and it helps in establishing and maintaining anticoagulant therapy.

Pruritus itching.

Self-limited denoting a disease that tends to cease after a definite period, e.g. pneumonia [2].

Seroconversion the production in a host of specific antibodies as a result of infection or immunisation. The antibodies can be detected in the host's blood serum following, but not preceding, infection or immunisation [1].

Serotype a subdivision of a species or subspecies distinguishable from other strains therein on the basis of antigenic character [2].

Serum is the clear, slightly yellow fluid which separates from blood when it clots. In composition it resembles blood plasma, but with fibrinogen removed. Sera containing antibodies and antitoxins against infections and toxins of various kinds (antisera) have been used extensively in prevention or treatment of various diseases [46].

Titre a measure of the concentration or activity of an active substance.

Urobilinogen also called stercobilinogen. A colourless, reduced form of stercobilin, in which the pyrrole rings are joined by $-CH_2-$ groups. It is oxidised by air to the coloured stercobilin [1].

Vaccine an antigenic preparation used to produce active immunity to a disease to prevent or ameliorate the effects of infection with the natural or "wild" organism. Vaccines may be living, attenuated strains of viruses or bacteria, which give rise to inapparent to trivial infections. Vaccines may also be killed or inactivated organisms or purified products derived from them. Formalin-inactivated toxins are used as vaccines against diphtheria and tetanus. Synthetically or genetically engineered antigens are currently being developed for use as vaccines. Some vaccines are effective by mouth, but most have to be given parenterally [1,46].

Vaccinee person receiving a vaccine.

Viremia the presence of viruses in the blood.

Virion a structurally complete virus, a viral particle [1].

Virosomes immunopotentiating reconstituted influenza virosomes (IRIVs) are safe, efficacious and easily prepared carrier systems for small virion particles, such as HAV. IRIVs are spherical, unilamellar vescicles, with a diameter of about 150 nm that combine components like haemagglutinin, neuraminidase and phospholipids of influenza virus and adsorbed HAV particles. The essential feature of IRIVs is that as well as the HAV antigen, their surface contains the fusion-inducing component haemagglutinin, which facilitates antigen delivery to immunocompetent cells. Most adults have been exposed to influenza haemagglutinin, and so IRIVs recruit primed cells leading to a rapid immune response [26].

Virus any of a number of small, obligatory intracellular parasites with a single type of nucleic acid, either DNA or RNA and no cell wall. The nucleic acid is enclosed in a structure called a capsid, which is composed of repeating protein subunits called capsomeres, with or without a lipid envelope. The complete infectious virus particle, called a virion, must rely on the metabolism of the cell it infects. Viruses are morphologically heterogeneous, occurring as spherical, filamentous, polyhedral, or pleomorphic particles. They are classified by the host infected, the type of nucleic acid, the symmetry of the capsid, and the presence or absence of an envelope [1].

References

1. Churchill's Illustrated Medical Dictionary. New York: Churchill Livingstone, 1989.
2. Stedman's Medical Dictionary, 26th edn. Baltimore: Williams & Wilkins, 1995.
3. Ambrosch F, et al. Immunogenicity and protectivity of a new liposomal hepatitis A vaccine. Vaccine 1997; 15(11): 1209–1213.
4. André FE. Approaches to a vaccine against hepatitis A: development and manufacture of an inactivated vaccine. J Infect Dis 1995; 171(Suppl 1): S33–S39.
5. Barzaga BN. Hepatitis A shifting epidemiology in South-East Asia and China. Vaccine 2000; 18(Suppl 1(2)): S61–S64.

6. Battegay M, Gust ID, Feinstone SM. Hepatitis A virus. In: Principles and Practice of Infectious Diseases (Mandell GL, Bennett JE, Dolin R, editors), 4th edn. New York: Churchill Livingstone; 1995; pp. 1636–1656.
7. Berge JJ, et al. The cost of hepatitis A infections in American adolescents and adults in 1997. Hepatology 2000; 31(2): 469–473.
8. Burke DS, Graham RR, Heisey GB. Hepatitis A virus in primates outside captivity. Lancet 1981; 2: 928.
9. Centers for Disease Control and Prevention. Prevention of hepatitis A through active or passive immunization: recommendations of the Advisory Committee on Immunization Practices. Morbidity Mortality Weekly Rep 1996; 45(RR15): 1–30.
10. Centers for Disease Control and Prevention, Epidemiology and Prevention of Viral Hepatitis A to E: An Overview 2000; http://www.cdc.gov/ncidod/diseases/hepatitis/slideset/httoc.htm.
11. Cianciara J. Hepatitis A shifting epidemiology in Poland and Eastern Europe. Vaccine 2000; 18(Suppl 1(2)): S68–S70.
12. Clemens R, et al. Clinical experience with an inactivated hepatitis A vaccine. J Infect Dis 1995; 171(Suppl 1): S44–S49.
13. Czeschinski PA, Binding N, Witting U. Hepatitis A and hepatitis B vaccinations: immunogenicity of combined vaccine and of simultaneously or separately applied vaccines. Vaccine 2000; 18(11–12): 1074–1080.
14. Emerson SU, et al. A simian strain of hepatitis A virus, AGM-27, functions as an attenuated vaccine for chimpanzees. J Infect Dis 1996; 173: 592–597.
15. Hillis WD. An outbreak of infectious hepatitis among chimpanzee handlers at a United States Air Force Base. Am J Hygiene 1961; 73: 316–328.
16. Hollinger FB. An overview of the clinical development of hepatitis A vaccine. Introduction. J Infect Dis 1995; 171(Suppl 1): S1.
17. Hollinger FB, et al. Discussion: who should receive hepatitis A vaccine? A strategy for controlling hepatitis A in the United States. J Infect Dis 1995; 171(Suppl 1): S73–S77.
18. Hollinger FB, Ticehurst JR. Hepatitis A virus. In: Fields Virology (Fields BN, Knipe DM, Howley PM, editors), 3rd edn. Philadelphia: Lippincott-Raven; 1996; pp. 735–782.
19. Innis B, et al. Protection against hepatitis A by an inactivated vaccine. J Am Med Assoc 1994; 271: 1328–1334.
20. Kessler H, et al. Hepatitis A and B at the London Zoo. J Infect 1982; 4: 63–67.
21. Koff RS. Hepatitis A. Lancet 1998; 341: 1643–1649.
22. Lemon SM. Hepatitis A virus. In: Encyclopedia of Virology (Webster RG, Granoff A, editors). London: Academic Press Ltd.; 1994; pp. 546–554.
23. Lemon SM. Type A viral hepatitis: epidemiology, diagnosis, and prevention. Clin Chem 1997; 43(8(B)): 1494–1499.
24. Lemon SM, et al. Genomic heterogeneity among human and nonhuman strains of hepatitis A virus. J Virol 1987; 61(3): 735–742.
25. Lemon SM, Robertson BH. Current perspectives in the virology and the molecular biology of hepatitis A virus. Semin Virol 1993; 4: 285–295.
26. Loutan L, et al. Inactivated virosome hepatitis A vaccine. Lancet 1994; 343: 322–324.
27. Maiwald H, et al. Long-term persistence of anti-HAV antibodies following active immunization with hepatitis A vaccine. Vaccine 1997; 15(4): 346–348.
28. Mao JS, et al. Further evaluation of the safety and protective efficacy of live attenuated hepatitis A vaccine (H2-strain) in humans. Vaccine 1997; 15(9): 944–947.

29. Mao JS, et al. Susceptibility of monkeys to human hepatitis A virus. J Infect Dis 1981; 144(1): 55–60.
30. McMahon BJ, et al. A program to control an outbreak of hepatitis A in Alaska by using an inactivated hepatitis A vaccine. Arch Pediatr Adolesc Med 1996; 150: 733–739.
31. Melnick JL. History and epidemiology of hepatitis A virus. J Infect Dis 1995; 171(Suppl 1): S2–S8.
32. Niu MT, et al. Two-year review of hepatitis A vaccine safety: data from the vaccine adverse event reporting system (VAERS). Clin Infect Dis 1998; 26: 1475–1476.
33. Poovorawan Y, et al. Safety, immunogenicity, and kinetics of the immune response to a single dose of virosome-formulated hepatitis A vaccine in Thais. Vaccine 1995; 13(10): 891–893.
34. Rajan E, Shattock AG, Fielding JF. Cost-effective analysis of hepatitis A prevention in Ireland. Am J Gastroenterol 2000; 95(1): 223–226.
35. Robertson BH, et al. Antibody response to nonstructural proteins of hepatitis A virus following infection. J Med Virol 1993; 40: 76–82.
36. Schulman AN, et al. Hepatitis A antigen particles in liver, bile, and stool in chimpanzees. J Infect Dis 1976; 134(1): 80–84.
37. Shapiro CN, Margolis HS. Worldwide epidemiology of hepatitis A virus infection. J Hepatol 1993; 18(Suppl 2): S11–S14.
38. Soucie JM, et al. Hepatitis A virus infections associated with clotting factor concentrate in the United States. Transfusion 1998; 38: 573–579.
39. Stapleton JT. Host immune response to hepatitis A virus. J Infect Dis 1995; 171(Suppl 1): S9–S14.
40. Stapleton JT, Lemon SM. In: Infectious Diseases (Hoeprich PD, Jordan MC, Ronald AR, editors), 5th edn. Philadelphia: Lippincott Co.; 1994; pp. 790–797.
41. Steffen R. Hepatitis A in travelers: the European experience. J Infect Dis 1995; 171(Suppl 1): S24–S28.
42. Tanaka J. Hepatitis A shifting epidemiology in Latin America. Vaccine 2000; 18(Suppl 1(2)): S57–S60.
43. Tormans G, Van Damme P, Van Doorslaer E. Cost-effectiveness analysis of hepatitis A prevention in travellers. Vaccine 1992; 10(Suppl 1): S88–S92.
44. Tufenkeji H. Hepatitis A shifting epidemiology in the Middle East and Africa. Vaccine 2000; 18(Suppl 1(2)): S65–S67.
45. Viral Hepatitis Prevention Board. News from the VHPB meeting in St. Julians, Malta. Viral Hepat 1997; 6(1).
46. Walton J, Barondess JA, Lock S. The Oxford Medical Companion. Oxford: Oxford University Press; 1994.
47. Wolfe MS. Hepatitis A and the American traveler. J Infect Dis 1995; 171(Suppl 1): S29–S32.

Viral Hepatitis
I.K. Mushahwar (editor)
© 2004 World Health Organisation. All rights reserved.

Hepatitis B

Nicoletta Previsani and Daniel Lavanchy
World Health Organization, Geneva, Switzerland

Arie J. Zuckerman
Royal Free and University College of Medical School, London, UK

Hepatitis B—an introduction

Hepatitis is a general term meaning inflammation of the liver and can be caused by a variety of different viruses such as hepatitis A, B, C, D and E. Since the development of jaundice is a characteristic feature of liver disease, a correct diagnosis can only be made by testing patients' sera for the presence of specific antiviral antigens or antibodies [15,23,31].

Of the many viral causes of human hepatitis few are of greater global importance than hepatitis B virus (HBV) [10,15,23,31].

Hepatitis B is a serious and common infectious disease of the liver, affecting millions of people throughout the world [6,10,15,23,31].

The severe pathological consequences of persistent HBV infections include the development of chronic hepatic insufficiency, cirrhosis, and hepatocellular carcinoma (HCC). In addition, HBV carriers can transmit the disease for many years [10,23,30,31].

Infection occurs very often in early childhood when it is asymptomatic and often leads to the chronic carrier state.

More than 2000 million people alive today have been infected with HBV at some time in their lives. Of these, about 350 million remain infected chronically and become carriers of the virus [6,15,23,38,51]. Three quarters of the world's population live in areas where there are high levels of infection.

Every year there are over 4 million acute clinical cases of HBV, and about 25% of carriers, 1 million people a year, die from chronic active hepatitis, cirrhosis or primary liver cancer [51].

Hepatitis B has also been called type B hepatitis, serum hepatitis, homologous serum jaundice [23,31].

What causes the disease?

Hepatitis B is caused by the HBV, an enveloped virus containing a partially double-stranded, circular DNA genome, and classified within the family hepadnavirus [10,15,23,30,31].

The virus interferes with the functions of the liver while replicating in hepatocytes (HC). The immune system is then activated to produce a specific reaction to combat and possibly eradicate the infectious agent. As a consequence of pathological damage, the liver becomes inflamed.

HBV may be the cause of up to 80% of all cases of HCC worldwide, second only to tobacco among known human carcinogens [15,38,51].

How is HBV spread?

One should not judge by appearance: most infected people look perfectly healthy and have no symptoms of disease, yet may be highly infectious. HBV is transmitted through percutaneous or parenteral contact with infected blood, body fluids, and by sexual intercourse [10,11,15,23]. HBV is able to remain on any surface it comes into contact with for about a week, e.g. table-tops, razor blades, blood stains, without losing infectivity [15,31].

HBV does not cross the skin or the mucous membrane barrier. Some break in this barrier, which can be minimal and insignificant, is required for transmission [31]. HBV is a large virus and does not cross the placenta, hence it cannot infect the fetus unless there have been breaks in the maternal–fetal barrier, e.g. via amniocentesis. Still, pregnant women who are infected with HBV can transmit their disease to their babies at birth. If not vaccinated at birth, many of these babies develop lifelong HBV infections, and many develop liver failure or liver cancer later in life [23].

Sexual intercourse with multiple partners or with persons who have multiple partners can be dangerous. Hepatitis B is the only sexually transmitted infection for which there is a protective vaccine [23].

All persons who are hepatitis B surface antigen (HBsAg) positive are potentially infectious. The many millions of people around the world who become HBV carriers are a constant source of new infections for those who have never contracted the virus [31].

Blood is infective many weeks before the onset of the first symptoms and throughout the acute phase of the disease. The infectivity of chronically infected individuals varies from highly infectious (HBeAg-positive) to often sparingly infectious (anti-HBe-positive).

Who is susceptible to infection?

Susceptibility is general. Only people who have been vaccinated successfully or those who have developed anti-HBs antibodies after HBV infection are immune to HBV infection.

Persons with congenital or acquired immunodeficiency including HIV infection, and those with immunosuppression including those with lymphoproliferative disease, and patients treated with immunosuppressive drugs including steroids and by maintenance haemodialysis are more likely to develop persistent infection with HBV.

Following acute HBV infection, the risk of developing chronic infection varies inversely with age. Chronic HBV infection occurs among about 90% of infants infected at birth, 25–50% of children infected at 1–5 years of age and about 1–5% of persons infected as older children and adults. Chronic HBV infection is also common in persons with immunodeficiency [10,15,23,31].

Where is HBV a problem, globally?

The world can be divided into three areas where the prevalence of chronic HBV infection is high (>8%), intermediate (2–8%), and low (<2%) [23,42].

High endemicity areas include South-east Asia and the Pacific Basin (excluding Japan, Australia, and New Zealand), sub-Saharan Africa, the Amazon Basin, parts of the Middle East, the central Asian Republics, and some countries in Eastern Europe. In these areas, about 70–90% of the population becomes HBV-infected before the age of 40, and 8–20% of people are HBV carriers [15].

In countries such as China, Senegal, Thailand, infection rates are very high in infants, and continue through early childhood. At that stage, the prevalence of HBsAg in serum may exceed 25%. In other countries such as Panama, Papua New Guinea, Solomon Islands, Greenland, and in populations such as Alaskan Indians, infection rates in infants are relatively low and increase rapidly during early childhood [15].

Low endemicity areas include North America, Western and Northern Europe, Australia, and parts of South America. The carrier rate here is less than 2%, and less than 20% of the population is infected with HBV [15,23].

The rest of the world falls into the intermediate range of HBV prevalence, with 2–8% of a given population being HBV carriers.

When is hepatitis B contagious?

The most important mode of HBV transmission globally is perinatal, from the mother to her newborn baby. If a pregnant woman is an HBV carrier and is also HBeAg-positive, her newborn baby has a 90% likelihood to be infected and become a carrier. Of these children, 25% will die later from chronic liver disease or liver cancer [15].

Another important mode of HBV transmission is from child to child during early life resulting from blood contact [11].

All patients with acute hepatitis B are HBeAg-positive, and therefore highly infectious and careless contact with their blood or body fluids can lead to HBV infection. HBeAg-positive specimens contain high concentrations of infectious virions and HBV DNA, in contrast to anti-HBe-positive samples, in which the number of hepatitis B virions is substantially reduced.

Why is there no treatment for the acute disease?

There is no specific treatment for acute viral hepatitis B [23]. Hepatitis B is a viral disease, and as such, antibiotics are of no value in the treatment of the infection.

The use of adrenocorticosteroids in the management of acute, uncomplicated hepatitis B is not indicated because they have no effect on the resolution of the underlying disease process, and may increase the rate of relapse. Early treatment of acute hepatitis B with steroids may result in the development of persistent infection. Corticosteroid therapy is only to be used in patients with chronic active hepatitis who are symptomatic, HBsAg-negative, and who have severe histologic lesions in liver biopsies [31].

The therapeutic effectiveness of interferon on the course and prognosis of acute hepatitis B is not known [23].

Haemodialysis, exchange transfusions, cross-perfusion, and immune globulin (IG) containing high titres of anti-HBs (HBIG) do not affect favourably the course of fulminant hepatitis.

Therapy for acute hepatitis B should be supportive and aimed at maintaining comfort and adequate nutritional balance [23]. Specific antiviral drugs such as lamivudine, a second generation nucleoside analogue, are available, and others are under development, but these drugs have not been evaluated for the treatment of acute hepatitis B.

The hepatitis B virus (HBV)

The HBV, a hepadnavirus, is a 42 nm partially double-stranded DNA virus, composed of a 27 nm nucleocapsid core (HBcAg), surrounded by an outer lipoprotein coat (also called envelope) containing the surface antigen (HBsAg) [10,11,15,23,30,31].

The family of hepadnaviruses comprises members recovered from a variety of animal species, including the woodchuck hepatitis virus (WHV), the ground squirrel hepatitis virus (GSHV), and the duck HBV. Common features of all of these viruses are enveloped virions containing 3–3.3 kb of relaxed circular, partially duplex DNA and virion-associated DNA-dependent polymerases that can repair the gap in the virion DNA template and have reverse transcriptase activities. Hepadnaviruses show narrow host ranges, growing only in species close to the natural host, like gibbons, African green monkeys, rhesus monkeys, and woolly monkeys [15,30,31].

HC infected *in vivo* by hepadnaviruses produce an excess of noninfectious viral lipoprotein particles composed of envelope proteins. Persistent infections display pronounced hepatotropism [15]. Mammalian hepadnaviruses fail to propagate in cell culture [23,30,31]. Intracellular HBV is non-cytopathic and causes little or no damage to the cell [6,10,15,23].

Electron microscopy (EM) picture and schematic representation of the hepatitis B virion

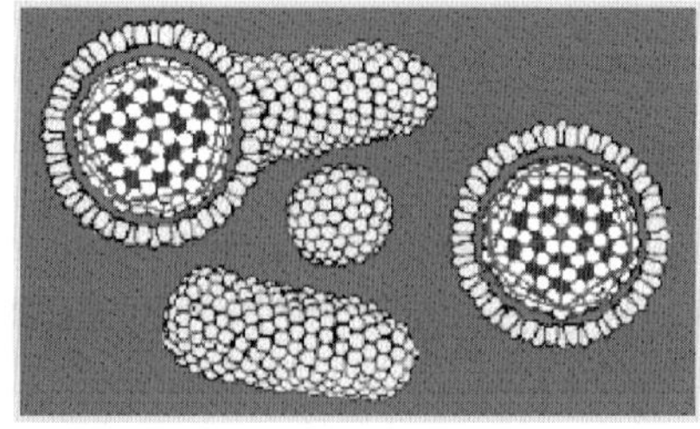

A diagrammatic representation of the hepatitis B virion and the surface antigen components.

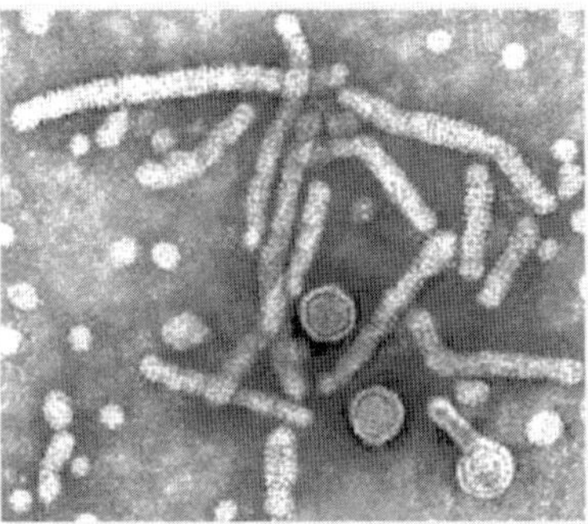

Virions are 42 nm in diameter and possess an isometric nucleocapsid or "core" of 27 nm in diameter, surrounded by an outer coat approximately 4 nm thick. The protein of the virion coat is termed "surface antigen" or HBsAg. It is sometimes extended as a tubular tail on one side of the virus particle. The surface antigen is generally produced in vast excess, and is found in the blood of infected individuals in the form of filamentous and spherical particles. Filamentous particles are identical to the virion "tails"—they vary in length and have a mean diameter of about 22 nm. They sometimes display regular, non-helical transverse striations.

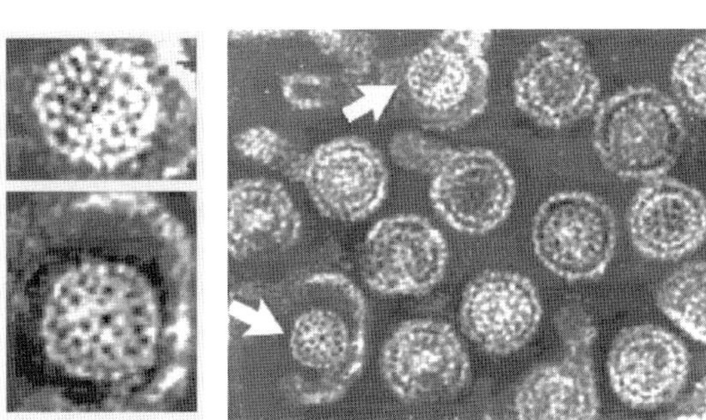

A group of hepatitis B virions (right) and enlargements of the two exposed cores (indicated by arrows).
From: University of Cape Town, South Africa, http://www.uct.ac.za/depts/mmi/ stannard/hepb.html

The hepatitis B virus life cycle

The HBV virion binds to a receptor at the surface of the hepatocyte [10]. A number of candidate receptors have been identified, including the transferrin receptor, the asialoglycoprotien receptor molecule, and human liver endonexin. The mechanism of HBsAg binding to a specific receptor to enter cells has not been established yet. Viral nucleocapsids enter the cell and reach the nucleus, where the viral genome is delivered [6,10,13,23]. In the nucleus, second-strand DNA synthesis is completed and the gaps in both strands are repaired to yield a covalently closed circular (ccc) supercoiled DNA molecule that serves as a template for transcription of four viral RNAs that are 3.5, 2.4, 2.1, and 0.7 kb long [6,10,23,31].

These transcripts are polyadenylated and transported to the cytoplasm, where they are translated into the viral nucleocapsid and precore antigen (C, pre-C), polymerase (P), envelope large (L), medium (M), small (S), and transcriptional transactivating proteins (X) [6,10,23,31].

The envelope proteins insert themselves as integral membrane proteins into the lipid membrane of the endoplasmic reticulum (ER).

The 3.5-kb species, spanning the entire genome and termed pregenomic RNA (pgRNA), is packaged together with HBV polymerase and a protein kinase into core particles where it serves as a template for reverse transcription of negative-strand DNA. The RNA to DNA conversion takes place inside the particles [10,23].

The new, mature, viral nucleocapsids can then follow two different intracellular pathways, one of which leads to the formation and secretion of new virions, whereas the other leads to amplification of the viral genome inside the cell nucleus [10,23]. In the virion assembly pathway, the nucleocapsids reach the ER, where they associate with the envelope proteins and bud into the lumen of the ER, from which they are secreted via the Golgi apparatus out of the cell [10,23]. In the genome amplification pathway, the nucleocapsids deliver their genome to amplify the intranuclear pool of covalently closed circular DNA (cccDNA) [10,23]. The precore polypeptide is transported into the ER lumen, where its amino- and carboxy-termini are trimmed and the resultant protein is secreted as precore antigen (eAg).

The X protein contributes to the efficiency of HBV replication by interacting with different transcription factors, and is capable of stimulating both cell proliferation and cell death [10,23].

The HBV polymerase is a multifunctional enzyme. The products of the P gene are involved in multiple functions of the viral life cycle, including a priming activity to initiate minus-strand DNA synthesis, a polymerase activity, which synthesizes DNA by using either RNA or DNA templates, a nuclease activity which degrades the RNA strand of RNA–DNA hybrids, and the packaging of the RNA pregenome into nucleocapsids [6,10,23]. Nuclear localisation signals on the polymerase mediate the transport of covalently linked viral genome through the nuclear pore [6,10].

Scheme of genome replication

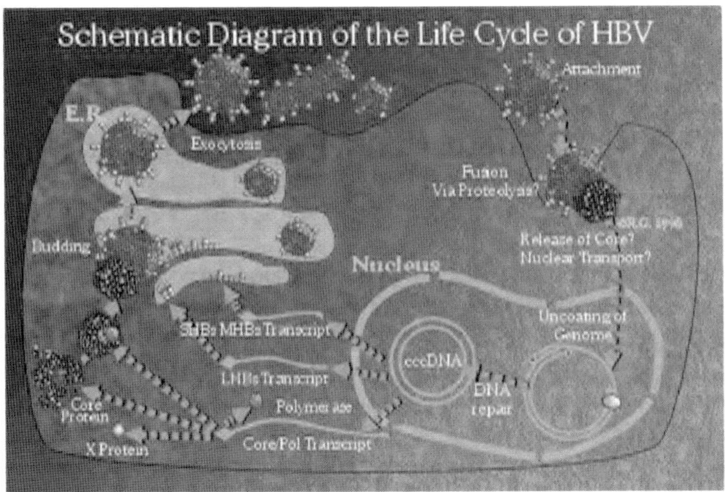

From http://www.globalserve.net/ ~ harlequin/HBV/hbvcycle.html

Morphology and physicochemical properties

Ultrastructural examination of sera from hepatitis B patients shows three distinct morphological forms [15].

The most abundant are small, spherical, noninfectious particles, containing HBsAg, that measure 17–25 nm in diameter. Concentrations of 10^{13} particles per ml or higher have been detected in some sera. These particles have a buoyant density of 1.18 g/cm^3 in CsCl, reflecting the presence of lipids, and a sedimentation coefficient that ranges from 39 to 54 S [15,31].

Tubular, filamentous forms of various lengths, but with a diameter comparable to that of the small particles, are also observed. They also contain HBsAg polypeptides [15,31].

The third morphological form, the 42 nm hepatitis B virion, is a complex, spherical, double-shelled particle that consists of an outer envelope containing host-derived lipids and all S gene polypeptides, the L, M, and S surface proteins, also known as pre-S1, pre-S2 and HBsAg. Within the sphere is an electron-dense inner core or nucleocapsid with a diameter of 27 nm. The nucleocapsid contains core proteins HBcAg, a 3.2-kb, circular, partially double-stranded viral DNA genome, an endogenous DNA polymerase (reverse transcriptase) enzyme, and protein kinase activity [15,23,31].

The sera of infected patients may contain as many as 10^{10} infectious virions per ml. The complete virion has a buoyant density of about 1.22 g/cm^3 in CsCl and a sedimentation coefficient of 280 S in sucrose gradients [15].

The polymerase protein is a DNA-dependent DNA polymerase, a reverse transcriptase, an RNAse H, and it binds to the $5'$ end of HBV DNA, acting thus as a primer for reverse transcription of the pregenome, an RNA intermediate, to form negative-strand DNA [31]. Furthermore, it plays important roles in the encapsidation of the viral pregenomic RNA. The polymerase protein is quite immunogenic during both acute and chronic infection [6].

ORF X encodes the protein X (17 kD), a transactivator for the viral core and S promoters. The X protein is the least-conserved protein among hepadnaviruses with only 33% amino acid homology between GSHV and HBV, and 71% between the two rodent viruses [6].

HBV coding organization

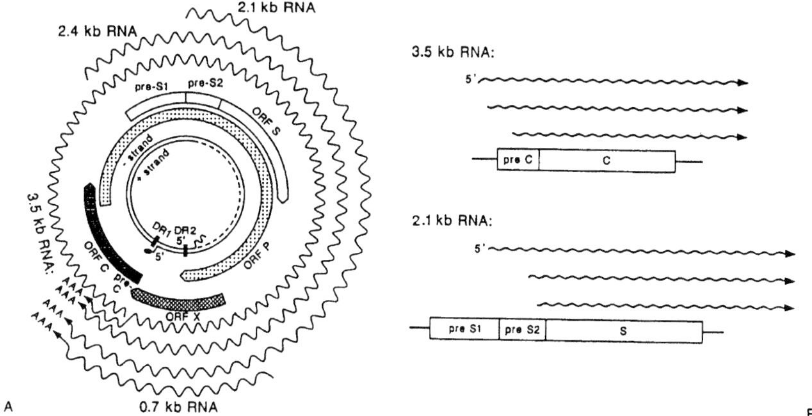

From Ref. [10], with permission (http://lww.com)

A: Diagrammatic representation of the HBV coding organization. *Inner circle* represents virion DNA, with dashes signifying the single-stranded genomic region; the locations of DR1 and DR2 sequence elements are as indicated. *Boxes* denote viral coding regions, with *arrows* indicating direction of translation. *Outermost wavy lines* depict the viral RNAs identified in infected cells, with arrows indicating direction of transcription. **B**: Fine structure of the $5'$ ends of the pre-C/C transcripts (*top*) and pre-S2/S transcripts (*bottom*) relative to their respective open reading frames [10].

Hepatitis B virus DNA and hepatocellular carcinoma

More than 85% of hepatocellular tumours examined harbor integrated HBV DNA, often multiple copies per cell. The viral DNA integrants are usually highly rearranged, with deletions, inversions, and sequence reiterations all commonly observed. Most of these rearrangements ablate viral gene expression, but the integrations alter the host DNA [10, 31,52].

Interestingly, tumours are clonal with respect to these integrants: every cell in the tumour contains an identical complement of HBV insertions. This implies that the integration event(s) preceded the clonal expansion of the cells. How integration is achieved is still not well understood. Since integration is not an obligatory step in the hepadnaviral replication cycle, and hepadnaviruses have no virus-encoded integration machinery, HBV DNA is probably assimilated into the nucleus by host mechanisms [10,23].

There is no similarity in the pattern of integration between different tumours, and variation is seen both in the integration site(s) and in the number of copies or partial copies of the viral genome [52]. The molecular mechanisms by which hepadnaviruses predispose to malignancy are still unknown [52].

Direct models

In the direct models, HBV DNA makes direct genetic contributions to the lesion by either providing cis-acting sequences deregulating host growth genes, or by providing trans-acting factors that interfere with cellular growth control [10].

Indirect models

In the indirect models, HBV genes and their products make no direct genetic contribution to the transforming event. Rather, HBV-induced liver injury triggering a series of host responses that lead to liver cell regeneration increases the probability of mutation and malignant transformation [10].

A better understanding of the immunologic mechanisms of liver cell injury could allow the development of therapeutic agents that would control these responses.

HBV mutants

Naturally occurring envelope, precore, core, and polymerase variants have been described [11,15,23].

Envelope antigenic variants may have a selective advantage over wild type under immune selection pressure, as observed in some cases after hepatitis B IG (HBIG) treatment or HBV vaccination. An epidemiological shift has not been observed yet.

A number of precore mutations preventing HBeAg synthesis have been identified in HBeAg-negative carriers. The most frequent variant has a G to A point mutation at nucleotide 83 (mutant HBV83, nucleotide 1896 of the genome, amino acid 144) in the precore region, introducing a stop codon at codon 28 [15,23]. The HBV83 mutant is predominantly found in Mediterranean and Asian countries but is uncommon in North America and Northern Europe. Precore mutants are found in patients with fulminant hepatitis or chronic active hepatitis, but also in asymptomatic carriers [11].

HBV core gene mutations have been reported in patients from Japan, Hong Kong, United States, and Italy. Most of the mutations are concentrated in the middle-third of the core gene, but although many of these mutations are located in regions that harbor B- and T-cell epitopes, they have not been proven to result in loss of immune recognition.

In rare patients where the function of the polymerase gene is impaired, additional compensatory mutations were found that minimized the impact of the impaired function of the polymerase.

HBV is far more heterogeneous than is generally thought. The HBV genome seems not to be characterized by a single representative genomic molecule, but by a pool of genomes which differ both in structure and function.

The public health importance of mutant HBVs is currently under debate. Further studies and a strict surveillance to detect the emergence of these viruses are crucial for a correct evaluation of the effectiveness of current immunization strategies [23,52,53].

Nomenclature of hepatitis B

HBV	Hepatitis B virus (complete infectious virion)	The 42 nm, double-shelled particle, originally called the Dane particle that consists of a 7-nm thick outer shell and a 27-nm inner core. The core contains a small, circular, partially double-stranded DNA molecule and an endogenous DNA polymerase. This is the prototype agent for the family Hepadnaviridae
HBsAg	Hepatitis B surface antigen (also called envelope antigen)	The complex of antigenic determinants found on the surface of HBV and of 22 nm particles and tubular forms. It was formerly designated Australia (Au) antigen or hepatitis-associated antigen (HAA)
HBcAg	Hepatitis B core antigen	The antigenic specificity associated with the 27 nm core of HBV
HBeAg	Hepatitis B e antigen	The antigenic determinant that is closely associated with the nucleocapsid of HBV. It also circulates as a soluble protein in serum
Anti-HBs, anti-HBc, and anti-HBe	Antibody to HBsAg, HBcAg, and HBeAg	Specific antibodies that are produced in response to their respective antigenic determinants

From Ref. [15] with permission (http://lww.com).

Antigenicity

All three coat proteins of HBV contain HBsAg, which is highly immunogenic and induces anti-HBs (humoral immunity). Structural viral proteins induce specific T-lymphocytes, capable of eliminating HBV-infected cells (cytotoxic T-lymphocytes; cellular immunity) [6,15].

HBsAg is heterogeneous antigenically, with a common antigen designated *a*, and two pairs of mutually exclusive antigens, *d* and *y*, and *w* (including several subdeterminants) and *r*, resulting in 4 major subtypes: *adw*, *ayw*, *adr* and *ayr* [23,30,31].

The distribution of subtypes varies geographically [30]. Because of the common determinants, protection against one subtype appears to confer protection to the other subtypes, and no difference in clinical features have been related to subtypes.

In the US, northern Europe, Asia, and Oceania, the *d* determinant is common, but the *y* determinant is found at lower frequency. The *d* determinant to the near exclusion of *y* is found in Japan. The *y* determinant, and rarely *d*, are found in Africa and in Australia aborigines. *y* is also frequently found in India and around the Mediterranean. In Europe, the US, Africa, India, Australia, and Oceania, the *w* determinant predominates. In Japan, China, and South-east Asia, the *r* determinant predominates. Subtypes adw, ady, and adr are each found in extensive geographic regions of the world. Subtype ayr is rare in the world, but it is commonly found in small populations in Oceania [23,52].

The c antigen (HBcAg) is present on the surface of core particles. HBcAg and core particles are not present in the blood in a free form, but are found only as internal components of virus particles [23,30]. The core antigen shares its sequences with the e antigen (HBeAg), identified as a soluble antigen, but no cross-reactivity between the two proteins is observed [30,31].

Viral oligopeptides of 8–15 amino acids are loaded on host cell MHC-class I molecules and are transported to the surface of the cell. HBV-specific T-lymphocytes can then detect infected cells and destroy them. This cell deletion triggered by inflammation cells may result in acute hepatitis. When the infection is self-limited, immunity results. If HBV is not eliminated, a delicate balance between viral replication and immunodefence prevails which may lead to chronic hepatitis and liver cirrhosis. In chronically infected cells the HBV DNA may integrate into the host cell DNA. As a long-term consequence, integration may lead to HCC [15,23,52].

Stability

The stability of HBV does not always coincide with that of HBsAg [15]. Exposure to ether, acid (pH 2.4 for at least 6 h), and heat (98°C for 1 min; 60°C for 10 h) does not destroy immunogenicity or antigenicity. However, inactivation may be incomplete under these conditions if the concentration of virus is excessively high [15]. Antigenicity and probably infectivity are destroyed after exposure of HBsAg to 0.25% sodium hypochlorite for 3 min [15]. Infectivity is lost after autoclaving at 121°C for 20 min or dry heat treatment at 160°C for 1 h [15,31].

HBV is inactivated by exposure to sodium hypochlorite (500 mg free chlorine per litre) for 10 min, 2% aqueous glutaraldehyde at room temperature for 5 min, heat treatment at 98°C for 2 min, Sporicidin (Ash Dentsply, York, PA) (pH 7.9), formaldehyde at 18.5 g/l (5% formalin in water), 70% isopropylalcohol, 80% ethyl alcohol at 11°C for 2 min, Wescodyne (a iodophor disinfectant, American Sterilizer Co., Erie, PA) diluted 1:213, or combined β-propriolactone and UV irradiation [15,45].

HBV retains infectivity when stored at 30–32°C for at least 6 months and when frozen at −15°C for 15 years. HBV present in blood can withstand drying on a surface for at least a week [15,31].

The disease

The course of hepatitis B may be extremely variable [31]. HBV infection has different clinical manifestations depending on the patient's age at infection and immune status, and the stage at which the disease is recognized.

During the incubation phase of the disease (6–24 weeks), patients may feel unwell with possible nausea, vomiting, diarrhoea, anorexia and headaches. Patients may then become jaundiced although low-grade fever and loss of appetite may improve. Sometimes HBV infection produces neither jaundice nor obvious symptoms [15,31].

The asymptomatic cases can be identified by detecting biochemical or virus-specific serologic alterations in their blood. They may become silent carriers of the virus and constitute a reservoir for further transmission to others.

Most adult patients recover completely from their HBV infection, but about 5–10%, will not clear the virus and will progress to become asymptomatic carriers or develop chronic hepatitis possibly resulting in cirrhosis and/or liver cancer [31]. Rarely, others may develop fulminant hepatitis and die.

People who develop chronic hepatitis may develop significant and potentially fatal disease [31]. In general, the frequency of clinical disease increases with age, whereas the percentage of carriers decreases.

Worldwide, about 1 million deaths occur each year due to chronic forms of the disease [39]. Persistent or chronic HBV infection is among the most common persistent viral infections in humans. More than 350 million people in the world today are estimated to be persistently infected with HBV. A large fraction of these are in eastern Asia and sub-Saharan Africa, where the associated complications of chronic liver disease and liver cancer are the most important health problems [31]. A small number of long-established chronic carriers apparently terminate their active infection and become HBsAg-negative (about 2% per year).

Survivors of fulminant hepatitis rarely become infected persistently, and HBsAg carriers frequently have no history of recognized acute hepatitis.

Spectrum of liver disease after HBV infection

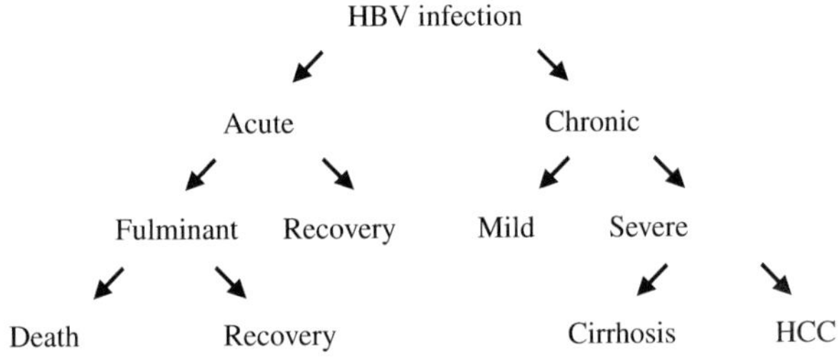

From Ref. [6], with permission (http://lww.com).

The infecting dose of virus and the age of the person infected are important factors that correlate with the severity of acute or chronic hepatitis B [23,31].

Only a small proportion of acute HBV infections are recognized clinically. Less than 10% of children and 30–50% of adults with acute HBV infection will have icteric disease [51]. Primary HBV infection may be associated with little or no liver disease or with acute hepatitis of severity ranging from mild to fulminant [31]. HBV infection is transient in about 90% of adults and 10% of newborn, and persistent in the remainder [23]. Most cases of acute hepatitis are subclinical, and less than 1% of symptomatic cases are fulminant [31].

Worldwide, about 350 million people are estimated to be infected chronically with HBV [39]. Persistent HBV infection is sometimes associated with histologically normal liver and normal liver function, but about one third of chronic HBV infections are associated with cirrhosis and HCC [31].

Clinical phases of acute hepatitis B infection

The acute form of the disease often resolves spontaneously after a 4–8 week illness. Most patients recover without significant consequences and without recurrence. However, a favourable prognosis is not certain, especially in the elderly who can develop fulminating, fatal cases of acute hepatic necrosis. Young children rarely develop acute clinical disease, but many of those infected before the age of seven will become chronic carriers [6,15,23,30,31].

The incubation period varies usually between 45 and 120 days, with an average of 60–90 days. The variation is related to the amount of virus in the inoculum, the mode of transmission and host factors [6,15,23,31].

The hallmark of acute viral hepatitis is the striking elevation in serum transaminase (aminotransferase) activity. The increase in aminotransferases, especially ALT, during acute hepatitis B varies from a mild/moderate increase of 3–10-fold to a striking increase of >100-fold. The latter does not necessarily imply a poor prognosis.

In patients with clinical illness, the onset is usually insidious with tiredness, anorexia, vague abdominal discomfort, nausea and vomiting, sometimes arthralgias and rash, often progressing to jaundice. Fever may be absent or mild [6,15,23, 31].

The icteric phase of acute viral hepatitis begins usually within 10 days of the initial symptoms with the appearance of dark urine followed by pale stools and yellowish discoloration of the mucous membranes, conjunctivae, sclerae, and skin. Jaundice becomes apparent clinically when the total bilirubin level exceeds 20–40 mg/l. It is accompanied by hepatomegaly and splenomegaly. About 4–12 weeks thereafter, the jaundice disappears and the illness resolves with the development of natural, protective antibodies (anti-HBs), in about 95% of adults [15].

46

The larger the virus dose, the shorter the incubation period and the more likely that icteric hepatitis will result. The largest virus doses received by patients may occur in transfusions of infectious blood [31].

In most cases, no special treatment or diet is required, and patients need not be confined to bed.

Acute hepatitis B is characterized by the presence of anti-HBc IgM serum antibodies converting to IgG with convalescence and recovery, and the transient ($<$6 months) presence of HBsAg, HBeAg, and viral DNA, with clearance of these markers followed by seroconversion to anti-HBsAg and anti-HBeAg. More than 90% of adult-onset infection cases fall into this category. The remaining 5–10% of adult-onset infection and over 90% of cases of neonatal infection become chronic, and may continue for the life span of the patient [23].

A small percentage of persons die from acute HBV.

Clinical features of chronic hepatitis B

Although most adult patients recover completely from an acute episode of hepatitis B, in a significant proportion, 5–10%, the virus persists in the body. This figure is much higher in children: 70–90% of infants infected in their first few years of life become chronic carriers of HBV [23,31].

Hepatitis B causes about 4 million acute infections worldwide each year. An estimated 350 million persons worldwide are chronic carriers of HBV, with 100 million carriers in China and 1 million carriers in the USA [40]. Of persistent HBsAg carriers, 70% have chronic persistent hepatitis (see below), and 30% have chronic active hepatitis (see below) [23].

Chronic hepatitis can cause serious destructive diseases of the liver and it contributes greatly to the worldwide burden of the disease [23]. Chronic hepatitis generally develops over many years during which individual patients will pass through a number of disease states. Surprisingly, some of the patients infected persistently may have no clinical or biochemical evidence of liver disease, while others may show signs of easy fatigability, anxiety, anorexia, and malaise [15,23].

Chronic hepatitis B is a prolonged ($>$6 months) infection with persistent serum levels of HBsAg and IgG anti-HBcAg and the absence of an anti-HBs response. HBV DNA and HBeAg are often detectable at high concentrations, but may disappear if viral replication ceases or if mutations occur that prevent the synthesis of the viral precore protein precursor of HBeAg. The associated inflammatory liver disease is variable in severity. It is always much milder than in acute hepatitis B, but it can last for decades and proceed to cirrhosis, and it is associated with a 100-fold increase in the risk of developing a HCC [15,31].

Three phases of viral replication occur during the course of HBV infection, especially in patients with chronic hepatitis B [11].

High replicative phase. In this phase HBsAg, HBeAg, and HBV DNA are present and detectable in the sera. Aminotransferase levels may increase, and moderate inflammatory activity is histologically apparent. The risk of evolving to cirrhosis is high.

Low replicative phase. This phase is associated with the loss of HBeAg, or a decrease or loss of the HBV DNA concentrations, and with the appearance of anti-HBe. Histologically, a decrease in inflammatory activity is evident. Serologic changes like the loss of HBV DNA and HBeAg are referred to as seroconversion.

Nonreplicative phase. Markers of viral replication are either absent or below detection level, and the inflammation is diminished. However, if cirrhosis has already developed, it persists indefinitely.

The laboratory abnormalities consist of elevation of the ALT, ranging from normal to 200 IU/l in up to 90% of patients. Transaminases, serum bilirubin, albumin, and gammaglobulin values are mild to markedly elevated, and autoimmune antibodies such as antinuclear antibody, anti-smooth muscle antibody and antimitochondrial antibody may be present [15].

Sustained increases in the concentrations of the aminotransferases together with the presence of HBsAg for >6 months is regarded as indicative of chronic hepatitis.

Up to 20% of the chronic persistent hepatitis cases progress to cirrhosis. This is a serious liver disease associated with chronic and often widespread destruction of liver substance occurring over a period of several years.

In cirrhosis, liver cells die and are progressively replaced with fibrotic tissue leading to nodule formation. The internal structure of the liver is deranged leading to the obstruction of blood flow and decrease in liver function. This damage is caused by recurrent immune responses stimulated by the presence of the virus. Because liver inflammation can be totally symptomless, progression of inflammation to cirrhosis can occur without the knowledge of the patient.

Therefore, most carriers are contagious but some are not. This is determined by the presence of HBV DNA.

Globally, HBV causes 60–80% of the world's primary liver cancers [38].

It is estimated that, in men, the lifetime risk of death from chronic disease which leads to cirrhosis and/or HCC is between 40 and 50%. In women the risk is about 15%, placing chronic hepatitis B infections among the 10 leading causes of death in men.

HBV and HCC

A number of HBV patients with chronic hepatitis will develop HCC [15,31]. Persons at increased risk of developing HCC include adult male and chronic hepatitis B patients with cirrhosis who contracted hepatitis B in early childhood [23]. Only about 5% of patients with cirrhosis develop HCC. On the other hand, between 60 and 90% of HCC patients have underlying cirrhosis [15,30,31].

The incidence of HCC varies with geography, race, age, and sex. HCC is responsible for 90% of the primary malignant tumours of the liver observed in adults. Worldwide, it is the seventh most frequent cancer in males and ninth most common in females.

Liver cancer is the cause of more than 500,000 deaths annually throughout the world, with a male:female ratio of 4:1. The frequency of HCC follows the same general geographic distribution pattern as that of persistent HBV infection. The age distribution of patients with clinically recognized tumours suggests that these tumours appear after a mean duration of about 35 years of HBV infection [15,31].

Patients who develop HCC as a result of malignant transformation of HC have a mean 5-year survival rate of 25–60% [15]. This variation depends on the size of the tumour, its resectability, and the presence or absence of α-fetoprotein (AFP). Non-resectable tumours have a mean survival rate of 5 months for AFP-positive tumours and of 10.5 months for AFP-negative tumours [15].

When serum AFP followed serially in HBsAg carriers rises significantly above the patient's own baseline ($>100\ \mu$g/ml), HCC can often be detected by liver scanning or ultrasound procedures at a stage when the tumour can be cured by surgical resection [31]. This suggests that HBsAg carriers should have regular serial serum AFP determinations and ultrasound examinations (at 6 months intervals for those above 40 years). Both these tests are recommended to be repeated regularly for all HBsAg carriers with cirrhosis [31].

HBV causes 60–80% of the world's primary liver cancer, and primary liver cancer is one of the three most common causes of cancer deaths in males in East and South-east Asia, the Pacific Basin, and sub-Saharan Africa [31].

Primary liver cancer is the eighth most common cancer in the world [31]. Up to 80% of liver cancers are due to HBV. When HCC presents clinically, the disease is fatal. The median survival frequency of HCC patients is less than 3 months. However, if the cancer is detected early, there is an 85% chance of a cure. Treatment involves surgery, hepatic irradiation, and anticancer drugs.

Progression to fulminant hepatitis B

Fulminant hepatitis B is a rare condition that develops in about 1% of cases. It is caused by massive necrosis of liver substance and is usually fatal [15,23].

Survival in adults is uncommon, prognosis for children is rather better. Remarkably, the few survivors usually recover completely without permanent liver damage and no chronic infection [15,31].

Patients infected with precore mutants often manifest severe chronic hepatitis, early progression with cirrhosis, and a variable response to interferon therapy. It may have an association with fulminant hepatic failure [52].

Genetic heterogeneity of HBV, coinfection or superinfection with other viral hepatitis agents, or host immunological factors, may be associated with the development of fulminant hepatitis B [15,31].

A rapid fall in ALT and AST in patients with fulminant hepatic failure may be erroneously interpreted as a resolving hepatic infection, when in fact HC are being lost and the outcome is fatal [11].

Extrahepatic manifestations of hepatitis B

Extrahepatic manifestations of hepatitis B are seen in 10–20% of patients as

- *Transient* serum sickness-like syndrome [15,23,31] with fever (<39°C), skin rash (erythematous, macular, macopapular, urticarial, nodular, or petechial lesions), polyarthritis (acute articular symmetrical inflammation, painful, fusiform swelling of joints of hand and knee, morning stiffness). Symptoms usually precede the onset of jaundice by a few days to four weeks and subside after onset of jaundice and may persist throughout the course of the disease. No recurrent or chronic arthritis occurs after recovery. Immune complexes (e.g. surface antigen–antibody) are important in the pathogenesis of other disease syndromes characterized by severe damage of blood vessels [31]:

- *Acute necrotizing vasculitis* (*polyarteritis nodosa*) [15,31] with high fever, anemia, leucocytosis, arthralgia, arthritis, renal disease, hypertension, heart disease, gastrointestinal disease, skin manifestations, neurologic disorders. Highly variable disease with mortality rate of 40% within 3 years unless treated. The diagnosis is established by angiography.

- *Membranous glomerulonephritis* [15,31] is present in both adults and children. Remission of nephropathy occurs in 85–90% of cases over a period of 9 years and is associated with clearance of HBeAg from serum.

- *Papular acrodermatitis of childhood* (Gianotti–Crosti syndrome) [15]: a distinctive disease of childhood. Skin lesions, lentil-sized, flat, erythematous, and papular eruptions localized to the face and extremities, last 15–20 days. The disease is accompanied by generalized lymphadenopathy, hepatomegaly, and acute anicteric hepatitis B of ayw subtype.

Immune complexes have been found in the sera of all patients with fulminant hepatitis, but are seen only infrequently in nonfulminant infections. Perhaps complexes are critical factors only if they are of a particular size or of a certain antigen-to-antibody ratio [52]. Why only a small proportion of patients with circulating complexes develop vasculitis or polyarteritis is still not clear.

Coinfection or superinfection with HDV

Hepatitis delta virus (HDV) is a defective virus that is only infectious in the presence of active HBV infection. HDV infection occurs as either coinfection with HBV or superinfection of an HBV carrier. Coinfection usually resolves. Superinfection, however, causes frequently chronic HDV infection and chronic active hepatitis. Both types of infections may cause fulminant hepatitis [3,31].

Routes of transmission are similar to those of HBV [3].

Preventing acute and chronic HBV infection of susceptible persons by vaccination will also prevent HDV infection [3,23].

Lamivudine, an inhibitor of HBV DNA replication, is not beneficial for the treatment of chronic hepatitis D [22].

Patterns of viral infection

Several patterns of infection define the spectrum of responses to HBV [31].
Self-limited HBsAg-positive primary HBV infection (see Fig. 1).
Self-limited primary infection without detectable serum HBsAg (see Fig. 2).
HBsAg-positive persistent HBV infection (see Fig. 3).

Diagnosis

Large-scale screening for HBV infection

Diagnosis of hepatitis is made by biochemical assessment of liver function. Initial laboratory evaluation should include: total and direct bilirubin, ALT, AST, alkaline phosphatase, prothrombin time, total protein, albumin, globulin, complete blood count, and coagulation studies [15,31].

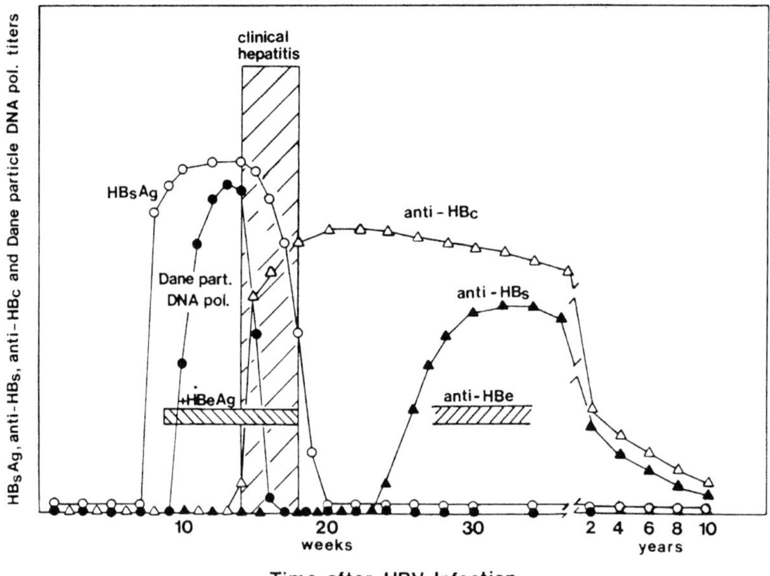

Fig. 1. Primary schematic representation of viral markers in the blood through a typical course of self-limited HBsAg-positive HBV infection [31]. This is the most common pattern of primary infection in adults. (From Ref. [31] with permission.)

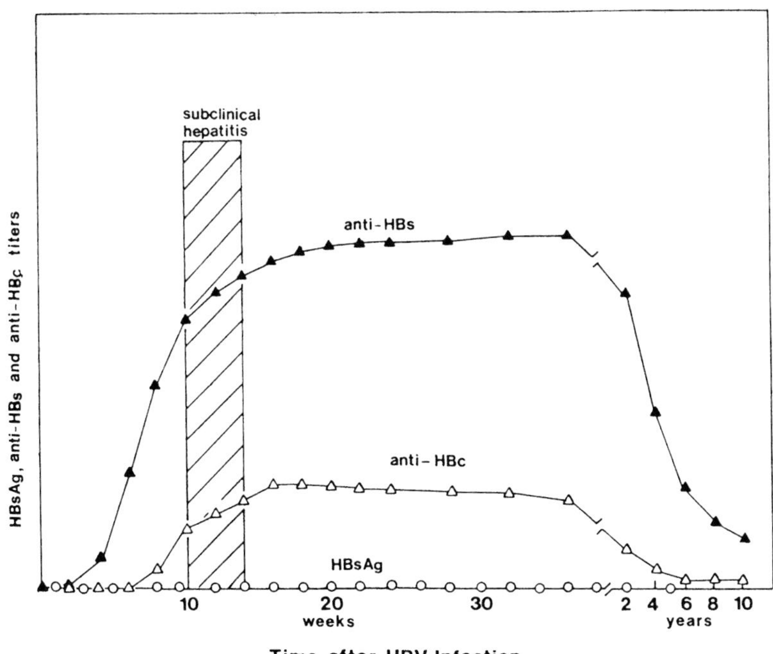

Fig. 2. Schematic representation of the serologic response through a typical course of HBsAg-negative primary HBV infection [31]. A significant number of patients with acute self-limited primary HBV infection never have detectable HBsAg in the blood. (From Ref. [31] with permission.)

Diagnosis is confirmed by demonstration in sera of specific antigens and/or antibodies. Three clinical useful antigen–antibody systems have been identified for hepatitis B:

HBsAg and antibody to HBsAg (anti-HBs).

Antibody (anti-HBc IgM and anti-HBc IgG) to hepatitis B core antigen (HBcAg).

Hepatitis B e antigen (HBeAg) and antibody to HBeAg (anti-HBe).

Tests specific for complete virus particles or DNA and DNA polymerase-containing virions, and for HDAg and HDV RNA in liver and serum are available only in research laboratories [31].

HBsAg can be detected in the serum from several weeks before onset of symptoms to months after onset. HBsAg is present in serum during acute infections and persists in chronic infections. The presence of HBsAg indicates that the person is potentially infectious [15,23,31].

Very early in the incubation period, pre-S1 and pre-S2 antigens are present. They are never detected in the absence of HBsAg. Hepatitis B virions, HBV DNA, DNA

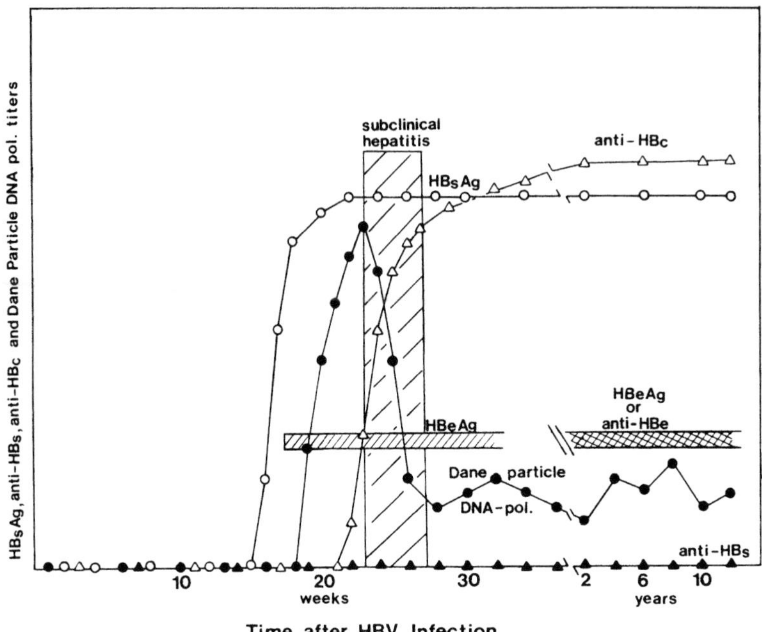

Fig. 3. Schematic representation of viral markers in the blood through a typical course of HBV infection that becomes persistent [31]. Patients who remain HBsAg-positive for 20 weeks or longer after primary infection are very likely to remain positive indefinitely and be designated chronic HBsAg carriers [31]. (From Ref. [31] with permission.)

polymerase, and HBeAg are then also detected. The presence of HBeAg is associated with relatively high infectivity and severity of disease [15,31].

Anti-HBc is the first antibody to appear. Demonstration of anti-HBc in serum indicates HBV infection, current or past. IgM anti-HBc is present in high titre during acute infection and usually disappears within 6 months, although it can persist in some cases of chronic hepatitis. This test may therefore reliably diagnose acute HBV infection. IgG anti-HBc generally remains detectable for a lifetime [15,23,31].

Anti-HBe appears after anti-HBc and its presence correlates to a decreased infectivity. Anti-HBe replaces HBeAg in the resolution of the disease [15,23,31].

Anti-HBs replaces HBsAg as the acute HBV infection is resolving. Anti-HBs generally persists for a lifetime in over 80% of patients and indicates immunity [15,23,31].

Acute hepatitis patients who maintain a constant serum HBsAg concentration, or whose serum HBeAg persists 8–10 weeks after symptoms have resolved, are likely to become carriers and at risk of developing *chronic liver disease* [15].

A complication in the diagnosis of hepatitis B is the rare identification of cases in which viral mutations change the antigens so they are not detectable.

Small-scale screening for HBV infection

Immunofluorescence studies, *in situ* hybridisation, immunohistochemistry, and thin-section EM are used to examine pathological specimens for the presence of HBV-associated antigens or particles, providing information about the relationship between HBV DNA replication and HBV gene expression [15].

Within the hepatocyte, HBsAg localizes in the cytoplasm, and HBcAg is seen in the nucleus and/or the cytoplasm. Detection of complete virions in the liver is uncommon [15].

DNA hybridisation techniques and RT-PCR assays have shown that almost all HBsAg/HBeAg-positive patients have detectable HBV DNA in their serum, whereas only about 65% of the HBsAg/anti-HBe-reactive patients are positive. All patients who recover from acute hepatitis B are negative for HBV DNA. On the other hand, some patients infected chronically who have lost their HBsAg remain HBV DNA-positive [15,31].

HBV serological markers in hepatitis patients

The three standard blood tests for hepatitis B can determine if a person is currently infected with HBV, has recovered, is a chronic carrier, or is susceptible to HBV infection [15,23,31].

Assay results			Interpretation
HBsAg	Anti-HBs	Anti-HBc	
+	−	−	Early acute HBV infection
+	+/−	+	Acute or chronic HBV infection. Differentiate with IgM-anti-HBc. Determine level of infectivity with HBeAg or HBV DNA
−	+	+	Indicates previous HBV infection and immunity to hepatitis B
−	−	+	Possibilities include: past HBV infection; low-level HBV carrier; time span between disappearance of HBsAg and appearance of anti-HBs; or false-positive or nonspecific reaction. Investigate with IgM anti-HBc, and/or challenge with HBsAg vaccine. When present, anti-HBe helps validate the anti-HBc reactivity
−	−	−	Another infectious agent, toxic injury to the liver, disorder of immunity, hereditary disease of the liver, or disease of the biliary tract
−	+	−	Vaccine-type response

From Ref. [15] with permission (http://lww.com).

Interpretation of HBV serologic markers in patients with hepatitis [15].

Host immune response

There is little evidence that humoral immunity plays a major role in the clearance of established infection. Cell-mediated immune responses, particularly those involving cytotoxic T-lymphocytes (CTLs), seem to be very important [30,31].

CD8-positive, class I major histocompatibility complex (MHC)-restricted CTLs directed against HBV nucleocapsid proteins are present in the peripheral blood of patients with acute, resolving hepatitis B. Such cells are barely detectable in the blood of patients with chronic HBV infection, suggesting that the inability to generate such cells may predispose to persistent infection, although their absence from the blood in chronic infection may be due to their sequestration elsewhere.

CTLs against envelope glycoprotein determinants, that are often CD4-positive, class II MHC-restricted, have also been detected.

Primary infection leads to an IgM and IgG response to HBcAg shortly after the appearance of HBsAg in serum, at onset of hepatitis. Anti-HBs and anti-HBe appear in serum only several weeks later, when HBsAg and HBeAg are no longer detected, although in many HBsAg-positive patients, HBsAg-anti-HBs complexes can be found in serum [23,30,31].

Serological markers of HBV infection

During HBV infection, the serological markers vary depending on whether the infection is acute or chronic [11,23,31].

Antigens	Antibodies
HBsAg Hepatitis B surface antigen is the earliest indicator of acute infection and is also indicative of chronic infection if its presence persists for more than 6 months. It is useful for the diagnosis of HBV infection and for screening of blood. Its specific antibody is anti-HBs	*Anti-HBs* This is the specific antibody to hepatitis B surface antigen. Its appearance 1–4 months after onset of symptoms indicates clinical recovery and sub-sequent immunity to HBV. Anti-HBs can neutralize HBV and provide protection against HBV infection
HBcAg Hepatitis B core antigen is derived from the protein envelope that encloses the viral DNA, and it is not detectable in the bloodstream. When HBcAg peptides are expressed on the surface of hepatocytes, they induce an immune response that is crucial for killing infected cells. The HBcAg is a marker of the infectious viral material and it is the most accurate index of viral replication. Its specific antibody is anti-HBc	*Anti-HBc* This is the specific antibody to hepatitis B core antigen. Antibodies to HBc are of class IgM and IgG. They do not neutralize the virus. The presence of IgM identifies an early acute infection. In the absence of HBsAg and anti-HBs, it shows recent infection. IgG with no IgM may be present in chronic and resolved infections. Anti-HBc testing identifies all previously infected persons, including HBV carriers, but does not differentiate carriers and non-carriers

(*continued*)

(continuation)

Antigens	Antibodies
HBeAg Hepatitis B e antigen appearing during weeks 3–6 indicates an acute active infection at its most infectious period, and means that the patient is infectious. Persistence of this virological marker beyond 10 weeks shows progression to chronic infection and infectiousness. Continuous presence of anti-HBe indicates chronic or chronic active liver disease. HBeAg is not incorporated into virions, but is instead secreted into the serum. Mutant strains of HBV exist that replicate without producing HBeAg. HBeAg's function is uncertain. Its specific antibody is anti-HBe	*Anti-HBe* This is the specific antibody to hepatitis B e antigen. During the acute stage of infection the seroconversion from e antigen to e antibody is prognostic for resolution of infection. Its presence in the patient's blood along with anti-HBc and in the absence of HBsAg and anti-HBs indicates low contagiousness and convalescence [31]
HBXAg Hepatitis BX antigen is detected in HBeAg positive blood in patients with both acute and chronic hepatitis. HBX Ag is a transcriptional activator. It does not bind to DNA. Its specific antibody is anti-HBX	*Anti-HBX* This is the specific antibody to hepatitis BX antigen. It appears when other virological markers are becoming undetectable
HBV DNA HBV DNA is detectable by PCR as soon as 1 week after initial infection, but the test is generally only performed for research purposes or to detect mutants that escape detection by current methods	
HBV DNA polymerase Tests for the presence of HBV DNA polymerase, detectable within 1 week of initial infection, are only performed for research purposes	

Serological test findings at different stages of HBV infection and in convalescence

Stage of infection	HBsAg	Anti-HBs	Anti-HBc		HBeAg	Anti-HBe
			IgG	IgM		
Late incubation period	+	−	−	−	+/−	−
Acute hepatitis B or persistent carrier state	+	−	+	+	+	−
HBsAg-negative acute hepatitis B infection	−	−	−	+	−	−
Recovery with loss of detectable anti-HBs	−	−	+	−	−	−
Healthy HBsAg carrier	+	−	+++	+/−	−	+

(*continued*)

56

(continuation)

Stage of infection	HBsAg	Anti-HBs	Anti-HBc IgG	IgM	HBeAg	Anti-HBe
Chronic hepatitis B, persistent carrier state	+	−	+++	+/−	+	−
HBV infection in recent past, convalescence	−	++	++	+/−	−	+
HBV infection in distant past, recovery	−	+/−	+/−	−	−	−
Recent HBV vaccination, repeated exposure to antigen without infection, or recovery from infection with loss of detectable anti-HBc	−	++	−	−	−	−

From Ref. [31] with permission.

HBV serological markers in different stages of infection and convalescence [23, 31,52].

Serological and clinical patterns of acute or chronic HBV infections

Acute HBV infection

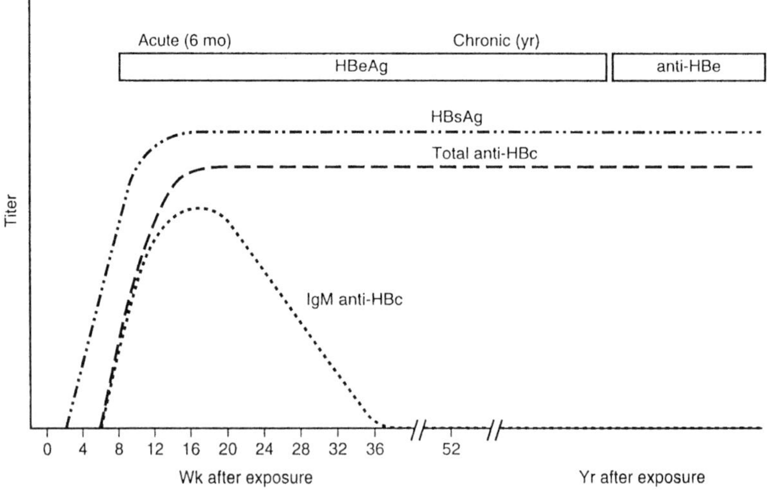

From Ref. [15] with permission.

Titre of HBsAg, antibody to hepatitis B core antigen (anti-HBc), IgM anti-HBc, and antibody to HBsAg (anti-HBs) in patients with acute hepatitis B with recovery [23].

Chronic HBV infection

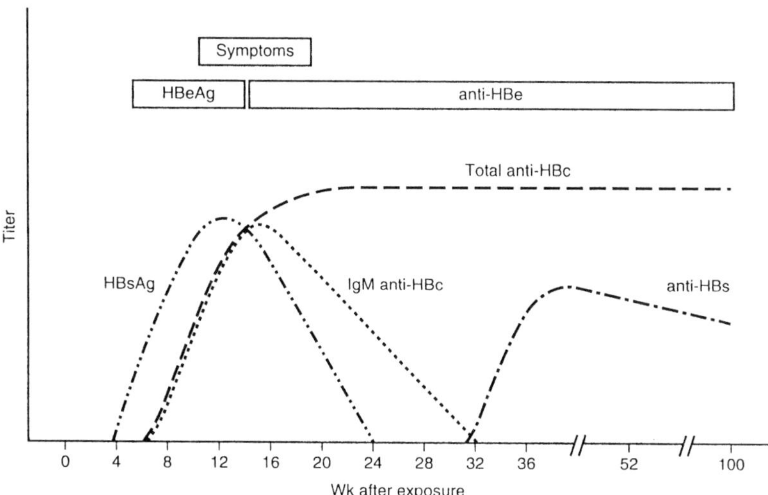

From Ref. [15] with permission.

Titre of HBsAg, antibody to hepatitis B core antigen (anti-HBc), and IgM anti-HBc during progression to chronic HBV infection [23].

Interpretation of hepatitis B markers

Marker	Infection		
	Acute	Chronic	Past
HBsAg	+	+	−
HBeAg	+ Early, then −	+/−	−
anti-HBs	−	−	+
anti-HBc IgM	+	−	−
anti-HBc IgG	+	+	+
anti-HBe	− Early, then +	+/−	+
HBV DNA	+ Early, then −	+/−	−
ALT	Increased (marked)	Increased (mild–moderate)	Normal

From Ref. [11], with permission.

Discordant or unusual hepatitis B serological profiles requiring further evaluation

Repeat testing of the same sample or possibly of an additional sample is advisable when tests yield discordant or unusual results [15].

HBsAg-positive/anti-HBc-negative	An HBsAg-positive response is accompanied by an anti-HBc-negative reaction only during the incubation period of acute hepatitis B, before the onset of clinical symptoms and liver abnormalities
HBsAg positive/anti-HBs positive/ anti-HBc positive	Uncommon, may occur during resolution of acute hepatitis B, in chronic carriers who have serious liver disease, or in carriers exposed to heterologous subtypes of HBsAg
Anti-HBc-positive only	Past infection not resolved completely
HBeAg-positive/HBsAg-negative	Unusual
HBeAg-positive/anti-HBe positive	Unusual
Anti-HBs positive only in a non-immunized person	It may be a result of passive transfer of anti-HBs after transfusion of blood from a vaccinated donor, in patients receiving clotting factors, after IG administration, or in newborn children of mothers with recent or past HBV infection. Passively acquired antibodies disappear gradually over 3 to 6 months, whereas actively produced antibodies are stable over many years. Apparently quite common when person has forgotten his/her immunization status!

Mutant proteins from mutant HBV strains may escape diagnostic detection. The presence of different serological markers should therefore be tested for a correct diagnosis. Diagnostic kits should contain antibodies against a variety of mutant proteins, if perfection is the goal.

Prevalence

HBV occurs worldwide [23,31].

The highest rates of HBsAg carrier rates are found in developing countries with primitive or limited medical facilities [23]. In areas of Africa and Asia, widespread infection may occur in infancy and childhood. The overall HBsAg carrier rates may be 10–15%.

The prevalence is lowest in countries with the highest standards of living, such as Great Britain, Canada, United States, Scandinavia, and some other European Nations. In North America infection is most common in young adults. In the USA and Canada, serological evidence of previous infection varies depending on age and socio-economic class. Overall, 5% of the adult USA population has anti-HBc, and 0.5% are HBsAg-positive.

In developed countries, exposure to HBV may be common in certain high-risk groups (see section on "Risk groups").

Adults infected with HBV usually acquire acute hepatitis B and recover, but 5–10% develop the chronic carrier state. Infected children rarely develop acute disease, but 25–90% become chronic carriers. About 25% of carriers will die from cirrhosis or primary liver cancer as adults [3,23].

In the past, recipients of blood and blood products were at high risk (for HBV infection). Over the last 25 years, testing blood donations for HBsAg has become a universal requirement. Testing procedures have made major progress in sensitivity in the last 15–20 years. However, 19% of countries reported that they were not testing all blood donations for HBsAg (WHO Global Database on Blood Safety, unpublished data). In the many countries where pretransfusion screening of blood donations for HBsAg is carried out systematically, the residual risk of HBV transmission is minimal. Moreover, plasma-derived medicinal products (including antihaemophilic factors) undergo additional viral inactivation and removal procedures resulting in greatly reduced or no transmission of HBV by these products.

However, the risk is still present in many developing countries. Contaminated and inadequately sterilized syringes and needles have resulted in outbreaks of hepatitis B among patients in clinics and physicians' offices. Occasionally, outbreaks have been traced to tattoo parlors and acupuncturists. Rarely, transmission to patients from HBsAg-positive health-care workers has been documented [41].

Reductions in the age-related prevalence of HBsAg in countries where hepatitis B is highly endemic and universal immunization of infants has been adopted suggest that it may be possible to eradicate HBV from humans.

Hepatitis B vaccines have been available since 1982 and have been used in hundreds of millions of individuals with an outstanding record of safety and impact on the disease. Carriage of HBV has already been reduced from high prevalence to low prevalence in immunized cohorts of children in many countries.

Prevalence of hepatitis B in various areas

Area	% of population positive for		Infection	
	HBsAg	Anti-HBs	Neonatal	Childhood
Northern, Western, and Central Europe, North America, Australia	0.2–0.5	4–6	Rare	Infrequent
Eastern Europe, the Mediterranean, Russia and the Russian Federation, South-west Asia, Central and South America	2–7	20–55	Frequent	Frequent
Parts of China, South-east Asia, tropical Africa	8–20	70–95	Very frequent	Very frequent

From Ref. [52], with permission.

World distribution map

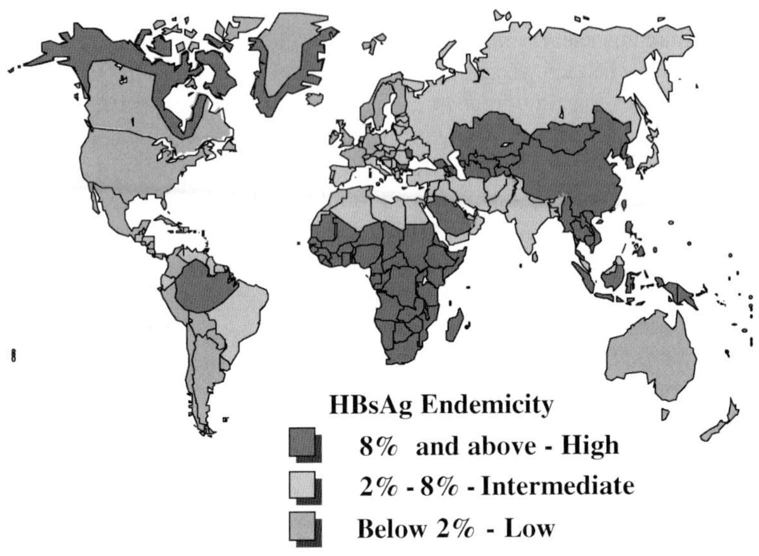

From Ref. [51].

Geographical distribution of chronic HBV infection.
(Note: *the boundaries and names shown and the designations used on this map do not imply the expression of any opinion whatsoever on the part of the World Health Organization concerning the legal status of any country, territory, city or area or of its authorities, or concerning the delimitation of its frontiers or boundaries. Dotted lines on maps represent approximate borderlines for which there may not yet be full agreement.*)

Pathogenesis

HBV infection contracted early in life may lead to chronic hepatitis, then to cirrhosis, and finally to HCC, usually after a period of 30–50 years. Once infected with HBV, males are more likely to remain persistently infected than women, who are more likely to be infected transiently and to develop anti-HBs.

It is possible that in man HBV is not carcinogenic by a direct viral mechanism. Instead, the role of HBV may be to cause chronic liver cell damage with associated host responses of inflammation and liver regeneration that continues for many years. This pathological process, especially when leading to cirrhosis, may be carcinogenic without involving a direct oncogenic action of the virus. No viral oncogene, insertional mutagenesis, or viral activation of oncogenic cellular genes has been demonstrated [30].

The expression of HBV proteins and the release of virions precedes biochemical evidence of liver disease. Moreover, large quantities of surface antigen can persist in liver cells of many apparently healthy persons who are carriers. HBV is therefore not directly cytopathic [6].

Three mechanisms seem to be involved in liver cell injury during HBV infections.

The first is an HLA class I restricted cytotoxic T-lymphocyte (CTL) response directed at HBcAg/HBeAg on HBV-infected HC [23,30,31]. A second possible mechanism is a direct cytopathic effect of HBcAg expression in infected HC [23,30,31]. A third possible mechanism is high-level expression and inefficient secretion of HbsAg [31].

Hypothetical course of immunopathogenesis of HBV

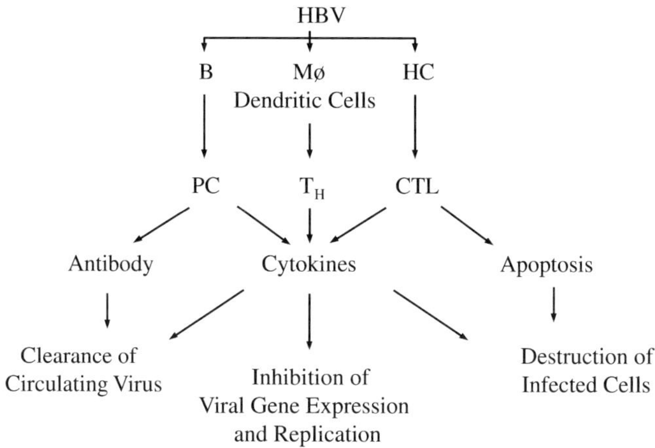

From Ref. [6], with permission (http://lww.com).

Eradication of HBV infection depends on the coordinate and efficient development of humoral and cell-mediated immune responses against HBV proteins. Antibodies secreted by plasma cells (PC) derived from antigen-specific B-cells (which usually recognize viral antigens in their native conformation) are mostly responsible for the neutralization of free circulating viral particles. Cytotoxic T-lymphocytes (CTL) that recognize endogenous viral antigens in the form of short peptides associated with human leukocyte antigen (HLA) class I molecules on the surface of the infected HC are the main effectors for the elimination of intracellular virus. They can do this by at least two different mechanisms: direct attachment to the cell membrane, causing the infected cell to undergo apoptosis; and the release of soluble cytokines that can downregulate viral gene expression, leading to the elimination of intracellular virus without destruction of the infected cell.

Both humoral and cytotoxic functions are more or less stringently regulated by the helper effect of the CD4 + T-cells (TH) that recognize exogenous viral antigens, released or secreted by liver cells, in the form of short peptides that associate with HLA class II molecules in the endosomal compartment of professional antigen-presenting cells such as B-cells, macrophages (Mø), and dendritic cells.

Transmission

Currently, there are four recognized modes of transmission [15,39].

1. From mother to child at birth (perinatal).

2. By contact with an infected person (horizontal).

3. By sexual contact.

4. By parenteral (blood-to-blood) exposure to blood or other infected fluids.

There is considerable variation between areas, countries and continents as to the age at which most transmission takes place.

There can be carriers with or without hepatitis [31].

There is no convincing evidence that airborne infections occur and faeces are not a source of infection, since the virus is inactivated by enzymes of the intestinal mucosa or derived from the bacterial flora. HBV is not transmitted by contaminated food or water, insects or other vectors [15,31].

HBsAg has been found in all body secretions and excretions. However, only blood, vaginal and menstrual fluids, and semen have been shown to be infectious [15,23,30,31].

Transmission occurs by percutaneous and permucosal exposure to infective body fluids. Percutaneous exposures that have resulted in HBV transmission include transfusion of unscreened blood or blood products, sharing unsterilized injection needles for i.v. drug use, haemodialysis, acupuncture, tattooing and injuries from contaminated sharp instruments sustained by hospital personnel [15,23,31].

Sexual and perinatal HBV transmission usually result from mucous membrane exposures to infectious blood and body fluids. Perinatal transmission is common in hyperendemic areas of South-east Asia and the far East, especially when HBsAg carrier mothers are also HBeAg-positive [15,23,31].

Infection may also be transmitted between household contacts and between sexual partners, either homosexual or heterosexual, and in toddler-aged children in groups with high HBsAg carrier rates [15,23].

IGs, heat-treated plasma protein fraction, albumin and fibrinolysin are considered safe when manufactured appropriately.

HBV is stable on environmental surfaces for at least 7 days, and indirect inoculation of HBV can occur via inanimate objects like toothbrushes, baby bottles, toys, razors, eating utensils, hospital equipment and other objects, by contact with mucous membranes or open skin breaks [31].

Infectious HBV can be present in blood without detectable HBsAg, so that the failure to detect antigen does not exclude the presence of infectious virus [31].

The source of infection cannot be identified in about 35% of cases.

The natural reservoir for HBV is man [38]. Closely related hepadnaviruses have been found in woodchucks and ducks, but they are not infectious for humans [10].

The reuse of the same, unsterilized needle and syringe for vaccination of many different children accounts for many unnecessary HBV infections [31,41].

People depending on repeated transfusion should be vaccinated against HBV.

HBV is about 100 times more infectious than HIV [38].

The role of non-human primates in the transmission of HBV

The only non-human primates that can develop productive HBV infection are the great apes (e.g. chimpanzees, orang-utans and gorillas). Chimpanzees have served as the model for the study of HBV infection for over 20 years [15].

Although chimpanzees may be infected in nature, there is no evidence that they are important sources for human infections because transmission from infected individuals requires specific patterns of intimate contact [31].

Gibbons are susceptible to HBV and have been infected successfully experimentally and also naturally by contact in captivity [35].

Experimental infections of woolly monkeys, tamarins, and other primate species have generally been unsatisfactory.

Risk groups

This is a list of groups of people who are at risk of contracting HBV [15,31].

- Infants born to infected mothers.
- Young children in day-care or residential settings with other children in endemic areas.
- Sexual/household contacts of infected persons.
- Health-care workers.
- Patients and employees in haemodialysis centres [4,41].
- Injection drug users sharing unsterile needles [41].
- People sharing unsterile medical or dental equipment.
- People providing or receiving acupuncture and/or tattooing with unsterile medical devices.
- Persons living in regions or travelling to regions with endemic hepatitis B [50].
- Sexually active heterosexuals.
- Men who have sex with men.

Frequent and routine exposure to blood or serum is the common denominator of health-care occupational exposure. Surgeons, dentists, oral surgeons, pathologists, operating room and emergency room staff, and clinical laboratory workers who handle blood are at the highest risk [31].

HBV infection is the major residual post-transfusion risk in developed countries because of the long window period, HBV mutants, the low viraemia (difficulties for PCR on pooled samples) and the very high infectivity.

Over one-third of patients with acute hepatitis B do not have readily identifiable risk factors [3].

Efforts to vaccinate persons in the major risk groups have had limited success because of the difficulties in identifying vaccination candidates belonging to high risk groups. Moreover, regulations have to be developed to ensure the implementation of vaccination programmes [3,37].

High risk persons should be post-tested within 1–2 months of receipt of the third dose of HBV vaccine, to identify good responders to vaccination. This policy is cost-saving since adequate responders do not need to be retested or given HBIG whenever they later are exposed to HBV. They also do not need to be offered booster doses of vaccine periodically.

Surveillance and control

Hepatitis B disease surveillance procedures should include:

- Monitoring disease incidence.
- Determination of sources of infection and modes of transmission by epidemiological investigation.
- Detection of outbreaks.
- Spread containment.
- Identification of contacts of case-patients for post-exposure prophylaxis.

Hepatitis B disease control measures should include:

- Immunization, the most effective and cost-saving means of prevention.
- Education of high risk groups and health-care personnel to reduce the risk of contracting the virus and to reduce the chances for transmission to others, as well as to promote acceptance of vaccination schemes.
- Screening of blood and blood products to reduce the chance that the blood supply system may contain pathogens like HBV.

Surveillance systems for hepatitis B vary in their methods and completeness. In many countries notification of HBV infections is mandatory. However, case definitions vary, laboratory confirmation is not always used, reporting systems differ, and distinctions are not always made between the types of viral hepatitis. In addition, under-reporting of HBV infection is commonplace. Surveillance systems need therefore to be strengthened and standardized.

For better standardization of surveillance systems, countries should follow the case definition of viral hepatitis B recommended by WHO:

- *A clinical case of acute viral hepatitis* is an acute illness that includes the discrete onset of symptoms and jaundice or elevated serum aminotransferase levels (>2.5 times the upper limit of normal).

- *A confirmed case of hepatitis B* is a suspected case that is laboratory confirmed: HBsAg-positive or anti-HBc-IgM-positive, and anti-HAV-IgM-negative.

The serological quality of the test used is crucial for firm diagnosis of infection. Countries without ready access to these tests may choose methods to detect HBsAg such as reverse passive haemagglutination (RPHA) or latex bead technology that are inexpensive. While not quite as sensitive as radioimmunoassay (RIA) antigen tests or enzyme-linked immunoabsorbent assay (ELISA), these tests are far better than not testing at all [15].

Regardless of the availability of serological tests, all countries are advised to report all cases of jaundice and suspected viral hepatitis. Countries with laboratory facilities can differentiate further between hepatitis A, B, C, and other types of hepatitis. Surveillance reports should be submitted on a regular basis.

Endemicity

There are no seasonal preferences for primary HBV infections [15,30].

Hepatitis B is highly endemic in all of Africa, some parts of South America, Alaska, northern Canada and parts of Greenland, Eastern Europe, the eastern Mediterranean area, South-east Asia, China, and the Pacific Islands, except Australia, New Zealand and Japan. In most of these areas, 5–15% of the population are chronically infected carriers of HBV, and in some areas may also carry HDV, which may lead to severe liver damage [23,42].

Even in low endemicity countries such as the USA, mortality from HBV was 5 times that from *Haemophilus influenzae* b (Hib) and 10 times that from measles before routine vaccination of children was introduced.

Incidence/epidemiology

The HBV is a ubiquitous virus with a global distribution [15,38]. Hepatitis B is one of the world's most common and serious infectious diseases. It is estimated that more than one third of the world's population has been infected with the HBV. About 5% of the population are chronic carriers of HBV, and nearly 25% of all carriers develop serious liver diseases such as chronic hepatitis, cirrhosis, and primary HCC. HBV infection causes more than one million deaths every year [15,23,30,39].

66

The HBsAg carrier rate varies from 0.1 to 20% in different populations around the world. The incidence of the HBsAg carrier state in populations is related most importantly to the incidence and age of primary infection [23].

In low-risk areas of the world, the highest incidence of the disease is seen in teenagers and young adults. Despite the low incidence of disease seen in the general population, certain groups who are sexually promiscuous or who have frequent contact with blood or blood products have a high rate of HBV infection. Nevertheless, the availability of an effective vaccine, optimized blood donor screening, and better sterilization procedures for blood derivatives have lowered substantially the infection risk [15].

In endemic areas of Africa and Asia, different epidemiological patterns are seen. In these regions, most infections occur in infants and children as a result of maternal–neonatal transmission or close childhood contact, although percutaneous exposure with contaminated needles or following unsafe injections is always a possibility in these countries [15,23].

The chronic liver disease and HCC associated with HBV infections are among the most important human health problems in high-prevalence regions.

Trends

There is no seasonal trend similar to that observed in hepatitis A infections [15].

Epidemics are unusual unless associated with contaminated blood or blood products, or the use of nonsterile injection equipment.

Evaluations of infant vaccination programmes need to compare vaccination coverage data with population-based serological analyses, since most HBV infection in young children are asymptomatic and are therefore not detected in surveillance studies of acute disease. A decline in the prevalence of chronic disease is on the other hand a major indicator of program success and infection reduction [23].

A reduction in the prevalence of chronic HBV infection after implementation of infant immunization programmes has been demonstrated in high endemicity areas like Alaska, Taiwan, Indonesia, Polynesia, and the Gambia [23].

The implementation of routine infant immunization will eventually achieve broad-population-based immunity to HBV infection and prevent HBV transmission among all age groups. However, it is only in the longer term that infant immunization in countries that have adopted the HBV vaccination programme will affect the incidence of hepatitis B and the severe consequences of chronic infections [37].

Costs

Hepatitis B is a significant health problem and vaccination saves both money and lives. Consideration of epidemiological and economic data shows that universal vaccination strategies are cost-effective even in countries with a low prevalence of hepatitis B. Hepatitis B prevention programmes incorporating universal immunization of newborns and/or adolescents have been highly successful in Spain and Italy, and their success offers an exemplary model for other countries [39].

Even in low HBV endemicity areas of the world it is more cost-saving for the society to follow prevention programmes against HBV infection for the younger age groups than to face an increase in chronic liver disease among adults [37].

The cost of vaccines has fallen dramatically since the early 1980s, to the point that paediatric-dose vaccine in quantities of several hundred thousand can be found for less than US$1 per dose in developing countries. However, even at US$0.5 per dose, a three-dose series costs more than the other six childhood vaccines recommended by the WHO Expanded Programme on Immunization (EPI) combined (BCG, three doses of DTP, four doses of OPV, and measles vaccines). Cost, therefore remains the primary obstacle to worldwide control of hepatitis B [23].

In the USA, the price of vaccination per dose is estimated at US$41 if given by a general practitioner, US$15 if administered through an existing childhood immunization programme, and US$17 if given through the school medical system [37].

Immune prophylaxis

In 1974, a special lot of high-titred human hepatitis B IG designated HBIG was introduced. HBIG is similar to conventional IG preparations except that it is prepared from plasma pre-selected for a high titre of anti-HBs (> 100,000 IU/ml of anti-HBs by RIA). The process used to prepare HBIG inactivates and eliminates HIV from the final product. There is no evidence that HIV can be transmitted by HBIG [3,15,23,31].

HBIG protects by passive immunization if given shortly before or soon after exposure to HBV. The protection is immediate, but it lasts only 3–6 months. HBIG is not recommended as pre-exposure prophylaxis because of high cost, limited availability, and short-term effectiveness. HBIG is generally not affordable in developing countries [15,23,31].

HBIG should be given to adults within 48 h of HBV exposure [23].

Maternal–neonatal transmission of HBV and the subsequent development of chronic hepatitis B in infected children has been reduced drastically, when HBIG was given to newborn babies of HBV carrier mothers in conjunction with the first dose of HB vaccine [15,23].

HBV vaccination and one dose of HBIG, administered within 24 h after birth, are 85–95% effective in preventing both HBV infection and the chronic carrier state. HB vaccine administered alone beginning within 24 h after birth, is 70–95% effective in preventing perinatal HBV infection [23].

Routine infant immunization programmes have shown that the currently available vaccines confer as much protection upon the infants as does a combination of vaccine and HBIG. Therefore, the additional expenses for the administration of HBIG can be avoided [23].

With the availability of a vaccine against hepatitis B and mandatory screening of blood donors for HBsAg and anti-HBc, there is little justification for the use of HBIG in pre-exposure prophylaxis, except for individuals failing to respond to vaccine, or in patients with disorders that preclude a response (e.g. agammaglobulinaemia) [15,23,31].

However, situations exist where post-exposure prophylaxis is essential or desirable. The effectiveness appears to diminish rapidly if administration is delayed for more than 3 days. Passive immunization is now generally combined with active immunization induced by vaccine, providing immediate protection and more durable immunity [23].

Safety of immune globulin

In 1972, routine screening of plasma donors for HBsAg was introduced, resulting in a sharp decline in the concentration of HBsAg inadvertently added to donor pools destined for IG production [15]. Since 1977, all tested lots of commercial IG contain anti-HBs at a titer of at least 1:100 by RIA [15]. Side effects associated with the administration of IG are rare [15].

Post-exposure prophylaxis

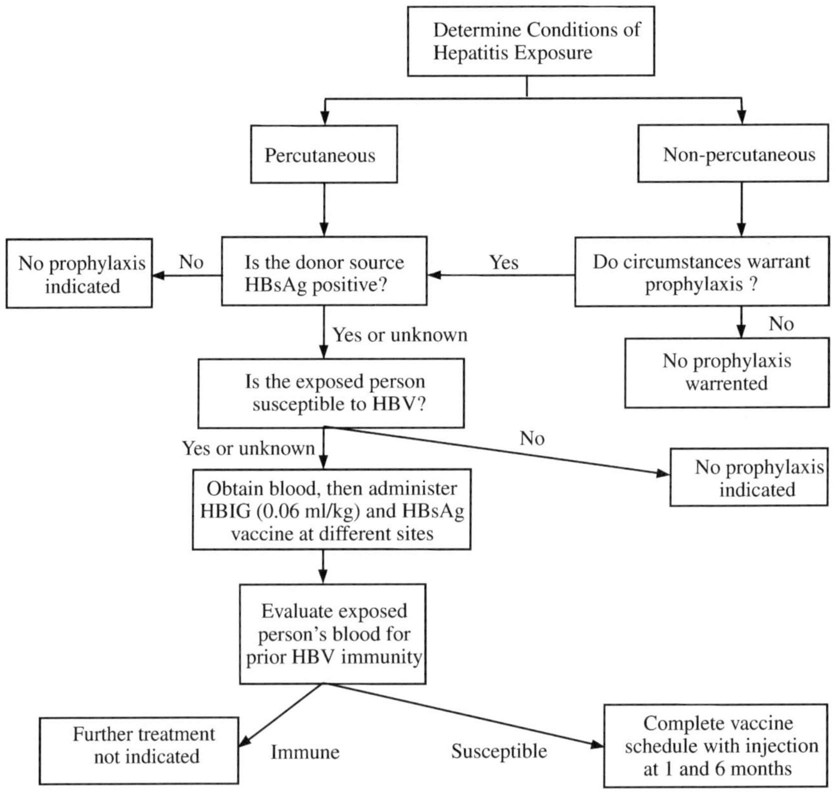

From Ref. [15], with permission (http://lww.com).

Algorithm for post-exposure prophylaxis of health-care personnel exposed to a potentially infectious source of HBV.

Abbreviation: HBIG, high-titred specific hepatitis B immune globulin [15].

Vaccines

Hepatitis B is a vaccine-preventable disease, but although global control of hepatitis B is achievable, it has not been attained yet [5,36,37]. In fact, a large pool of carriers and the burden of their disease remains, so that efforts must necessarily continue to treat the various stages of disease.

HB vaccine is the first and currently the only vaccine against a major human cancer. Vaccination is the most effective tool in preventing the transmission of HBV and HDV. Vaccines are composed of the surface antigen of HBV (HBsAg), and are produced by two different methods: plasma derived or recombinant DNA. When administered properly, hepatitis B vaccine induces protection in about 95% of recipients [5].

A safe and effective vaccine against HBV infection has been available for 20 years. HB vaccine is effective in preventing HBV infections when it is given either before exposure or shortly after exposure, At least 85–90% of HBV-associated deaths are vaccine-preventable.

Despite the availability of a vaccine, worldwide infection persists.

Systematic hepatitis B vaccination of newborns renders the screening of pregnant women for HBsAg-status before delivery superfluous [23].

WHO recommends that hepatitis B vaccine be included in routine immunization services in all countries. The primary objective of hepatitis B immunization is to prevent chronic HBV infections, which result in chronic liver disease later in life. By preventing chronic HBV infections, the major reservoir for transmission of new infections is also reduced.

Plasma-derived vaccines

These vaccines, derived from the plasma of HBsAg-positive donors, consist of highly purified, formalin-inactivated and/or heat-inactivated, alum-adsorbed, hepatitis B subvirion particles (22 nm) of HBsAg that are free of detectable nucleic acid, and, therefore, noninfectious [15,23].

The first plasma-derived hepatitis B vaccines manufactured in the USA and in France were licensed in 1981–1982 (Heptavax B®, Merck & Co., Hevac B®, Institut Pasteur). They contain 20 μg/ml HBsAg and the preservative thimerosal at a concentration of 1:20,000 [3,15,31].

Plasma-derived HB vaccines are no longer produced in North America or western Europe, but several hundred million doses are produced in the Republic of Korea, China, Vietnam, Myanmar, India, Indonesia, Iran and Mongolia [15,23,31].

More than 200 million doses of plasma-derived vaccines have been distributed globally, and the safety record is impressive. Local reactions are generally insignificant clinically and are limited to mild pain or discomfort at the injection site in up to 25% of the vaccine recipients [15].

Recombinant DNA yeast-derived or mammalian cell-derived vaccines

In the mid-1980s, an alternative, genetically engineered vaccine became available. The new technologies offer manufacturers a shorter production cycle (12 instead of 65 weeks), batch-to-batch consistency, and continuous supply of material, allowing the replacing of plasma-derived vaccines available on the market [15,31].

In recombinant DNA technology, the S gene (pre-S1, pre-S2, S) is cloned and isolated, inserted into an expression plasmid and introduced into yeast (*S. cerevisiae*) or mammalian (Chinese hamster ovary, CHO) cells. The desired protein(s) is(are) expressed and assembled into 22 nm antigenic particles [15,23,31].

As on natural HBsAg particles, the a epitope that elicits the most important immune response is exposed on the surface of artificial particles. Natural and artificial particles differ in the glycosylation of HBsAg [15,23].

The only mammalian cell-derived vaccine available is GenHevac B$^®$ (Pasteur Mérieux Connaught, 1993). GenHevac B$^®$ contains both preS2 and S proteins [15].

The two major yeast-derived hepatitis B vaccines that are licensed in most countries are Engerix-B$^®$ (SmithKline Beecham, 1992) and Recombivax HB$^®$ (Merck & Co.). Both recombinant products contain nonglycosylated HBsAg particles (only S protein) that have been physicochemically purified, adsorbed on aluminium hydroxide, and preserved with thimerosal. Only Recombivax HB$^®$ is treated with formaldehyde [15].

The yeast-derived HB-VAX DNA$^®$ (Pasteur Mérieux MSD), containing only the S protein, is produced in France.

Recombinant HB vaccines are produced in Belgium, China, Cuba, France, India, Israel, Japan, the Republic of Korea, Switzerland, the USA and Vietnam [23].

India has developed an indigenous yeast-derived, recombinant DNA vaccine, Shanvac-B$^®$ (Santha Biotechnics, 1997). At about US$ 14 for three doses, Shanvac-B$^®$ is within the reach of the EPI [2].

A licence application in both Europe and the USA has been filed in 1998 for Hepagene$^®$ (Medeva), the first recombinant hepatitis B vaccine to incorporate significant levels of HBV's pre-S1 and pre-S2 epitopes, and S protein. Further, like the surface of the virus itself, Hepagene$^®$'s surface proteins are glycosylated. The result is that Hepagene$^®$ closely mimics the surface of HBV and produces a better immune response than that of other recombinant HB vaccines. Hepagene$^®$ has also been studied as an immunotherapy for the treatment of hepatitis B. Results are comparable with results reported after treatment with lamivudine [29].

Cross-protection by different serotype vaccines against different HBV subtypes has been observed in chimpanzees [15]. Post-exposure immunization after an HBV challenge has also been effective in chimpanzees [15,28].

Vaccination of HBV carriers is safe but ineffective in eliminating HBsAg from chronically infected individuals [15]. The HBV vaccine produces neither therapeutic nor adverse effects for individuals who possess antibodies against HBV from a previous infection. Passively acquired antibody will not interfere with active immunization.

Combination vaccines

The HBsAg vaccines (HB) can be combined with other vaccines such as Calmette-Guérin bacillus (BCG), measles, mumps, and rubella (MMR), Hib, and diphtheria, tetanus and pertussis combined with polio (DTP–polio). SmithKline Beecham offers a tetravalent DTP–HB vaccine, and a combined hepatitis A–hepatitis B vaccine [15].

The combined hepatitis A and B vaccine (Twinrix®, SmithKline Beecham) has been introduced in Australia, Canada and some countries in Europe in 1997. In its adult formulation it contains 720s EL.U. of hepatitis A antigen (Havrix®) and 20 μg of HBsAg (Engerix®-B) adsorbed onto aluminium salts [15,34].

Neonates born to mothers who are HBeAg-positive should be given a combination of passive and active immunization to provide immediate protection with HBIG in the first 6 h after delivery, followed by long-term immunity with the vaccine. At the currently recommended doses, HBIG does not interfere with the active immune response of the vaccine. When concurrent administration of HBIG and vaccine are contemplated, different sites should be used [15].

The vaccines are to be administered by intramuscular injection in the anterolateral aspect of the thigh of newborns and infants or the deltoid (arm) muscle of children and adults in order to achieve optimal protection [3,15].

The recommendation for universal infant vaccination neither precludes vaccinating adults identified to be at high risk of infection nor alters previous recommendations for post-exposure prophylaxis for hepatitis B [3].

Vaccine batches should be stored at 2–8°C but not frozen. Freezing destroys the potency of the vaccine since it dissociates the antigen from the adjuvant alum interfering with the immunogenicity of the preparation [15].

The vaccine is thermostable and neither reactogenicity nor immunogenicity is altered after heating at 45°C for 1 week or 37°C for 1 month [23].

Factors that may reduce the immunogenicity of hepatitis vaccines include age (>40 years), gender, weight, genetics, haemodialysis, HIV infection, immunosuppression, tobacco smoking, subcutaneous injection, injection into the buttocks, freezing of vaccine, and accelerated schedule [15,31].

An initial anti-HBs titre of >10 IU/l is regarded as being protective. Although the initial anti-HBs titre is followed by a decline of antibody, a rapid anamnestic response develops after exposure to the virus [23,31].

The duration of vaccine-induced immunity is uncertain but it is definitely long term (>15 years). At present there is no recommendation for the administration of booster doses, although future studies could demonstrate a need for boosters [15,23,31].

A recent study designed to determine the safety and immunogenicity of a DNA vaccine consisting of a plasmid encoding HBsAg delivered into human skin suggests that this gene delivery system may induce a booster response, but that the vaccine at the dose used (0.25 ug) did not induce primary immune responses [32].

Hepatitis B vaccines available internationally

Manufacturer	Brand name[a]	Country	Type
Centro de Ingenieria Genetica Y Biotecnologia	Enivac-HB	Cuba	Recombinant DNA
Chiel Jedang	Hepaccine-B	South Korea	Plasma-derived
Korea Green Cross	Hepavax B	South Korea	Plasma-derived
Korea Green Cross	Hepavax-Gene	South Korea	Recombinant DNA
LG Chemical	Euvax B	South Korea	Recombinant DNA
Merck Sharp & Dohme	Recompivax H-B-Vax II	United States	Recombinant DNA
Merck Sharp & Dohme	Comvax	United States	Combined Hib and (recombinant)
Pasteur Mérieux Connaught	Genhevac B	France	Recombinant DNA (mammalian cell)
SmithKline Beecham	Engerix-B	Belgium	Recombinant DNA
SmithKline Beecham	Twinrix	Belgium	Combined hepatitis A and B (recombinant)
SmithKline Beecham	Tritanrix-HB	Belgium	Combined DTP and recombinant
SmithKline Beecham	Infanrix-HB	Belgium	Combined DTP (acellular P) and HB (recombinant)
Swiss Serum and Vaccines Institute	Heprecombe	Switzerland	Recombinant DNA (mammalian cell)

DTP, diphteria, tetanus and pertussis; HB, hepatitis B; Hib, *Haemophilus influenza* type b.
From Ref. [23], with permission.
[a]Brand names may vary in different countries.

Numerous producers who sell only in country of production are not listed. Presence on this list does not imply endorsement of these products by the World Health Organization.

Recommendations for pre-exposure immunization with hepatitis B vaccine

Here is a list of groups for whom pre-exposure vaccination is recommended. If all members of these groups were immunized, the incidence of hepatitis B would decrease rapidly [3,5,15].

- Infants (universal immunization).

- Infants and adolescents not vaccinated previously (catch-up vaccination).

- Persons with occupational risk (exposure to blood or blood-contaminated environments) and students of health-care professions before they have blood contact.

- Clients and staff of institutions for the developmentally disabled and susceptible contacts in day-care programmes who are at increased risk from HBV carrier clients with aggressive behaviour or special medical problems that increase the risk of exposure.

- Haemodialysis patients. Vaccination before dialysis treatment is recommended.

- Recipients of frequent and/or large volumes of blood or blood components.

- Susceptible injecting drug abusers.

- Sexually active men or women (homosexual and bisexual men; persons with recently acquired sexually transmitted disease; prostitutes; promiscuous hetero-sexuals).

- Susceptible inmates of long-term correctional facilities who have a history of high risk behaviour.

- Household contacts and sex partners of HBV carriers.

- Populations with a high incidence of disease.

- International travellers to areas of high HBV endemicity if specific at-risk circumstances exist [50].

- Transplant candidates before transplantation.

In 1991 the WHO/EPI recommended that HB vaccine be included in national immunization programmes in all countries with an HBV carrier rate of 8% or over by 1995, and in all other countries (regardless of HBsAg prevalence) by 1997. Countries with a low prevalence may consider immunization of all adolescents (before age of 13) as an addition or alternative to infant immunization.

So far (March 2002) 151 countries [Albania, American Samoa, Andorra, Anguilla, Antigua and Barbuda, Argentina, Armenia, Australia, Austria, Azerbaijan, Bahamas, Bahrain, Barbados, Belarus, Belgium, Belize, Bermuda, Bhutan, Bolivia, Bosnia and Herzegovina, Botswana, Brazil, British Virgin Islands, Brunei Darussalam, Bulgaria, Cambodia, Canada, Cayman Islands, China, C.N. Mariana Islands, Colombia, Cook Islands, Costa Rica, Côte d'Ivoire, Cuba, Cyprus, D. People's R. of Korea, Dominica, Dominican Republic, Ecuador, Egypt, El Salvador, Eritrea, Estonia, Fiji, France, French Guiana, French Polynesia, Gambia, Georgia, Germany, Ghana, Greece, Grenada, Guadeloupe, Guam, Guyana, Honduras, Indonesia, Iran (Islamic Republic of), Iraq, Israel, Italy, Jamaica, Jordan, Kazakhstan, Kenya, Kiribati, Kuwait, Kyrgyzstan, Lao People's D. R., Latvia, Lebanon, Libyan Arab Jamahiriya, Lithuania, Luxembourg, Madagascar, Malawi, Malaysia, Maldives, Marshall Islands, Martinique, Mauritius, Mexico, Micronesia (Federated States of), Monaco, Mongolia, Montserrat, Morocco, Mozambique, Nauru, Netherlands Antilles, New Caledonia, New Zealand, Nicaragua, Niue, Oman, Pakistan, Palau, Panama, Papua New Guinea, Paraguay, Peru, Philippines, Poland, Portugal, Puerto Rico, Qatar, Republic of Korea, Romania, Rwanda, Saint Kitts and Nevis, Saint Vincent and the Grenadines, Samoa, San Marino, Saudi Arabia, Seychelles, Singapore, Slovakia, Slovenia, Solomon Islands, South Africa, Spain, Suriname, Swaziland, Syrian Arab

80

Immunization strategies

Routine infant immunization: HB immunization of all infants as an integral part of the national immunization schedule should be the highest priority in all countries.

Additional immunization strategies that should be considered depending on the epidemiology of HBV transmission in a particular country are given below

- Prevention of perinatal HBV transmission

  In order to prevent HBV transmission from mother to infant, the first dose of HB vaccine needs to be given as soon as possible after birth (preferably within 24 h). In countries where a high proportion of chronic infections is acquired perinatally (e.g. South-east Asia), a birth dose should be given to infants. It is usually most feasible to give HB vaccine at birth when infants are born in hospitals. Efforts should also be made in these countries to give HB vaccine as soon as possible after delivery to infants delivered at home. In countries where a lower population of chronic infections is acquired perinatally (e.g. Africa), the highest priority is to achieve high DTP3 and HB3 vaccine coverage among infants. In these countries, use of a birth dose may also be considered after disease burden, cost-effectiveness, and feasibility are evaluated.

- Catch-up vaccination of older persons

  In countries with a high prevalence of chronic HBV infection (HBsAg prevalence $> 8\%$), catch-up immunization is not usually recommended because most chronic infections are acquired among children < 5 years of age, and thus, routine infant vaccination will rapidly reduce HBV transmission. In countries with lower endemicity of chronic HBV infection, a higher proportion of chronic infections may be acquired among older children, adolescents and adults; catch-up immunization for these groups may be considered.

Vaccine formulations

Hepatitis B vaccine is available in monovalent formulations that protect only against HBV infection and also in combination formulations that protect against HBV and other diseases.

- Monovalent hepatitis B vaccines *must be used* to give the birth dose of hepatitis B vaccine.

- Combination vaccines that include hepatitis B vaccine must not be used to give the birth dose of hepatitis B vaccine because DTP and Hib vaccines are not recommended to be given at birth.

- Either monovalent or combination vaccines may be used for later doses in the hepatitis B vaccine schedule. Combination vaccines can be given whenever all of the antigens in the vaccine are indicated.

Schedule

Hepatitis B vaccine schedules are very flexible; thus, there are multiple options for adding the vaccine to existing national immunization schedules without requiring additional visits for immunization.

Practically, it is usually easiest if the three doses of hepatitis B vaccine are given at the same time as the three doses of DTP (Option I). This schedule will prevent infections acquired during early childhood, which account for most of the HBV-related disease burden in high endemic countries, and also will prevent infections acquired later in life.

However, this schedule will not prevent perinatal HBV infections because it does not include a dose of hepatitis B vaccine at birth. Two schedule options can be used to prevent perinatal HBV infections: a three-dose schedule of monovalent hepatitis B vaccine, with the first dose given at birth and the second and third doses given at the same time as the first and third doses of DTP vaccine (Option II); or a four-dose schedule in which a birth dose of monovalent HepB vaccine is followed by three doses of a combination vaccine, e.g. DTP hepatitis B (Option III). The three-dose schedule (Option II) is less expensive, but may be more complicated to administer because infants receive different vaccines at the second immunization visit than at the first and third visits. The four-dose schedule (Option III) may be easier to administer in practice, but is more costly, and vaccine supply issues may make it unfeasible.

Administration

Hepatitis B vaccine is given by intramuscular injection in the anterolateral aspect of the thigh (infants) or deltoid muscle (older children). It can be given safely at the same time as other vaccines (e.g. DTP, Hib, measles, OPV, BCG, and yellow fever). If the hepatitis B vaccine is given on the same day as another injectable vaccine, it is preferable to give the two vaccines in different limbs.

Injection equipment

The injection equipment for hepatitis B vaccine is the same type as that for all other EPI vaccines (except for BCG vaccine).

- 0.5 ml Auto-disable (AD) syringes are recommended.

- If AD syringes are not available, standard disposable syringes (1.0 or 2.0 ml) must be used ONCE ONLY, and safely disposed of after use.

- A 25 mm, 22 or 23 gauge needle is recommended.

Dosage

The standard paediatric dose is 0.5 ml.

Vaccine procurement

In most countries, hepatitis B vaccine procured through The Vaccine Fund will be supplied through the UNICEF procurement mechanism. The number of hepatitis B vaccine doses required is estimated using the size of the birth cohort, the coverage rate for DTP and the number of doses in the immunization schedule. These calculations should also include wastage and the size of the reserve stock.

Presentation

Hepatitis B vaccines are available in liquid single-dose and multi-dose glass vials, and in pre-filled single-dose injection devices (e.g. Uniject™).

Storage and shipping volume

Storage volumes (vial plus packet containing vial plus other packaging) for hepatitis B vaccines supplied through UNICEF are as per the figure below: for comparison, the total storage volume for other EPI vaccines (BCG, DTP, measles, OPV, TT) is about 11.0 cm^3 per dose.

Cold chain issues

The storage temperature for hepatitis B vaccine is the same as for DTP vaccine, from 2 to 8°C. *Hepatitis B vaccine should never be frozen.* If frozen, hepatitis B vaccine loses its potency.

Adding hepatitis B vaccine to the national immunization schedule will require cold chain assessments at all administrative levels:

- To assure adequate storage capacity is available.

- To assure policies and procedures are in place to prevent freezing of hepatitis B vaccine.

Reducing vaccine wastage

Since hepatitis B vaccines are more expensive than the traditional EPI vaccines, it is important to monitor vaccine wastage and to develop and implement strategies to reduce wastage.

Strategies to reduce wastage include:

- Careful planning of vaccine ordering and distribution.

- Implementation of WHO's multi-dose vial policy.

- Appropriate use of single-dose and multi-dose vials.

- Careful maintenance of the cold chain.

- Attention to vaccine security; and

- Reducing missed opportunities for immunization.

Injection safety

Hepatitis B vaccine should be supplied with AD syringes and safety boxes.

Managers at each level are responsible for ensuring that adequate supplies are available at all times so that each injection is given with a sterile injection device. Attention should also be given to proper use and disposal of safety boxes to collect these materials.

Revision of immunization forms and materials

An important element of integrating hepatitis B vaccine into national immunization programmes is to revise training and informational materials, immunization cards and forms used to monitor and evaluate immunization services.

Training

Training for health-care staff is essential because these staff are responsible for handling and administering hepatitis B vaccine and they are a major source of information for parents and others in the general public.

Advocacy and communication

Advocacy and communication efforts are important in order to generate support and commitment for the new vaccine. The primary target audiences are decision-makers/opinion leaders, health-care staff, and the general public (including parents).

What information is needed to assess hepatitis B disease burden?

Adequate seroprevalence data needed to assess hepatitis B disease burden are generally available in all countries, or from adjacent countries with similar HBV endemicity. Thus, additional seroprevalence studies are usually not needed.

How should hepatitis B vaccine be phased into the existing infant immunization services?

A strategy in which hepatitis B vaccine is given to infants who have not yet completed the DTP vaccine series at the time hepatitis B vaccine is introduced is generally the most feasible to implement.

Combination prophylaxis with lamivudine and HBIG prevents hepatitis B recurrence following liver transplantation [15,24]. Subjecting hepatitis B patients who develop end-stage liver disease to liver transplantation is very controversial because the graft is inevitably reinfected, especially if the patient is HBV DNA-positive. To counteract this problem, the long-term i.v. administration of HBIG to these patients before the operation and continuously thereafter, helps maintain a minimal level of anti-HBs in the serum at all times. Some patients relapse, however, when therapy is interrupted [15].

Adoptive transfer of immunity to hepatitis B has been a novel approach to terminating HBV infection in the carrier after bone marrow transplantation from a hepatitis B immune donor [11,15].

Several new agents (e.g. Ritonavir, Adefovir, Dipivoxil, Lobucavir, Famvir, FTC, N-acetyl-cysteine (NAC), PC1323, Theradigm-HBV, Thymosin-alpha, Ganciclovir [14]) are in development, and some encouraging data are available.

Chronic hepatitis B: potential drug therapy

Agent	Effective	Ineffective	Toxic	Under evaluation
Interferon	Interferon-α	Interferon-γ		Interferon-β
Antiviral	Lamivudine	Acyclovir	Fialuridine	Ribavirin
	Famciclovir	Dideoxyinosine	Adenine	Lamivudine (long term)
		Azidothymidine	Arabinoside	Famciclovir (long term)
		Foscarnet		Adefovir
				Entecavir
Immunomodulatory		Prednisone		Adoptive immune transfer
		Interleukin-2		
		Thymosin		
		Levamisole		

From Ref. [11], with permission.

Goals of interferon therapy

Goal	Implication
Loss of HBeAg	Significant decrease of infectious potential accompanied by clinical benefit
Loss of HBV DNA	Loss of ability of HBV to replicate
Return to normal ALT levels	Cessation of hepatic inflammation and interruption of progression of liver injury
Loss of HBsAg	Eradication of HBV

Contraindications for interferon therapy for chronic hepatitis B

Hepatic decompensation	Albumin <3.0 g/l
	Bilirubin >51.3 μmol/l (30 mg/l)
	Prolonged prothrombin time >3.0 s
Portal hypertension	Variceal bleed
	Ascites
	Encephalopathy
Hypersplenism	Leukopenia ($< 2 \times 10^9 l^{-1}$)
	Thrombocytopenia ($< 7 \times 10^7 l^{-1}$)
Psychiatric depression	Severe, suicide attempt
Autoimmune disease	Polyarteritis nodosa, rheumatoid arthritis, thyroiditis
Major system impairment	Cardiac failure
	Obstructive airways disease
	Uncontrolled diabetes
Pregnancy	
Current intravenous drug abuse	

From Ref. [11], with permission.

Side-effects of interferon therapy

Constitutional	Flu-like illness
	Fever
	Rigors
	Arthralgia
	Myalgia
	Fatigue
Hematologic	Leukopenia
	Thrombocytopenia
Alopecia	
Neuropsiachiatric	Depression
	Insomnia
	Irritability
Weight loss	
Ocular	
Autoimmune	Hypothyroidism
	Diabetes

From Ref. [11], with permission.

Guidelines for epidemic measures

1. When two or more cases occur in association with some common exposure, a search for additional cases should be conducted.

2. Introduction of strict aseptic techniques. If a plasma derivative like antihaemophilic factor, fibrinogen, pooled plasma or thrombin is implicated, the lot should be withdrawn from use.

3. Tracing of all recipients of the same lot in search for additional cases.

4. Relaxation of sterilization precautions and emergency use of unscreened blood for transfusions may result in increased number of cases.

Future considerations

Attaining global immunization coverage is a goal still unmet.

The development of a better and cheaper antiviral therapy should be pursued intensively for chronic HBV infections [30].

Strategies to activate appropriate immune responses during chronic virus infections may offer the best approach for terminating such infections.

Attempts at protecting the whole community by vaccinating only high-risk individuals have not been successful [37]. Universal vaccination is necessary to control and possibly eradicate hepatitis B. The next step is finding strategies for meeting that goal in countries with different health-care structures and financial resources.

WHO Goals

WHO aims at controlling HBV worldwide to decrease the incidence of HBV-related chronic liver disease, cirrhosis, and HCC by integrating HB vaccination into routine infant (and possibly adolescent) immunization programmes [3,23,36].

Persons infected with HBV during infancy or early childhood are more likely to become infected chronically and to develop life-shortening chronic liver disease such as cirrhosis or even liver cancer than adults. This is one important reason why emphasis should be placed upon preventing HBV among the youngest age groups.

In 1991, the Global Advisory Group of EPI set 1997 as the target for integrating the hepatitis B vaccination into national immunization programmes worldwide. The group recommended strategies for implementation and delivery that vary according to epidemiology: advocating integration of the vaccine into immunization programmes by 1995 in countries with a HBV carrier prevalence of 8% or higher, and setting 1997 as the target date for all other countries. WHO endorsed the recommendation in May 1992, and the World Health Assembly added a disease reduction target for hepatitis B in 1994, calling for an 80% decrease in new HBV child carriers by 2001.

Commitment of public health resources to eliminate the spread of HBV requires recognition of the importance of hepatitis B, persistent efforts to ensure that populations are protected, and patience to achieve the goals of disease reduction [23].

Glossary

Alopecia loss of hair occurring at any site and from any cause.

ALT alanine aminotransferase an enzyme that interconverts L-alanine and D-alanine. It is a highly sensitive indicator of hepatocellular damage. When such damage occurs, ALT is released from the liver cells into the bloodstream, resulting in abnormally high serum levels. Normal ALT levels range from 10 to 32 U/l; in women, from 9 to 24 U/l. The normal range for infants is twice that of adults.

Amino acids the basic units of proteins, each amino acid has a $NH-C(R)-COOH$ structure, with a variable R group. There are altogether 20 types of naturally occurring amino acids.

Antibody a protein molecule formed by the immune system which reacts specifically with the antigen that induced its synthesis. All antibodies are immune globulins [1].

Antigen any substance which can elicit in a vertebrate host the formation of specific antibodies or the generation of a specific population of lymphocytes reactive with the substance. Antigens may be protein or carbohydrate, lipid or nucleic acid, or contain elements of all or any of these as well as organic or inorganic chemical groups attached to protein or other macromolecule. Whether a material is an antigen in a particular host depends on whether the material is foreign to the host and also on the genetic makeup of the host, as well as on the dose and physical state of the antigen [1].

Arthralgia joint pain with objective findings of heat, redness, tenderness to touch, loss of motion, or swelling.

AST aspartate aminotransferase the enzyme that catalyses the reaction of aspartate with 2-oxoglutarate to give glutamate and oxaloacetate. Its concentration in blood may be raised in liver and heart diseases that are associated with damage to those tissues. Normal AST levels range from 8 to 20 U/l. AST levels fluctuate in response to the extent of cellular necrosis [1].

B-cells also known as B lymphocytes. A class of white blood cells, which carry out humoral immune response. They mature in the bone marrow.

Bilirubin is the chief pigment of bile, formed mainly from the breakdown of haemoglobin. After formation it is transported in the plasma to the liver to be then excreted in the bile. Elevation of bile in the blood (>30 mg/l) causes jaundice [43].

Carcinoma a malignant epithelial tumour. This is the most frequent form of cancer.

Carrier is a person who has HBV (HCV, HDV) in his or her blood for longer than 6 months even if all symptoms have disappeared. Because the virus is present in the blood, it can be transmitted to others. The HBV carrier can be recognized by a specific blood test.

Cirrhosis a chronic disease of the liver characterized by nodular regeneration of HC and diffuse fibrosis. It is caused by parenchymal necrosis followed by nodular proliferation of the surviving HC. The regenerating nodules and accompanying fibrosis interfere with blood flow through the liver and result in portal hypertension, hepatic insufficiency, jaundice and ascites.

Codon the smallest unit of genetic material that can specify an amino acid residue in the synthesis of a polypeptide chain. The codon consists of three adjacent nucleotides.

Complete blood count chemical analysis of various substances in the blood performed with the aim of (i) assessing the patient's status by establishing normal levels for each individual patient, (ii) preventing disease by alerting to potentially dangerous levels of blood constituents that could lead to more serious conditions, (iii) establishing a diagnosis for already present pathologic conditions, and (iv) assessing a patient's progress when a disturbance in blood chemistry already exists.

Cytopathic that kills the cells.

Cytoplasm the protoplasm of the cell which is outside of the nucleus. It consists of a continuous aqueous solution and the organelles and inclusions suspended in it. It is the site of most of the chemical activities of the cell.

Endemic prevalent continuously in some degree in a community or region [43].

Endoplasmic reticulum a network or system of folded membranes and interconnecting tubules distributed within the cytoplasm of eukaryotic cells. The membranes form enclosed or semi-enclosed spaces. The endoplasmic reticulum functions in storage and transport, and as a point of attachment of ribosomes during protein synthesis.

Enzyme any protein catalyst, i.e. substance which accelerates chemical reactions without itself being used up in the process. Many enzymes are specific to the substance on which they can act, called substrate. Enzymes are present in all living matters and are involved in all the metabolic processes upon which life depends.

Epidemic an outbreak of disease such that for a limited period a significantly greater number of persons in a community or region suffer from it than is normally the case. Thus an epidemic is a temporary increase in prevalence. Its extent and duration are determined by the interaction of such variables as the nature and infectivity of the casual agent, its mode of transmission and the degree of pre-existing and newly acquired immunity [43].

Epitope also known as antigenic determinant. A localized region on the surface of an antigen which antibody molecules can identify and bind.

Fulminant describes pathological conditions that develop suddenly and are of great severity [1].

Genome the total genetic information present in a cell. In diploid cells, the genetic information contained in one chromosome set [1].

Golgi apparatus a cytoplasmic organelle which is composed of flattened sacs resembling smooth endoplasmic reticulum. The sacs are often cup-shaped and located near the nucleus, the open side of the cup generally facing toward the cell surface. The function of the Golgi apparatus is to accept vesicles from the endoplasmic reticulum, to modify the contents, and to distribute the products to other parts of the cell or to the cellular environment.

Hepadnavirus family of single-stranded DNA viruses of which hepatitis B virus (HBV) and woodchuck hepatitis virus (WHV) are members.

Hepatocytes are liver cells [1].

Humoral pertaining to the humors, or certain fluids, of the body [1].

Icterus see jaundice.

IgG antibodies IgG is the most abundant of the circulating antibodies. It readily crosses the walls of blood vessels and enters tissue fluids. IgG also crosses the placenta and confers passive immunity from the mother to the fetus. IgG protects against bacteria, viruses, and toxins circulating in the blood and lymph.

IgM antibodies IgMs are the first circulating antibodies to appear in response to an antigen. However, their concentration in the blood declines rapidly. This is diagnostically useful, because the presence of IgM usually indicates a current infection by the pathogen causing its formation. IgM consists of five Y-shaped monomers arranged in a pentamer structure. The numerous antigen-binding sites make it very effective in agglutinating antigens. IgM is too large to cross the placenta and hence does not confer maternal immunity.

Immune globulin (IG) is a sterile preparation of concentrated antibodies (immune globulins) recovered from pooled human plasma processed by cold ethanol fractionation. Only plasma that has tested negative for (i) HBsAg, (ii) antibody to human immunodeficiency virus (HIV), and (iii) antibody to hepatitis C virus (HCV) is used to manufacture IG. IG is administered to protect against certain diseases through passive transfer of antibody. The IGs are broadly classified into five types on the basis of physical, antigenic and functional variations, labeled respectively IgM, IgG, IgA, IgE and IgD.

Immune system our body's natural defence system, involving antibodies and a class of white blood cells called lymphocytes.

Incidence the number of cases of a disease, abnormality, accident, etc. arising in a defined population during a stated period, expressed as a proportion, such as x cases per 1000 persons per year [1].

Interferon a protein produced in organisms infected by viruses, and effective at protecting those organisms from other virus infections. Interferons exert virus-nonspecific but host-specific antiviral activity by inducing the transcription of cellular genes coding for antiviral proteins that selectively inhibit the synthesis of viral DNA and proteins. Interferons also have immunoregulatory functions. Production of interferon can be stimulated by viral infection, especially by the presence of double-stranded RNA, by intracellular parasites, by protozoa, and by bacteria and bacterial products. Interferons have been divided into three distinct types (α, β, and γ) associated with specific producer cells and functions, but all animal cells are capable of producing interferons, and certain producer cells (leukocytes and fibroblasts) produce more than one type (both α and β).

Jaundice is a yellow discolouration of the skin and mucous membranes due to excess of bilirubin in the blood, also known as icterus [43].

Leukopenia an abnormal decrease in the number of leukocytes in the blood.

Lumen the cavity or channel between a tube or tubular structure.

Lymphocyte a leukocyte of blood, bone marrow and lymphatic tissue. Lymphocytes play a major role in both cellular and humoral immunity, and thus several different functional and morphologic types must be recognized, i.e. the small, large, B-, and T-lymphocytes, with further morphologic distinction being made among the B-lymphocytes and functional distinction among T-lymphocytes [1].

Lymphoproliferative disease a neoplastic or systemic tumor-like proliferation of lymphocytes, as in lymphoid leukemia, malignant lymphomas, or in Waldenström's macroglobulinemia [1].

Major histocompatibility complex (MHC) originally defined as the genetic locus coding for those cell surface antigens presenting the major barrier to transplantation between individuals of the same species. Now known to be a cluster of genes on human chromosome 6 or mouse chromosome 17 that encodes the MHC molecules. These are the

MHC class I molecules or proteins that present peptides generated in cytosol to CD8 T-cells, and the MHC class II molecules or proteins that present peptides degraded in cellular vesicles to CD4 T-cells. The MHC also encodes proteins involved in antigen processing and host defence. The MHC is the most polymorphic gene cluster in the human genome, having large numbers of alleles at several different loci. Because this polymorphism is usually detected using antibodies or specific T-cells, the MHC proteins are often called major histocompatibility antigens.

Myalgia pain in the muscles.

Nucleotide a molecule formed from the combination of one nitrogenous base (purine or pyrimidine), a sugar (ribose or deoxyribose) and a phosphate group. It is a hydrolysis product of nucleic acid [1].

Nucleus a membrane-bounded compartment in an eukaryotic cell which contains the genetic material and the nucleoli. The nucleus represents the control center of the cell. Nuclei divide by mitosis or meiosis.

Plasma the liquid matrix in which the blood cells and blood proteins are suspended in. It contains an extensive variety of solutes dissolved in water. Water accounts for about 90% of blood plasma.

Plasmid a small, circular DNA molecule, separate from the bacterial chromosome, capable of independent replication.

Polymerase an enzyme which catalyses the replication of DNA (DNA polymerase) or RNA (RNA polymerase).

Prevalence is the number of instances of infections or of persons ill, or of any other event such as accidents, in a specified population, without any distinction between new and old cases [43].

Promoter a region of DNA usually occurring upstream from a gene coding region and acting as a controlling element in the expression of that gene. It serves as a recognition signal for an RNA polymerase and marks the site of initiation of transcription.

Prophylaxis is the prevention of disease, or the preventive treatment of a recurrent disorder [43].

Protein large molecule made up of many amino acids chemically linked together by amide linkages. Biologically important as enzymes, structural protein and connective tissue.

Reverse transcriptase an enzyme that catalyses the formation of DNA using an RNA template, and is thus an RNA-dependent DNA polymerase. The name refers to the fact that the enzyme transcribes nucleic acids in the reverse order from the usual DNA-to-RNA transcription.

Rigors stiffness.

RT-PCR reverse transcriptase-polymerase chain reaction. A technique commonly employed in molecular genetics through which it is possible to produce copies of DNA sequences rapidly.

Seroconversion the production in a host of specific antibodies as a result of infection or immunization. The antibodies can be detected in the host's blood serum following, but not preceding, infection or immunization [1].

Serum is the clear, slightly yellow fluid which separates from blood when it clots. In composition it resembles blood plasma, but with fibrinogen removed. Sera containing antibodies and antitoxins against infections and toxins of various kinds (antisera) have been used extensively in prevention or treatment of various diseases [43].

T-cells also known as T-lymphocytes. White blood cells, which function in cell-mediated response. They originate from stem cells in the bone marrow but mature in the thymus.

Thrombocytopenia a fewer than normal number of platelets per unit volume of blood, i.e. fewer than 130×10^9 platelets per liter.

Titre a measure of the concentration or activity of an active substance.

Transcription the process by which a strand of RNA is synthesized with its sequence specified by a complementary strand of DNA, which acts as a template. The enzymes involved are called DNA-dependent RNA polymerases.

Translation the process of forming a specific protein having its amino acid sequence determined by the codons of messenger RNA. Ribosomes and transfer RNA are necessary for translation [1].

Tumour a lump due to uncontrolled cell division, may be benign or malignant. Malignant tumours cause cancer. Tumours are able to spread to other parts of the body (metastasize) and begin secondary growths at these other sites.

Vaccine an antigenic preparation used to produce active immunity to a disease to prevent or ameliorate the effects of infection with the natural or "wild" organism. Vaccines may be living, attenuated strains of viruses or bacteria, which give rise to inapparent to trivial infections. Vaccines may also be killed or inactivated organisms or purified products derived from them. Formalin-inactivated toxins are used as vaccines against diphtheria and tetanus. Synthetically or genetically engineered antigens are currently being developed for use as vaccines. Some vaccines are effective by mouth, but most have to be given parenterally [1,43].

Vaccinee person receiving a vaccine.

Virion a structurally complete virus, a viral particle [1].

Virus any of a number of small, obligatory intracellular parasites with a single type of nucleic acid, either DNA or RNA and no cell wall. The nucleic acid is enclosed in a structure called a capsid, which is composed of repeating protein subunits called capsomeres, with or without a lipid envelope. The complete infectious virus particle, called a virion, must rely on the metabolism of the cell it infects. Viruses are morphologically heterogeneous, occurring as spherical, filamentous, polyhedral, or pleomorphic particles. They are classified by the host infected, the type of nucleic acid, the symmetry of the capsid, and the presence or absence of an envelope [1].

References

1. Churchill's Illustrated Medical Dictionary. New York: Churchill Livingstone, 1989.
2. Abraham P, et al. Evaluation of a new recombinant DNA hepatitis B vaccine (Shanvac-B). Vaccine 1999; 17: 1125–1129.
3. Centers for Disease Control and Prevention. Hepatitis B virus: a comprehensive strategy for eliminating transmission in the United States through universal childhood vaccination: recommendations of the immunization practices advisory committee (ACIP). Morbidity Mortality Weekly Rep 1991; 40: 1–19. http://www.cdc.gov/ncidod/diseases/hepatitis/v40rr13.htm.
4. Centers for Disease Control and Prevention. Control measures for hepatitis B in dialysis centers, http://www.cdc.gov/ncidod/hip/control.htm, 1998.
5. Centers for Disease Control and Prevention. Hepatitis B vaccine, http://www.cdc.gov/ncidod/diseases/hepatitis/b/hebqafn.htm, 1998.
6. Chisari FV, Ferrari C. Viral hepatitis. In: Viral Pathogenesis (Nathanson N, et al. editors). Philadelphia: Lippincott-Raven; 1997; pp. 745–778.
7. Clemens R, et al., Booster immunization of low- and non-responders after a standard three dose hepatitis B vaccine schedule—results of a post-marketing surveillance. Vaccine 1997; 15: 349–352.
8. Dienstag JL, et al. Extended lamivudine retreatment for chronic hepatitis B: maintenance of viral suppression after discontinuation of therapy. Hepatology 1999; 30: 1082–1087.
9. Fattovich G, et al. Long-term outcome of hepatitis B e antigen-positive patients with compensated cirrhosis treated with interferon alfa. European Concerted Action on Viral Hepatitis (EUROHEP). Hepatology 1999; 26: 1338–1342.
10. Ganem D, Schneider RJ. Hepadnaviridac: The Viruses and Their Replication. In: Fields Virology (Knipe DM, et al. editors), 4th edn. Philadelphia: Lippincott Williams & Wilkins; 2001; pp. 2923–2969.
11. Gitlin N. Hepatitis B: diagnosis, prevention, and treatment. Clin Chem 1997; 43(8B): 1500–1506.
12. Goldwater PN. Randomized, comparative trial of 20 μg vs 40 μg Engerix B vaccine in hepatitis B vaccine non-responders. Vaccine 1997; 15: 353–356.

13. Guidotti LG, et al., Hepatitis B virus nucleocapsid particles do not cross the hepatocyte nuclear membrane in transgenic mice. J Virol 1994; 68: 5469–5475.
14. Hadziyannis SJ, Manesis EK, Papakonstantinou A. Oral ganciclovir treatment in chronic hepatitis B virus infection: a pilot study. J Hepatol/title > 1999; 31: 210–214.
15. Hollinger FB, Liang TJ, Hepatitis B virus. In: Fields Virology (Knipe DM, et al. editors), 4th edn. Philadelphia: Lippincott Williams & Wilkins; 2001; pp. 2971–3036.
16. Ikeda K, et al. Interferon decreases hepatocellular carcinogenesis in patients with cirrhosis caused by the hepatitis B virus: a pilot study. Cancer 1998; 82: 827–835.
17. Iwarson S. Why the Scandinavian countries have not implemented universal vaccination against hepatitis B. Vaccine 1998; 16: S56–S57.
18. Janssen HL, et al. Interferon alfa for chronic hepatitis B infection: increased efficacy of prolonged treatment. Hepatology 1999; 30: 238–243.
19. Kane MA. Status of hepatitis B immunization programmes in 1998. Vaccine 1998; 16(Suppl): S104–S108.
20. Lai CL, et al. A one-year trial of lamivudine for chronic hepatitis B. Asia Hepatitis Lamivudine Study Group. N Engl J Med 1998; 339: 61–68.
21. Lai CL, et al. Lamivudine is effective in suppressing hepatitis B virus DNA in Chinese hepatitis B surface antigen carriers: a placebo-controlled trial. Hepatology 1997; 25: 241–244.
22. Lau DT, et al. Lamivudine for chronic delta hepatitis. Hepatology 1999; 30: 546–549.
23. Mahoney FJ, Kane M. Hepatitis B vaccine. In: Vaccines (Plotkin SA, Orenstein WA, editors), 3rd edn. Philadelphia: W.B. Saunders Company; 1999; pp. 158–182.
24. Markowitz JS, et al. Prophylaxis against hepatitis B recurrence following liver transplantation using combination lamivudine and hepatitis B immune globulin. Hepatology 1998; 28: 585–589.
25. Marwick C, Mitka M. Debate revived on hepatitis B vaccine value. J. Am. Med. Assoc. 1999; 282: 15–17.
26. Nevens F, et al. Lamivudine therapy for chronic hepatitis B: a six-month randomized dose-ranging study. Gastroenterology 1997; 113: 1258–1263.
27. Niederau C, et al. Long-term follow-up of HBeAg-positive patients treated with interferon alfa for chronic hepatitis B. N Engl J Med 1996; 334: 1422–1427. http://www.nejm.org/content/1996/0334/0022/1422.asp.
28. Ogata N, et al. Licensed recombinant hepatitis B vaccines protect chimpanzees against infection with the prototype surface gene mutant of hepatitis B virus. Hepatology 1999; 30: 779–786.
29. Pride MW, et al. Evaluation of B and T-cell responses in chimpanzees immunized with Hepagene, a hepatitis B vaccine containing pre-S1, pre-S2 gene products. Vaccine 1998; 16: 543–550.
30. Robinson WS. Hepatitis B viruses. General features (human). In: Encyclopedia of Virology (Webster RG, Granoff A, editors). London: Academic Press Ltd; 1994; pp. 554–569.
31. Robinson WS. Hepatitis B virus and hepatitis D virus. In: Principles and Practice of Infectious Diseases (Mandell GL, Bennett JE, Dolin R, editors), 4th edn. New York: Churchill Livingstone; 1995; pp. 1406–1439.
32. Tacket CO, et al. Phase 1 safety and immune response studies of a DNA vaccine encoding hepatitis B surface antigen delivered by a gene delivery device. Vaccine 1999; 17: 2826–2829.
33. Tassopoulos NC, et al. Recombinant interferon-alpha therapy for acute hepatitis B: a randomized, double-blind, placebo-controlled trial. J Viral Hepat 1997; 4: 387–394.

34. Thoelen S, et al. The first combined vaccine against hepatitis A and B: an overview. Vaccine 1999; 17: 1657–1662.
35. Thornton SM, Walker S, Zuckerman JN. Management of hepatitis B virus infections in two gibbons and a western lowland gorilla in a zoological collection. Vet Record 2001; 149: 113–115.
36. Van Damme P, Kane M, Meheus A. Integration of hepatitis B vaccination into national immunization programmes. Br Med J 1997; 314: 1033–1037.
37. Viral Hepatitis Prevention Board. Antwerp VHPB Report. Editorial. Control of viral hepatitis in Europe. Viral Hepat 1996; 4(2), http://hgins.uia.ac.be/esoc/VHPB/vhv4n2.html.
38. Viral Hepatitis Prevention Board. Prevention and control of hepatitis B in the community. Commun Dis Ser 1996, 1.
39. Viral Hepatitis Prevention Board. The clock is running 1997: deadline for integrating hepatitis B vaccinations into all national immunization programmes, Fact Sheet VHPB/ 1996/1, http:// hgins.uia.ac.be/esoc/VHPB/vhfs1.html, 1996.
40. Viral Hepatitis Prevention Board. News from the VHPB meeting in St. Julians, Malta. Viral Hepat 1997; 6, http://hgins.uia.ac.be/esoc/VHPB/maltatxt.html.
41. Viral Hepatitis Prevention Board. Ensuring injection safety and a safe blood supply. Fact Sheet VHPB/ 1998/3, http://hgins.uia.ac.be/esoc/VHPB/vhfs3.html, 1998.
42. Viral Hepatitis Prevention Board. Universal HB immunization by 1997: where are we now? Fact Sheet VHPB/ 1998/2, http://hgins.uia.ac.be/esoc/VHPB/vhfs2.html, 1998.
43. Walton J, Barondess JA, Lock S. The Oxford Medical Companion. Oxford: Oxford University Press; 1994.
44. Wilson JN, Nokes DJ, Carman WF. The predicted pattern of emergence of vaccine-resistant hepatitis B: a cause for concern? Vaccine 1999; 17: 973–978.
45. World Health Organization. 2nd edn. Laboratory Biosafety Manual. Geneva: WHO; 1993.
46. World Health Organization. Expanded Programme on Immunization (EPI); lack of evidence that hepatitis B vaccine causes multiple sclerosis. Weekly Epidemiol Record 1997; 72: 149–152.
47. World Health Organization. Hepatitis B immunization; WHO position. Weekly Epidemiol Record 1998; 73: 329–329.
48. World Health Organization. No Scientific Justification to Suspend Hepatitis B Immunization. Geneva: WHO; 1998 Press Release WHO/67.
49. World Health Organization. Children's Vaccines—Safety First. Geneva: WHO; 1999 Note for the press N°18.
50. World Health Organization. Health risks and their avoidance—hepatitis B. International Travel and Health. Vaccination Requirements and Health Advice. Geneva: WHO; 1999; p. 67.
51. World Health Organization. Introduction of Hepatitis B Vaccine into Childhood Immunization Services. Geneva: World Health Organization; 2001 unpublished document WHO/V&B/ 01.31; available on request from Department of Vaccines and Biologicals, World Health Organization, 1211 Geneva 27, Switzerland.
52. Zuckerman AJ. Hepatitis viruses. In: Medical Microbiology (Baron S, editor), 4th edn. Galveston, TX: The University of Texas Medical Branch at Galveston; 1996; pp. 849–863.
53. Zuckerman AJ. Effect of hepatitis B virus mutants on efficacy of vaccination. Lancet 2000; 355: 1382–1384.

Viral Hepatitis
I.K. Mushahwar (editor)
© 2004 Elsevier B.V. All rights reserved.

Hepatitis B virus mutants: emergence and impact on diagnostic detection

Paul F. Coleman
Core Infectious Disease R&D, Abbott Diagnostics Division, Abbott Laboratories, Abbott Park, Illinois, USA

The following article will address recent information concerning the emergence of hepatitis B virus (HBV) variants. The mechanism by which these variants arise and their impact on viral antigens and diagnostic detection will be discussed with particular emphasis on the HBV surface antigen mutants. For more extensive review of the clinical impact of hepatitis B mutants and their effects on the management of viral infection the reader is referred to other reviews [1–4].

Hepatitis B replication infidelity

The replication of HBV is specifically addressed in another review in this monograph (see Birkenmeyer, Chapter 4, Hepatitis B virus: Life cycle and morphogenesis); however, two aspects of HBV replication influence the spawning of viable mutants. One, that viral replication proceeds through a reverse transcriptase step and second, that the hepatitis B genome consists of four major open reading frames coding for the core, surface, X, and polymerase genes which partially overlap. More specifically, the polymerase gene overlaps the envelope gene and also partially overlaps the X and core genes. HBV reverse transcriptase has a notoriously poor proofreading function that leads to the generation of multiple variant transcripts from a single RNA template. These transcription errors lead to the initial generation of a quasi-species pool [5]. This initial pool is further selected for either silent or compatible mutations in the overlapping reading frame especially in the surface antigen/polymerase gene overlap [6]. As a result, a wild-type HBV infection provides a fertile ground for mutant occurrence when a phenotypic selection pressure is applied. Correspondingly, all four genes have had mutant or variants identified for the surface, polymerase, core, and X genes.

Surface antigen structure

The product of the envelope gene consists of three proteins that have different initiation sites but the same termination site. The most significant of these proteins, in a diagnostic sense, is the small hepatitis B surface antigen (HBsAg) protein of 226 amino acids. Between positions 100 and 160 of this sequence is a hydrophilic domain termed the 'a' determinant, which represents the immunodominant antigen of HBsAg. The 'a' determinant is the major antigenic recognition site of current HBsAg diagnostic assays.

The HBsAg is a highly conformational structure stabilized by conserved disulfide-bonded cysteine residues. Alteration of these cysteine residues by site-directed mutagenesis resulted in a significant reduction in antigenicity and reduced levels of protein expression [7,8]. Therefore, the conserved disulfide bonds form a backbone that is necessary for the proper three-dimensional presentation of the 'a' determinant antigen. Utilizing a combination of monoclonal antibody competition experiments [9], synthetic peptide mapping experiments [10,11], and phage display experiments [12,13] a proposed schematic of the 'a' determinant was constructed as shown in Fig. 1.

The key features of this model include a large laminar loop stabilized by bonding between cysteine residues 108–138 with a finger-like projection stabilized by disulfide-bonded 121–124 cysteine residues, and a second projecting loop stabilized by bonding between cysteine pairs 136–149, and 139–147. Interestingly, the human immune response to HBsAg is primarily directed against disulfide-bonded conformational epitopes of the 'a' determinant and can be classified into one of three distinct groups [14]. Alteration of these conformational epitopes presumably results in failure to neutralize viral infection.

Surface antigen mutants

The initial description of an HBsAg mutant was made in the breakthrough HBV infection of a child (subsequently vaccinated and passively immunized) born to a HBV-positive mother [15]. DNA sequencing of the child's vaccine escape virus was shown to contain a substitution mutation of glycine to arginine at HBsAg amino acid position 145 (Gly/Arg

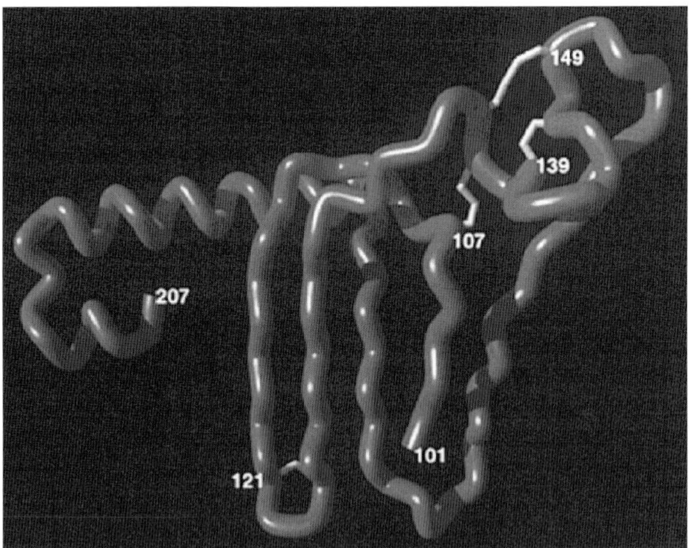

Fig. 1. Proposed conformational model of the 'a' determinant displaying amino acids 101–207 of HBsAg (see Ref. [13]).

145) [16]. The child subsequently remained positive for this mutant for a period of over 12 years despite having protective anti-HBs titer. The Gly/Arg 145 substitution alters the second projecting loop of the 'a' determinant such that neutralizing antibody for wild-type antigen no longer recognizes the altered sequence, hence the term vaccine escape mutant. While the Gly/Arg 145 substitution by far remains the predominant mutant described in the literature, a full array of substitution mutants (see Ref. [1]), some restricted insertion mutants [17–19], and a deletion mutant [20] have been described within the hydrophilic amino acid domain at positions 100–160 of HBsAg. Some of these substitution mutants appear to be of academic interest as they occur at very low levels in long-term HBV carriers and have only been elucidated by the use of highly amplified polymerase chain reaction (PCR) techniques. Care must be taken that the amplification technology itself is not responsible for artificial PCR-introduced mutations [21]. Other mutant isolates may be infrequently occurring natural variants [22].

Some investigators have proposed that with the increased acceptance of HBV vaccination, mutants will become the predominant HBV strains in several years time [23]. Others argue that HBsAg mutants are replication defective and lack the necessary viral transmissibility [24,25]. In either case, a study to assess the susceptibility of 9 HBsAg assay configurations to 25 defined and quantitated HBsAg recombinant mutants generated by site-directed mutagenesis was performed [26] as summarized in Tables 1 and 2. As the level of protein expression varies greatly for each recombinant HBsAg mutant depending on the nature of the mutation, it is extremely important to quantitate each recombinant protein to a standard antigen concentration prior to immunoassay testing or incorrect false negative results can be obtained. This lack of quantitation can account for some of the conflicting results for immunoassay detection of HBsAg mutants published in subsequent studies [27]. Table 1 shows the immunoassay detection of HBsAg mutants of known prevalence. In a study in Singapore, the Gly/Arg 145 mutation was present alone or in combination with other mutations in 70% of the isolated HBsAg mutants [28]. Of importance is that mutant constructs containing the commonly occurring Gly/Arg 145 mutation are not detected by some commercial assays and this false negative reaction is influenced somewhat by assay configuration but more so by the selection of assay reagents.

Table 2 shows the immunodetection of an expanded selection of less prevalent mutants. The data clearly show that substitution in and around the small projecting loop (amino acids 138–147) of the 'a' determinant reduces immunoreactivity in some diagnostic assays. Substitution at positions 40 and 196 outside of the 'a' determinant are readily detected by all assays. Position 120 appears somewhat unique in that it is near the proposed finger projection on the 'a' determinant but can still reduce immunoreactivity in Auszyme and Commercial assay A. In two cases, sufficient serum sample was available to demonstrate parallel reactivity of the recombinant protein and the original clinical sample.

The true challenge for HBsAg assays is detection of the rarely occurring insertion mutants shown at the bottom of Table 2. Here the HBsAg sequence derived from the clinical specimen itself was used to generate recombinant antigen. Only the polyclonal capture, polyclonal detection, assay configuration of Ausria was capable

Table 1

Immunoassay detection of 1 ng/ml recombinant HBsAg of known mutation prevalence by Abbott assays and commercial assays A, B, and C. All positive samples confirmed in their respective assays

Mutants/configuration	Ausria Poly/poly	Auszyme Mono/mono	IMx Mono/poly	AxSYM Mono/poly	PRISM Mono/poly	Architect Mono/poly	Assay A Mono/mono	Assay B Mono/mono	Assay C Poly/mono
Wild-type	++	++	++	++	++	++	++	++	++
Thr/Ser 126	++	+	+	++	++	++	+	++	++
Gln/His 129	++	+	+	++	++	++	+	++	++
Met/Leu 133	++	++	++	++	++	++	++	++	++
Asp/Ala 144	++	++	++	++	++	++	−	++	++
Gly/Arg 145	++	++	++	++	++	++	−	−	−
Thr/Ser126 + Gly/Arg145	++	++	+	++	++	++	−	−	−
Pro/Leu142 + Gly/Arg145	++	++	++	++	++	++	−	−	−
Pro/Ser142 + Gly/Arg145	++	++	++	++	++	++	−	−	−
Asp/Ala144 + Gly/Arg145	++	++	++	++	++	+	−	−	−

Key: ++ = detection equivalent to wild-type antigen, + = detection less than wild-type but still positive, − = not detected. [26].

Table 2

Immunoassay detection of 1 ng/ml recombinant HBsAg of lesser prevalent mutations by Abbott assays and other commercial assays. All positive samples confirmed in their respective assays

	Ausria Poly/poly	Auszyme Mono/mono	IMx Mono/poly	AxSYM Mono/poly	PRISM Mono/poly	Architect Mono/poly	Assay A Mono/mono	Assay B Mono/mono	Assay C Poly/mono
Configuration/mutants									
Wild-type	++	++	++	++	++	++	++	++	++
Asn/Ser 40	++	++	++	++	++	++	++	++	++
Pro/Thr 111	++	++	+	++	++	++	++	++	++
Thr-Thr/Ile-Ile 115,116	++	++	++	++	++	++	++	+	++
Thr/Ser 118	++	++	++	++	++	++	++	++	++
Pro/Gln 120	++	−	++	++	++	++	−	++	++
Serum	++	−	++				−		
Thr/Ile 131	++	++	++	++	++	++	++	++	++
Pro/Ser 135	++	++	++	++	++	++	−	−	++
Lys/Glu 141	++	−	++	++	++	++	−	−	++
Pro/Leu 142	++	++	++	++	++	++	−	−	++
Pro/Ser 142	++	++	++	++	++	++	++	−	++
Gly/Ala 145	++	++	++	++	++	++	+	++	++
Gly/Lys 145	++	++	++	++	++	++	−	−	−
Thr/His 148	++	+	++	++	++	++	−	++	++
Ser/Trp 154	++	+	+	++	++	++	−	−	−
MetMetMet/SerSerSer 196−198	++	++	++	++	++	++	++	++	++
Clinical seqs									
123 insert Arg + Ala (adw2)	++	−	−	−	−	+	−	−	++
123 insert Asp + Thr	++	−	−	−	−	−	−	−	−
Serum (ayw1)	++	−	−				−		

Key: ++ = detection equivalent to wild-type antigen, + = detection less than wild-type but still positive, − = not detected [26].

104

of detecting both insertion mutants at low levels, indicating that conserved epitopes with wild-type antigen exist even in the highly rearranged insertion mutants. Monoclonal antibodies against these conserved epitopes are being included in later generation assays as the Architect assay demonstrates by its detection of the Arg + Ala 123 insertion mutant.

These studies with recombinant surface antigen mutants underscore the usefulness of mapping the epitope susceptibility of HBsAg assays and their component reagents to enable the development of more robust assay formats for the future.

Polymerase mutants

As immune response presents a selective pressure for mutants in the surface antigen gene, so does nucleoside analogue therapy select for variants in the polymerase gene. Long-term treatment with lamivudine (LMV) has been associated with the selection of HBV strains that have reduced response to LMV [29]. RNA-dependent polymerases have a conserved catalytic domain composed of the amino acid sequence Tyr_{551}-Met-Asp-Asp_{554} or abbreviated as the YMDD motif [43]. Mutation of the active site methionine to either isoleucine or valine has been shown to be the primary mutation responsible for LMV resistance [29]. This mutation in the hepatitis B polymerase active site renders the virus defective in replication [30]. Secondary compensating mutations in the polymerase gene 'finger' region have been described which restores replication efficacy [31].

The effect of LMV polymerase mutations on HBsAg detection is of some concern as the same nucleotide (nt) sequence in different reading frames is utilized to produce both proteins. The YMDD motif mutants correspond to surface antigen substitution mutants at positions 195 or 196. These mutants occur well outside of the "a" determinant and they would be expected to have little impact on HBV serodetection as the 196–198 mutant in Table 2 demonstrates. However, the secondary LMV finger mutations occur in regions of the HBV genome that do overlap the "a" determinant and, therefore, could impact immunoassay detection. The Pro120 to Ser mutant was detected by HBsAg assays in our hands (unpublished results). Another secondary LMV mutation was shown to produce a stop codon the S-gene resulting in defective viral secretion [32].

In contrast to LMV, famciclovir therapy appears to induce mutations only upstream of the YMDD motif and preferentially in immunocompromised individuals [33]. In an interesting parallel, a famciclovir mutant isolated from an immunocompetent host was also shown to produce a stop codon in the S-gene producing a replication compromised virus.

With the advent of new nucleoside analogue treatments for HBV such as adefovir dipivoxil and entecavir, the option of using alternate therapy regimens is now available to treat YMDD mutants generated by LMV therapy [34,35]. These regimens will have to be monitored to assess their impact on S-gene function.

Precore/core mutants

The core gene is utilized in a single reading frame with two start codons to produce both the necessary structural hepatitis B core antigen (HBcAg) nucleocapsid protein and the hepatitis B e antigen (HBeAg) secretory protein. Secretion of HBeAg is not essential for viral disease progression [36]. HBeAg initially appeared to have a key immune modulation role in HBV-induced liver damage in fulminant hepatitis (Ogata et al., 1993) but subsequent studies have cast some doubt on the true biological role of this viral antigen [2]. The relatively common precore mutation of guanosine to adenine in nt position 1896 of the prototype HBV sequence, causes a tryptophan codon to mutate to a stop codon which prevents the synthesis of HBeAg [37]; [44]. Alternately, mutations in the basal core promoter region (adenine 1762 to threonine or guanosine 1764 to adnine) result in impaired secretion of HBeAg. Here the antigenicity of HBeAg is not altered and does not present a challenge to immunodetection, the effect is specifically on extent of antigen secretion. Most reports of the HBeAg-negative precore mutant have been from southern Europe and Asia. This variant is uncommon in the United States and northern Europe.

Mutations in the immunodominant epitopes of HBcAg have been described as a result of immune selective pressure on B- and T-cell epitopes [38]; however, since the major immunoassays utilize a polyclonal antibody detection step for HBcAg, these mutations would be predicted to have little effect on antigen detection.

X gene mutants

Several studies have shown that mutations within the X gene coding region can effect the detection of HBV diagnostic markers [39,40]. Most of these mutations seem to originate in the immunological epitopes of the X protein [41] which in itself is not commonly used as a marker for HBV infection. Investigators have found three kinds of X gene mutations which have important clinical significance. The first type of mutation is an eight nt deletion between nt 1770 and nt 1777, which truncates 20 amino acids from the carboxyl terminus of the X protein and probably damages the enhancer II/core promoter elements. This deletion mutation suppresses both the replication and expression of HBV DNA, which results in the absence of HBV markers in these patients (HBsAg, HBcAg, HBeAg, and their corresponding antibody response) despite the presence of confirmed viral infection. The second type of described mutation was substitution mutations and they seem to relate to fulminant hepatitis [41]. The third group of mutations was reported to be other HBV X gene deletion mutants that seem to be involved in hepatocellular carcinoma [42].

Conclusion

The occurrence of mutations within the HBV genome is a function of both a unique replicative cycle employing a RNA-dependant polymerase and the utilization of overlapping reading frames to produce multiple viral proteins. These mutants present a challenge to immunoassay manufacturers when alterations occur in immunodominant epitopes for the mutations may abrogate viral antigen detection depending on the immunoassay configuration. Surveillance of HBV mutants is an ongoing and seemingly open-ended undertaking as new treatment regimens are initiated in an effort to control

106

infection. Areas endemic for HBV infection would be a key location for these surveillance studies, as prevalence of infection would increase the chance that HBV mutants that challenge diagnostic detection would be isolated.

References

1. Carman W. The clinical significance of surface antigen variants of hepatitis B virus. J Viral Hepat 1997; 4: 11–20.
2. Brunetto M, Rodriguez U, Bonino F. Hepatitis B virus mutants. Intervirology 1999; 42: 69–80.
3. Lok A. Hepatitis B infection: pathogenesis and management. J Hepatol 2000; 32: 89–97.
4. Hunt M, McGill J, Allen M, Conndreay L. Clinical relevance of hepatitis B viral mutations. Hepatology 2000; 31: 1037–1044.
5. Ngui S, Hallet R, Teo C. Natural and iatrogenic variation in hepatitis B virus. Rev Med Virol 1999; 9: 183–209.
6. Weinberger K, Bauer T, Bohm S, Jilg W. High genetic variability of the group-specific a-determinant of hepatitis B virus surface antigen (HBsAg) and the corresponding fragment of the viral polymerase in chronic virus carriers lacking detectable HBsAg in serum. J Gen Virol 2000; 81: 1165–1174.
7. Mangold C, Streeck R. Mutational analysis of the cysteine residues in the hepatitis B virus small envelope protein. J Virol 1993; 67: 4588–4597.
8. Antoni B, Rodriguez-Crespo I, Gomez-Gutierrez J, Nieto M, Peterson D, Gavilanes F. Site-directed mutagenesis of cysteine residues of hepatitis B surface antigen. Eur J Biochem 1994; 222: 121–127.
9. Peterson D, Paul D, Lam J, Tribby I, Achord D. Antigenic structure of hepatitis B surface antigen: identification of the 'd' subtype determinant by chemical modification and use of monoclonal antibodies. J Immunol 1984; 132: 920–927.
10. Ohnuma H, Takai E, Machida A, Tsuda F, Okamoto H, Tanaka T, Naito M, Munekata E, Miki K, Miyakawa Y, Mayumi M. Synthetic oligopeptides bearing a common or subtypic determinant of hepatitis B surface antigen. J Immunol 1990; 145: 2265–2271.
11. Qiu X, Schroeder P, Bridon D. Identification and characterization of a C(K/R)TC motif as a common epitope present in all subtypes of hepatitis B surface antigen. J Immunol 1996; 156: 3350–3356.
12. Folgori A, Tafi R, Meola A, Felici F, Galfre G, Cortese R, Monaci P, Nicosia A. A general strategy to identify mimotopes of pathological antigens using only random peptide libraries and human sera. EMBO J 1994; 13: 2236–2243.
13. Chen Y-C, Delbrook K, Dealwis C, Mimms L, Mushahwar I, Mandecki W. Discontinuous epitopes of hepatitis B surface antigen derived from a filamentous phage peptide library. Proc Natl Acad Sci USA 1996; 93: 1997–2001.
14. Maillard P, Pillot J. At least three epitopes are recognized by the human repertoire in the hepatitis B virus group a antigen inducing protection; possible consequences for seroprevention and serodiagnosis. Res Virol 1998; 149: 153–161.
15. Zanetti A, Tanzi E, Manzillo G, Maio G, Sbreglia C, Caporaso N, Thomas H, Zuckerman A. Hepatitis B variant in Europe. Lancet 1988; 8620: 1132–1133.
16. Carman W, Sanetti A, Karayiannis P, Waters J, Manzillo G, Tanzi E, Zuckerman A, Thomas H. Vaccine-induced escape mutant of hepatitis B virus. Lancet 1990; 336: 325–329.
17. Yamamoto K, Horikita M, Tsuda F, Itoh K, Akahane Y, Yotsumoto S, Okamoto H, Miyakawa Y, Mayumi M. Naturally occurring escape mutants of hepatitis B virus with various mutations

in the S gene in carriers seropositive of antibody to hepatitis B surface antigen. J Virol 1994; 68: 2671–2676.

18. Carman W, Korula J, Wallace L, MacPhee R, Mimms L, Decker R. Fulminant reactivation of hepatitis B due to envelope protein mutant of HBV that escaped detection by monoclonal HBsAg ELISA. Lancet 1995; 345: 1406–1407.

19. Hou J, Karayiannis P, Waters J, Lou K, Liang C, Thomas H. A unique insertion in the S gene of surface antigen-negative hepatitis B virus Chinese carriers. Hepatology 1995; 21: 273–278.

20. Weinberger K, Zoulek G, Bauer T, Bohm S, Jilg W. A novel deletion mutant of hepatitis B virus surface antigen. J Med Virol 1999; 58: 105–110.

21. Gunther S, Sommer G, von Breunig F, Iwanska A, Kalinina T, Sterneck M, Will H. Amplification of full-length hepatitis B virus genomes from samples from patients with low levels of viremia: frequency and functional consequences of PCR-introduced mutations. J Clin Microbiol 1998; 36: 531–538.

22. Carman W, Van Deursen F, Mimms L, Hardie D, Coppola R, Decker R, Sanders R. The prevalence of surface antigen variants of hepatitis B virus in Papua New Guinea, South Africa, and Sardinia. Hepatology 1997; 26: 1658–1666.

23. Wilson J, Nokes D, Carman W. Predictions of the emergence of vaccine-resistant hepatitis B in The Gambia using a mathematical model. Epidemiol Infect 2000; 124: 295–307.

24. Ogata N, Zanetti A, Yu M, Miller R, Purcell R. Infectivity and pathogenicity in chimpanzees of a surface gene mutant of hepatitis B virus that emerged in a vaccinated infant. J Infect Dis 1997; 175: 511–523.

25. Ogata N, Cote P, Zanetti A, Miller R, Shapiro M, Gerin J, Purcell R. Licensed recombiant hepatitis B vaccines protect chimpanzees against infection with the prototype surface gene mutant of hepatitis B virus. Hepatology 1999; 30: 779–786.

26. Coleman P, Chen Y-C, Mushahwar I. Immunoassay detection of hepatitis B surface antigen mutants. J Med Virol 1999; 59: 19–24.

27. Ireland J, O'Donnell B, Basuni A, Kean J, Wallace L, Lau G, Carman W. Reactivity of 13 in vitro expressed hepatitis B surface antigen variants in 7 commercial diagnostic assays. Hepatology 2000; 31: 1176–1182.

28. Oon C, Lim G, Ye Z, Goh K, Tan K, Yo S, Hopes E, Harrison T, Zuckerman A. Molecular epidemiology of hepatitis B virus vaccine variants in Singapore. Vaccine 1995; 13: 699–702.

29. Ling R, Mutimer D, Ahmed M, Boxall E, Elias E, Dusheiko G, Harrison T. Selection of mutations in the hepatitis B virus polymerase during therapy of transplant recipients with lamivudine. Hepatology 1996; 24: 711–713.

30. Melegari M, Scaglioni P, Wands J. Hepatitis B virus mutants associated with 3TC and famcicloir administration are replication defective. Hepatology 1998; 27: 628–633.

31. Bock T, Tillman H, Torresi J, Klempnauer J, Locarnini S, Manns M. Selection of hepatitis B virus polymerase mutants with enhanced replication by lamivudine treatment after liver transplantation. Gastroenterology 2002; 122: 264–273.

32. Yeh C, Chien R, Chu C, Liaw Y. Clearance of the original hepatitis B virus YMDD-motif mutants with emergence of distinct lamivudine-resistant mutants during prolonged lamivudine therapy. Hepatology 2000; 31: 1318–1326.

33. Aye T, Bartholomeusz A, Shaw T, Bowden S, Breschkin A, McMillan J, Angus P, Locarnini S. Hepatitis B virus polymerase mutations during antiviral therapy in a patient following liver transplantation. J Hepatol 1997; 26: 1148–1153.

34. Tassopoulos N, Hadziyannis S, Cianciara J. Entecavir is effective in treating patients with chronic hepatitis B who have failed lamivudine therapy. Hepatology 2001; 34: 340A.

35. Peters M, Hann H, Martin P. Adefovir dipivoxil alone and in combination with lamivudine suppresses lamivudine-resistant hepatitis B virus replication, 16 week interim analysis. J Hepatol 2002; 36: S6–S7.
36. Tong S, Li J, Vitvitski L, Trepo C. Replication capacities of natural and artificial precore stop codon mutants of hepatitis B virus: relevance of pregenome encapsidation signal. Virology 1992; 191: 237–245.
37. Carman W, Fagan E, Hadziyannis S, Karayiannis P, Tassopoulos N, Williams R, Thomas H. Association of a precore genomic variant of hepatitis B virus with fulminant hepatitis. Hepatology 1991; 14: 219–222.
38. Chuang W, Omata M, Ehata T. Precore and core clustering mutations in chronic hepatitis B virus infection. Gastroenterology 1993; 104: 263–271.
39. Uchida T, Shimojima M, Gotoh K, Shikata T, Tanaka E, Kiyosana K. "Silent" hepatitis B virus mutants are responsible for non-A, non-B, non-C, non-D, non-E hepatitis. Microbiol Immunol 1994; 38: 281–285.
40. Uchida T, Gotoh K, Shikata T. Complete nucleotide sequences and the characteristics of two hepatitis B virus mutants causing serologically negative acute or chronic hepatitis. J Med Virol 1995; 45: 247–252.
41. Uchida T, Takafumi S, Shiuzana H. Mutations of the X region of hepatitis B virus and their clinical implications. Pathol Int 1997; 47: 183–193.
42. Hsia C, Nakashima Y, Tabor E. Deletion mutants of the hepatitis B virus X gene in human hepatocellular carcinoma. Biochem Biophys Res Commun 1997; 241: 726–729.
43. Poch O, Sauvaget I, Delarue M, Tordo N. Identification of four conserved motifs among the RNA dependent polymerase encoding elements. EMBO J 1989; 8: 3867–3874.
44. Ogata N, Miller R, Ishak K, Purcell R. The complete nucleotide sequence of a pre-core mutant of hepatitis B virus implicated in fulminant hepatitis and its biological characterization in chimpanzees. Virology 1993; 194: 263–276.

Viral Hepatitis
I.K. Mushahwar (editor)
© 2004 Abbott Laboratories. All rights reserved.

Hepatitis B virus: life cycle and morphogenesis

Larry G. Birkenmeyer
Core Research and Development, Abbott Diagnostics Division, Abbott Laboratories, Abbott Park, Illinois, USA

This chapter will cover the life cycle of the hepatitis B virus from infection of the host cell to secretion of mature virions, with emphasis on research published over the last few years. The proteins and nucleic acids directly involved in this process, and how they interact with one another to mediate infection, replication and viral morphogenesis will be discussed. In addition, an update on splicing of viral transcripts, the novel proteins that result and their potential significance will be presented. Other topics including genomic variability, viral gene expression, antiviral therapy, epidemiology and vaccines have been covered by recent reviews [1–10] and will not be addressed here.

General virology

The hepadnaviridae family of viruses [11] contains members that infect birds (ducks [12], herons [13] and geese [14]), rodents (ground squirrels [15,16] and woodchucks [17,18]) and apes (woolly monkey [19], chimps [20,21], gibbons [22,23], orang-utans [24], gorillas [25] and humans [26]). Phylogenetic analysis supports the presence of indigenous strains in each of these ape species [23–25], and the recently reported woolly monkey virus may exhibit an evolutionary relationship intermediate between the rodent and human viruses [19]. Although hepatocytes are the primary site of viral infection and replication in these species, other possible extra-hepatic sites of replication have been reported, including the pancreas, kidney, bile duct epithelial cells, spleen and peripheral blood mononucleocytes (PBMCs) [27,28]. In some of these cases the evidence might be explained by virion adsorption at extra-hepatic sites without subsequent infection [29]. However, for PBMCs, the presence of HBV sequences different than those found in the patient serum [30], and the detection of viral DNA replication intermediates [31] supports these cells as a site of productive viral infection.

The structure of the mature virus has been covered in recent reviews [32,33]. Briefly, the virion is a spherical particle approximately 42 nm in diameter consisting of an outer envelope comprising lipids [34] (up to 30% of the particle mass [35]) and three related envelope proteins, large (L), middle (M) and small (S) [36]. Individual virions contain approximately 40–80 copies each of L and M and 360–400 copies of S [37,38]. Inside the envelope is an icosahedral capsid (32–36 nm in diameter) constructed of 180 or 240 core proteins [39,40]. Within the capsid is a single partially double-stranded DNA genome [41] approximately 3.2 kb in length, and the viral polymerase [42,43] (1–2 copies [44,45]). At least two host-derived proteins, a kinase [46,47] and a chaperone protein [48], are contained within, or are tightly associated with, the capsid. Subviral serum particles [26] (20–22 nm in width), consisting of the three envelope proteins in various ratios [37] but containing no capsid or nucleic acid, are present in large excess vs. mature virus.

There are 7 viral proteins encoded by 4 major overlapping open reading frames (ORFs) present in the human HBV genome [32,33] (see Fig. 1). The three envelope glycoproteins (S, M and L), encoded by the preS/S gene, are initiated from three separate in-frame start codons and terminate at a common stop codon. S, also known as the major surface antigen of HBV (HBsAg), is 226 amino acids long. In addition to the complete S region, M and L possess adjacent N-terminal extensions of 55 (preS2 region) and, depending on subtype, 108–119 (preS1 region) amino acids, respectively, forming proteins 281 (M) and 389–400 (L) residues in length. Core protein (183–185 amino acids) and the multi-functional viral polymerase (up to 845 amino acids) are also associated with the virion and are derived from the C and P ORFs, respectively. A domain involved in self-assembly of core proteins into capsids comprises most of the molecule

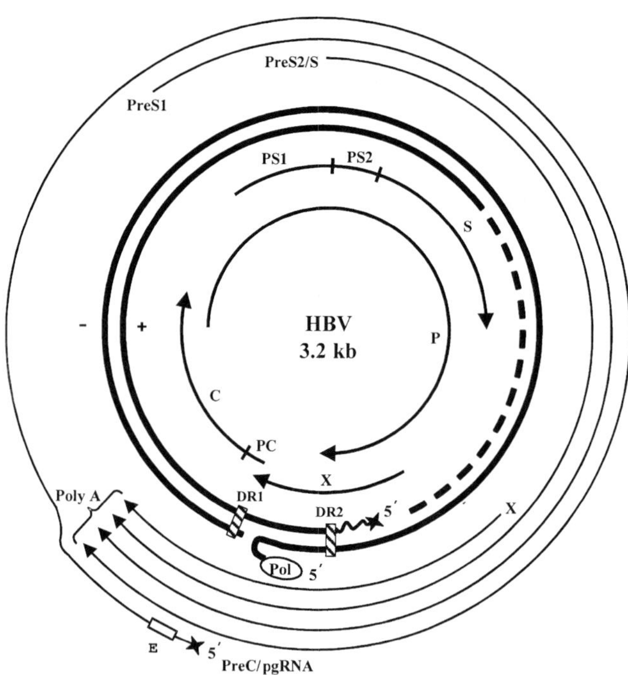

Fig. 1. Genomic organization of HBV. The minus (−) and plus (+) viral DNA strands (bold lines) are depicted associated with their 5′-modifications [polymerase (Pol), RNA primer (wavy line) and 5′ cap structure (star)]. The gapped nature of the DNA plus strand is indicated by the dashed line. Major viral transcript classes and open reading frames are shown around the perimeter and in the center, respectively. Abbreviations are as follows: PreC/pgRNA, precore/pre-genomic RNA; E, epsilon sequence; DR, direct repeat; PC, precore; C, core; PS, preS; P, polymerase.

while the extreme C-terminal portion resembles a protamine-like domain involved in nucleic acid binding. Four functionally distinct regions make up the polymerase protein, (i) a terminal protein domain (TP) involved in replication, (ii) a spacer region, (iii) the reverse transcription domain (RT), and (iv) the RNaseH domain.

The two remaining viral proteins, X and precore (also known as the HBV "e" antigen or HBeAg) are not part of the mature virion X (154 amino acids), a pleiotropic trans-activator of both viral and host genes, is encoded by the X gene. Recent reviews [4,9,49] covering various aspects of X protein expression and function have been published. HBeAg translation is initiated from an upstream start codon in-frame with core. It is initially identical to core except for an additional 29 N-terminal amino acids. Post-translational cleavage at both ends leads to the mature secretable form of HBeAg (157–170 amino acids) [50]. HBeAg is typically thought to be involved in maintenance of immune tolerance in the host, although other roles are possible.

Post-transcriptional processing

There are four major size classes of polyadenylated HBV transcripts, all of which share a common $3'$-end [33,51,52]. The largest (3.5 kb) is longer than genome length and thus terminally redundant [53]. It contains the pre-genomic RNA (pgRNA) which, in addition to serving as a template for genomic replication, functions as a bi-cistronic mRNA for translation of the viral core and polymerase proteins [54]. A slightly longer transcript is initiated about 30 nucleotides upstream of pgRNA and is utilized solely for translation of the precore protein [55,56]. The L envelope protein is encoded by a 2.4-kb transcript while the M and S envelope proteins are translated from a set of 2.1-kb transcripts with heterogeneous $5'$-ends. Finally, a low level transcript of 0.8 kb likely serves as message for translation of the X protein.

Splicing of viral transcripts is not thought to be necessary for successful completion of the viral life cycle [57]. A post-transcription regulatory element (PRE) needed for export from the nucleus of intronless mRNAs has been identified [58,59]. It is bipartite in nature and is present in a region covered, at least partially, by all four classes of viral transcripts [60]. Orientation, but not position, of PRE within a transcript is critical for its proper function which is 3-fold: (i) promotion of cleavage/polyadenylation of intronless mRNAs, (ii) inhibition of splicing activity, and (iii) promotion of nuclear export of unspliced/intronless mRNAs [61].

Although not apparently required, an *in vivo* role for splicing of HBV transcripts has been suggested by several studies. For example, a spliced L envelope protein transcript was reported to be essential for replication of duck HBV (DHBV) in infected primary hepatocytes and in duck livers [62]. In humans, multiple spliced viral RNAs and their derived genomes have been detected [63], particularly in chronic carriers [64]. For HBV genotypes A, C, D and E, viral particles with DNA derived from a single dominant splice variant was detected in two different hepatoma cell lines and in patient sera, suggesting a conserved biological function [65]. Distribution analysis of viral particles containing this variant indicated that they were much more likely to be retained intracellularly than to be secreted, consistent with an intracellular role for the variant.

Proteins generated as a result of splicing of viral transcripts have been reported. Antibodies to a novel protein (HBSP) generated by expression of a singly spliced HBV RNA, was detected in a third of the chronic carriers tested [66]. The HBSP protein, a fusion of part of the viral polymerase with a different open reading frame, was also found to be present in infected liver samples. In another case, splicing of the pgRNA transcript led to expression of a novel 43 kd polymerase/S fusion glycoprotein which was detected in cell lysates, viral particles and 22-nm subviral particles [67]. Clearly, splicing of HBV RNA transcripts is much more prevalent than previously thought, resulting in protein fusion products both within the cell and in viral particles. As our understanding increases of the interactions between virus and host in natural infections, it will not be surprising to find that splicing is involved, adding yet another layer of complexity to a rapidly expanding area of study.

Sequences associated with splicing may affect other aspects of RNA processing. A conserved 30 nucleotide region, encompassing both a splice donor and acceptor site within the overlapping polymerase and S genes, appears to be essential for maintaining stability of viral transcripts in cis [68]. Its elimination results in a nearly complete loss of the 2.1 kb preS2/S transcript in transfected cells even though transcriptional initiation remains unchanged. In contrast, a 150 nucleotide region covering part of the enhancerII/ core promotor complex behaves as a negative cis-regulator of transcript stability [69]. The orientation dependency for its function suggests that it is involved in destabilization of RNAs transcribed from the viral DNA plus strand, opposite to that of all other viral transcripts. Thus, it may serve to eliminate aberrant transcripts of the wrong orientation which could interfere with normal expression.

Virus uptake

Identification of the primary host cell receptor required for binding to, and infection of, human hepatocytes has been elusive [70]. In fact, it seems likely that several host proteins will be involved in the process which is both tissue and species restricted [71,72]. This appears to be the case for DHBV in which binding and uptake of the virus involves at least two host proteins, duck carboxypeptidase D (DCPD) [73–75] and the p protein component of the duck glycine decarboxylase (DGD) complex [76,77]. Both proteins bind to the same general region of the DHBV preS domain, and mutations within this region, or monoclonal antibodies specific for this region of preS, curtail DHBV binding and/or infection of duck hepatocytes. The p protein is present principally in tissues infected by DHBV, and is located both on the cell surface and in the cytoplasm [76]. Conversely, DCPD, which also appears on the cell surface, exhibits a widespread tissue distribution yet still confers species specificity [74]. Inactivation of the enzymatic carboxypeptidase domains of DCPD does not result in a loss of receptor function, indicating a structural role with regards to DHBV uptake [78].

Numerous candidates for the human HBV hepatocyte receptor(s) involving interaction with the envelope proteins have been reported [70]. It was hypothesized that polymerized human serum albumin, which binds to the preS2 domain of HBV and also to the cell surface of hepatocytes, forms a potential bridge leading to infection.

However, the binding is not species specific, and does not seem to be relevant under physiological conditions [79]. The preS2 domain has also been implicated in hepatocyte binding through its N-linked glycan [80] and by association with fibronectin found in the extra-cellular matrix of human liver sinusoids [81]. Interaction between transferrin and the preS2 domain has also been implied [82], but there is no direct evidence supporting this association.

Participation of the HBV L protein in uptake by hepatocytes is supported on several fronts. N-terminal myristylation of L, which is unique to the L envelope protein [83], is required for HBV infectivity, but not virion formation, in both humans [84,85] and in ducks [86,87]. Binding has been demonstrated between preS1 and isolated hepatocyte membranes [88–90], although it was not clear to which side of the membrane this occurred. Studies involving mutagenesis, synthetic peptides and antibodies suggest that amino acid residues 21–47 of L are needed for hepatocyte binding [89,91]. However, mutation analysis indicates that most of the preS1 domain is needed for infection of primary human hepatocytes [92].

Potential host cell receptor proteins with reported binding to preS1 include glyceraldehyde-3-phosphate-dehydrogenase (GAPD) [93], IgA receptor [94], IL-6 [95], asialoglycoprotein receptor [96] and several unidentified proteins including a cellular surface protein (p80) present on multiple cell types [97], a soluble 50-kd serum glycoprotein [98] and a 31-kd liver cell protein [99].

Finally, the S envelope protein is almost certainly involved in the initial steps of infection. Hepatitis B surface antigen (HBsAg) binds to hepatocytes [90,100], and this binding is inhibited by antibody to HBsAg and by competing recombinant HBsAg [101, 102]. Binding of HBsAg by two liver-associated plasma membrane proteins, apolipo-protein H (apo-H) [101] and annexin V (also known as endonexin II) [103], has suggested them as potential host cell receptors. Apo-H is primarily a serum glycoprotein which can undergo a membrane-induced conformational shift [104]. This shift may mimic that reported when apo-H is complexed with a monoclonal antibody, resulting in a 100-fold increase in affinity for HBsAg binding [102]. Thus, the shift in conformation could specifically prepare membrane-bound apo-H for binding to HBsAg. Apo-H has also been shown to associate with the gag and env proteins of HIV [105].

Annexin-V, a Ca^{2+}-dependent phospholipid-binding protein, binds HBsAg in a species-specific manner [103]. Rat hepatoma cells expressing [106], or simply coated with [107], human annexin-V are productively infected with HBV, but the rat annexin-V cannot be substituted even though it has 90% amino acid sequence homology with the human protein. HBsAg amino acid residues 125–131 and 158–169, in conjunction with phosphatidylserine, seem to participate in formation of a conformational epitope important for annexin-V binding [108,109]. Annexins have also been reported to be associated with early steps of cytomegalovirus [110] and influenza virus [111] infections. The large number of host proteins that appear to associate with the HBV envelope proteins suggest that most such interactions are the result of a general stickiness of the envelope proteins, and so may not be relevant to infections *in vivo*. Definitive identification of the HBV receptor(s) and understanding of how it mediates infection remains to be determined.

114

Following binding to the receptor, capsid release of enveloped viruses into the host cell requires fusion of the viral envelope with a host membrane, either at the cell surface or after an energy-dependent internalization into an intracellular compartment. Nearly all of the information pertaining to this process comes from the DHBV animal model. DHBV, complexed with carboxypeptidase D, appears to be endocytosed in an energy-dependent fashion [75]. Instead of the normal targeting to the golgi or lysosome, it is maintained endosomally [78] where, it is proposed, binding to a putative secondary receptor initiates nucleocapsid release into the cytoplasm by membrane fusion. Carboxypeptidase D mutants, deficient in endocytosis, were blocked in their ability to be infected by DHBV [78].

Properties of the viral envelope must also be considered when studying release of the nucleocapsid into the cell. A mixed membrane topology of the L protein, formed post-translationally, has been described for both human and duck HBV, wherein approximately half of the preS domains are displayed on the virus surface and the other half are sequestered inside the envelope [112]. Mature DHBV envelope particles exhibit some L proteins in a partially translocated or translocation-ready conformation [113]. Furthermore, a normally hidden hydrophobic portion of the L protein S domain apparently becomes exposed on the virus surface of DHBV particles exposed to conditions which cause envelope proteins of viruses to shift from a metastable to a fusion-competent state [114]. Thus, at least a portion of the viral particle L proteins are deemed to exist in a "spring loaded" conformation which may facilitate virus uncoating during infection. It is interesting to note that the hepadnaviruses have an amino acid sequence at the N-terminus of the S domain that resembles membrane fusion sequences of other viruses [115]. Synthetic peptides corresponding to this region from avian and mammalian HBV strains promote fusion of lipid bilayers via membrane insertion [116–118].

Replication

Following release into the cytoplasm, the nucleocapsid is routed to the nucleus (to be discussed below) and the viral DNA is introduced into the nucleoplasm. Although the associated viral polymerase has been reported to be involved [119,120], the details of this latter step are largely unknown. The relaxed circular DNA (rcDNA) of the virion is converted from its partially double-stranded nicked form (see Fig. 1) into a covalently closed circular DNA (cccDNA). This must entail removal of the covalently attached polymerase from the 5′-end of the minus strand, and extension of the incomplete plus strand to full-length along with excision of its capped 5′-terminal RNA primer. Inhibitor studies show that completion of the plus strand and formation of cccDNA apparently does not require the viral polymerase activity [121], although it may help perform this function under normal conditions. Both strands are then ligated to generate cccDNA. Further studies are needed to clarify the details of cccDNA formation. The host cell RNA polymerase II enzyme uses the cccDNA as a template for transcription of the terminally redundant 3.5 kb pgRNA which is polyadenylated and transported into the cytoplasm for translation [32].

As stated previously, the pgRNA serves as template for translation of the viral core and polymerase proteins [54]. A stem-loop structure (termed epsilon) present at the $5'$-end of pgRNA is co-translationally bound by the nascent polymerase resulting in cis-binding by each polymerase molecule to its own mRNA [122]. Epsilon is also present at the $3'$-end of the terminally redundant pgRNA but is somehow prevented from forming a complex with the polymerase. Properties of epsilon that are important for generating a functional pgRNA/polymerase complex include both its primary sequence and secondary structure [122] and the nearby $5'$ cap structure which may explain why only the $5'$-copy of epsilon is utilized [123]. Surprisingly, only specific positions (less than 10% of the residues) in epsilon are required to be ribonucleotides for function *in vitro* based on a study using DHBV [124].

Simple binding of epsilon by the polymerase is not sufficient for progression to subsequent steps [125,126]. An induced fit model [126] has been proposed for generation of an active complex, consistent with the binding-induced conformational changes observed both in the epsilon sequence [127] and in the RT domain of the polymerase [128]. Formation of the functional pgRNA/polymerase complex appears to be facilitated by binding to cellular chaperon proteins [129]. Involvement of the avian hsp90/p23 chaperon complex requiring ATP hydrolysis [48,130] and two human chaperons proteins, the ER glycoprotein GRP94 [131], and human hsp90 [132] has been reported. The C-terminal regions of both the TP and RT polymerase domains interact with hsp90 independently [132].

RT of the pgRNA by the associated polymerase occurs cytoplasmically, most likely within the assembled capsid. Neither the polymerase nor the pgRNA can be encapsidated without first forming the epsilon/polymerase complex [122,133]. Further details of the nucleocapsid assembly are provided below. Interestingly, a majority of the viral polymerase molecules in DHBV is not encapsidated but instead is associated with a large cytoplasmic structure in post-translationally modified forms, only the smallest of which appears to be encapsidated [134]. The same may be true for human HBV. Further studies are needed to determine the relevance of this observation.

Minus strand synthesis is initiated by the hydoxyl group of the HBV polymerase amino acid residue Tyr63 [135], which is part of a GxY domain common to picornavirus and other hepadnavirus polymerases. A 6 base bulge in the stem portion of epsilon serves as template for the synthesis of the first 4 minus strand nucleotides [122,133]. *In vitro* experiments were done to define the polymerase regions involved in this priming reaction. They are amino acid residues 20–175 (optimally 1–199) of the TP domain which overlaps the spacer domain, and amino acid residues 300–775 (optimally 300–800) which contains the RT and most of the RNaseH domains [136]. These two polymerase sections separated by the spacer domain are tightly associated with one another [135]. Amino acid regions required for this interaction are still ill-defined but are a significantly smaller subset of those required for priming [136].

The primer along with the covalently attached polymerase is transferred to a sequence complementary to the primer near the $3'$-end of the pgRNA at direct repeat sequence 1 (DR1) [122,133] (see Fig. 1). This transfer reaction directly involves the polymerase and marks a transition of the polymerase from a phosphonoformic acid (PFA) resistant priming mode to a PFA sensitive elongation mode [137]. However, primer transfer is not required for the switch in modes. Other factors that affect the transfer are the length and/ or sequence of the primer produced from the bulge template [138], and the transfer distance between donor and acceptor sites with a genome length of 3.2 kb being favored [139]. A model has been proposed in which the transferred polymerase/primer complex scans bi-directionally for a suitable acceptor site starting approximately 3.2 kb from the priming site [139]. Following transfer, the polymerase extends the primer to the $5'$-end of the pgRNA template, completing RT of the minus strand. Full-length DNA/RNA hybrids have been detected in virions isolated from plasma and liver, indicating that the polymerase-mediated digestion of the pgRNA template occurs after synthesis of the DNA minus strand [140]. The RNaseH activity of the viral polymerase degrades the pgRNA except for the capped $5'$-terminal 157–18 nucleotides, the $3'$-end of which overlaps the $5'$-copy of DR1 [122,133].

Plus strand DNA synthesis is initiated from within direct repeat sequence 2 (DR2) near the $5'$-end of the minus strand template, using as a primer the residual 15–18 nucleotide long-capped $5'$-end of pgRNA. This requires directed transfer of the RNA oligonucleotide from DR1 to DR2 [122,133]. Kinetically, it seems likely that the RNA primer transfer is governed by the polymerase, as appears to be the case for the minus strand primer [137], since both the RNA primer and the minus strand template exist as single copies within the capsid. The viral polymerase extends the RNA primer the short distance to the $5'$-end of the minus strand template. A third and final switch now occurs. The $3'$-end of the nascent plus strand is transferred to a short complementary terminal redundancy (r) overlapping DR1 at the extreme $3'$-end of the minus strand, thereby circularizing the genome [141]. The length of the plus strand prior to this switch affects the efficiency with which the switch occurs [142]. A short cis-acting sequence, the upstream binding sequence (UBS), located at the $5'$-end of pgRNA between epsilon and DR1, is essential for the generation or elongation of the plus strand primer [143]. Binding of a host nuclear 65-kd protein to the UBS RNA sequence positively correlates with successful plus strand priming [143]. Since this priming step typically takes place in cytoplasmic capsids, the 65-kd protein may be exported from the nucleus in a complex with pgRNA. It should be noted, however, that an intact capsid is not required for plus strand synthesis [144]. Finally, the transferred plus strand $3'$-end is incompletely extended, resulting in a partially double-stranded rcDNA with a nicked minus strand covalently bound to the polymerase at its $5'$-end, and a gapped plus strand with a capped RNA primer still attached at its $5'$-end [32,122]. Once replication is complete, the new nucleocapsid must be routed back to the nucleus to repeat the cycle of non-semi-conservative replication, or to the membrane-bound envelope proteins for envelopment and secretion.

Capsid formation/structure

Core proteins, which make up the capsid, are divided into two major regions based on function: (i) a highly alpha helical capsid assembly domain (amino acids 1–149), and (ii) a basic protamine-like domain involved in RNA packaging into the capsid (amino acids 150–183) [145]. The core proteins spontaneously form dimers [146,147], a process involving amino acid residues 78–117 [148], and the dimers then self-assemble into capsid particles, via interactions between the alpha helical regions [149]. Specifically, two regions participate in the multimerization of dimers into capsids, amino acids from the N-terminus to residue 73 and amino acids 120–143 [148,150]. In fact, a single mutation, wherein Pro138 is changed to Gly, can prevent the full-length core protein from self-assembling into particles [151]. Although disruption of the region involved in dimerization often prevents capsid self-assembly, this is not always the case [148]. Deletion of amino acid residues 77–93 disrupts the dimerization region but does not prevent particle self-assembly [150].

Truncated core proteins lacking the protamine domain, form "T"-shaped dimer structures in which the tips of the cross bar make the inter-dimer contacts needed for further polymerization, and the stem is a bundle of four amphipathic alpha-helical hairpins (two from each subunit) constituting the dimer interface [152–154]. This differs from most other viral capsid structures. However, a similar structure was determined for HIV-1 capsid protein, suggesting a common evolutionary origin [155]. The immunodominant loop of the core protein (amino acid residues 78–83) lies at the tip of the 30 Å long stems which form outward pointing spikes on the capsid surface [152,154,156].

Capsid phosphorylation

The arginine-rich C-terminal domain of HBV core contains three identified phosphorylation sites at serine residues 155, 162, and 170 (subtype ayw) [157], and the phosphorylation state of core has been reported to influence encapsidation, replication and targeting within the cell [157–159]. Since no difference was observed in the overall phosphorylation level between core dimers and assembled particles containing both polymerase and viral RNA, it was reported that core protein is phosphorylated prior to assembly into capsids [160]. However, other studies indicate that phosphorylation by a phosphokinase (PK) occurs within the assembled capsid [159]. Either way, the phosphorylation state of core has no marked effect on formation of the capsid shell. In contrast, encapsidation of pgRNA requires phosphorylation of Ser162 and is further optimized by phosphorylation at Ser170 while the state of Ser155 is of little consequence [160,161]. This is consistent with the observation that phosphorylation occurs prior to initiation of DNA synthesis.

All three serine phosphorylation sites mentioned above form part of a three-times repeated SPRRR sequence (Ser-Pro-Arg-Arg-Arg), which overlaps the nuclear location motif of core [162,163]. Localization of capsids to the nucleus is, in fact, dependent on the repeated SPRRR motifs. An *in vitro* study demonstrated that phosphorylation of Ser170 was required for targeting of core particles to the nuclear core complex [164]. In a proposed model, phosphorylation of the core protein promotes a conformational change allowing the nuclear targeting portion of the normally sequestered C-terminus of core [165] to be exposed on the capsid surface, thus promoting transport to the nucleus [164]. Once located to the nuclear membrane, the capsid is too large [166], or is otherwise

118

barred from entering the nucleus through the nuclear pores [167]. Exactly how the viral DNA subsequently enters the nucleoplasm is unknown, although the viral polymerase may be sufficient for this function [119,120].

Reported inhibition of core nuclear localization by phosphorylation [157] is in conflict with the above observations, unless unassembled non-phosphorylated core proteins are preferentially taken into the nucleus. Alternatively, core sequences detected inside the nucleus may actually represent non-phosphorylated precore proteins, which have had the signal peptide cleaved off, but retain an intact C-terminus. In fact, in the absence of precore expression, core sequences were detected only in the cytoplasm and not the nucleus [168]. Consistent with these observations are reports of signal cleaved precore proteins being released into the cytoplasm and localizing to the nucleus [169–171].

In light of the apparently important role of core phosphorylation in the viral life cycle, the association of a PK activity with the virion is not surprising. PK-C, apparently associated with the C-terminal region of the core protein, has been detected in purified capsids by Western blot [159]. Another PK activity, resembling that associated with the liver form of glyceraldehyde-3-phosphate dehydrogenase (GAPD), was detected in similar preparations. It was further demonstrated that, GAPD phosphorylates purified capsids *in vitro*. However, the site(s) of phosphorylation by GAPD on the core protein was not reported and the protein could not be detected by Western blot [172]. A 46 kd serine-PK, capable of specifically binding and phosphorylating the core protein C-terminus, was also reported to be associated with purified capsids [173]. As a whole, these studies suggest the possibility of encapsidation of multiple PK activities.

Envelopment of capsid

Secretion of a mature virion requires that nucleocapsid particles, instead of being targeted to the nucleus, be routed to the membranes of the compartment between the endoplasmic reticulum (ER) and the golgi apparatus, which contains the viral surface proteins needed to initiate envelopment [174]. Since nuclear localization of the capsid may be regulated through phosphorylation (see above), it is possible that an alternative phosphorylation state might be involved in routing of the capsid to the ER membrane. Whether this or some other specific mechanism controls targeting, or if the intracellular distribution of capsids simply reflects a dynamic equilibrium between nuclear and ER localization is not known.

Envelopment proceeds through specific interactions between the membrane-bound envelope proteins and the capsid. The three envelope proteins are inserted co-translationally into the ER membrane, but only the L and S forms are required for virion formation [175,176]. Trimming, by glycosidase, of the *N*-glycan of the envelope glycoproteins is required for proper routing of these proteins, and in fact, inhibition of glycosidase prevents secretion of mature virions [177]. The N-terminal preS domain of L is initially maintained in the cytoplasm and not inserted co-translationally into the membrane. As mentioned earlier, about 50% of L proteins subsequently undergo a post-translational translocation of the preS domain to the ER lumenal space [112]. Those on the lumenal side are thought to participate in host cell binding while those that remain on the cytoplasmic side likely serve as a matrix-like protein (not found in the hepadnaviridae) to promote interaction with the capsid during envelopment. Prevention of co-translational translocation of the preS domain is dependent on a cytosolic anchorage determinant (preS1 amino acid residues 70–94) [178]. The hsc70 chaperon specifically binds preS1 in an ATP-dependent fashion, and deletion of the cytosolic anchorage determinant abolishes hsc70 binding of L [178]. PreS1 amino acid residues 63–107 have been mapped as the site of interaction with hsc70, which encompasses the cytosolic anchorage determinant [178].

Two regions within the envelope proteins have been identified as potential sites of interaction with the capsid [179–181]. Deletion of most of the preS1 domain (N-terminal amino acid residues up through 102) does not prevent virion formation [182]. However, amino acid residues 103–124 are important [183], consistent with the observation that a synthetic peptide overlapping this region, identical to the 13 C-terminal amino acids of preS1, binds efficiently to core particles [179]. This region, highly conserved among the mammalian hepadnaviruses [183], most likely interacts with the spike tip formed by the core dimers [180]. A second region, amino acid sequence 56–80 in the cytosolic loop of S, also binds efficiently to core particles and is involved in virion formation, but is not critical for secretion of subviral particles. Although this second region is also present in the L protein, it seems to act primarily in the context of S [181]. The two regions act synergistically with regard to capsid binding [180,181].

Regions of the core protein involved in envelopment of the nucleocapsid to form mature virions are not the same as those described for core protein dimerization. However, there is significant overlap with regard to those regions needed for capsid particle formation. N-terminal extensions of core protein do not inhibit capsid formation, but do prevent production of secretable mature virions, suggesting a defect in the envelopment process [184]. The same is true for mutations centered around core amino acid residues Thr 12, Pro 134 and Ser 14 1, all of which are situated within the body of the capsid and are not present in the spike [185]. A synthetic peptide that binds to the spike tip formed by dimerization of the core protein, blocks binding between the core protein and the L protein *in vitro*, and reduces production of mature virions *in vivo* [186]. Thus, both the capsid body and its protruding spikes participate in envelopment.

Finally, not all nucleocapsids are enveloped equally. There seems to be a strong preference for envelopment of those in which replication has progressed at least through synthesis of the DNA minus strand. Newly formed capsids still containing only the pgRNA, or capsids blocked for RT of the minus strand, are not secreted in mature virions [187,188], suggesting that the replication state is somehow reflected on the surface of the nucleocapsid. In contrast, binding of synthetic envelope peptides to capsid particles does not seem to be affected by whether or not the particles contain encapsidated DNA [189]. This implies that the stage at which secretion of replication deficient or immature nucleocapsids is blocked involves transport of the nucleocapsid to the ER membrane, or budding of the membrane-bound particle.

120

Future directions

Given the compact multi-functional nature of its genome, the complexity of its life cycle and the intricacies of its gene regulation, it is not surprising that after more than 30 years of study, the workings of the hepatitis B virus are still only partially revealed. Its impact on the lives of hundreds of millions of people around the world makes a complete understanding of all aspects of the virus imperative. In particular, definitive identification of the human HBV receptor(s) and a more comprehensive explanation for the wide-ranging effects of the viral X protein are of high priority. With regard to the latter, an association of the X protein with the host proteasome complex has been recently reported [190–193]. Altering the function of a hub of cellular activity such as the proteasome is consistent with the observed pleiotropic effects of the X protein.

Other important areas include development of a permissive and an efficient permanent cell culture system or a practical animal model that more closely mimics HBV infection in humans. Also, a greater understanding is needed of the interactions between the host immune system and the defenses deployed by the virus to avoid elimination, and how they can be exploited to improve treatment.

References

1. Brunetto MR, Rodriguez UA, Bonino F. Intervirology 1999; 42: 69–80.
2. Hussain M, Lok AS. J. Viral Hepat. 1999; 6: 183–194.
3. Kramvis A, Kew MC. J. Viral Hepat. 1999; 6: 415–427.
4. Murakami S. Intervirology 1999; 42: 81–99.
5. Ngui SL, Hallet R, Teo CG. Rev. Med. Virol. 1999; 9: 183–209.
6. Hunt CM, McGill JM, Allen MI, Condreay LD. Hepatology 2000; 31: 1037–1044.
7. Maddrey WC. J. Med. Virol. 2000; 61: 362–366.
8. Torresi J, Locarnini S. Gastroenterology 2000; 118: S83–S103.
9. Yeh CT. J. Gastroenterol. Hepatol. 2000; 15: 339–341.
10. Zuckerman AJ. Lancet 2000; 355: 1382–1384.
11. Gust ID, Burrell CJ, Coulepis AG, Robinson WS, Zuckerman AJ. Intervirology 1986; 25: 14–29.
12. Mason WS, Seal G, Summers J. J. Virol. 1980; 36: 829–836.
13. Sprengel R, Kaleta EF, Will H. J. Virol. 1988; 62: 932–937.
14. Chang SF, Netter HJ, Bruns M, Schneider R, Frolich K, Will H. Virology 1999; 262: 39–54.
15. Marion PL, Oshiro LS, Regnery DC, Scullard GH, Robinson WS. Proc. Natl Acad. Sci. USA 1980; 77: 2941–2945.
16. Testut P, Renard CA, Terradillos O, Vitvitski-Trepo L, Tekaia F, Degott C, Blake J, Boyer B, Buendia MA. J. Virol. 1996; 70: 4210–4219.
17. Summers J, Smolec J, Snyder R. Proc. Natl Acad. Sci. USA 1978; 75: 4533–4537.
18. Galibert F, Chen TN, Mandart E. J. Virol. 1982; 41: 51–65.
19. Lanford RE, Chavez D, Brasky KM, Burns RB III, Rico-Hesse R. Proc. Natl Acad. Sci. USA 1998; 95: 5757–5761.
20. Vaudin M, Wolstenholme AJ, Tsiquaye KN, Zuckerman AJ, Harrison TJ. J. Gen. Virol. 1988; 69: 1383–1389.

21. Takahashi K, Brotman B, Usuda S, Mishiro S, Prince AM. Virology 2000; 267: 58–64.
22. Mimms LT, Solomon LR, Ebert JW, Fields H. Biochem. Biophys. Res. Commun. 1993; 195: 186–191.
23. Norder H, Ebert JW, Fields HA, Mushahwar IK, Magnius LO. Virology 1996; 218: 214–223.
24. Warren KS, Heeney JL, Swan RA, Heriyanto, Verschoor EJ. J. Virol. 1999; 73: 7860–7865.
25. Grethe S, Heckel JO, Rietschel W, Hufert FT. J. Virol. 2000; 74: 5377–5381.
26. Dane DS, Cameron CH, Briggs M. Lancet 1970; i(7649): 695–698.
27. Seeger C, Ganem D, Varmus HE. J. Virol. 1984; 51: 367–375.
28. Michalak TI. Immunol. Rev. 2000; 174: 98–111.
29. Kock J, Theilmann L, Galle P, Schlicht HJ. Hepatology 1996; 23: 405–413.
30. Laskus T, Wang LF, Radkowski M, Vargas H, Cianciara J, Poutous A, Rakela J. J. Gen. Virol. 1997; 78: 649–653.
31. Cabrerizo M, Bartolome J, Caramelo C, Barril G, Carreno V. Hepatology 2000; 32: 116–123.
32. Nassal M. Curr. Top. Microbiol. Immunol. 1996; 214: 297–337.
33. Seeger C, Mason WS. Microbiol. Mol. Biol. Rev. 2000; 64: 51–68.
34. Gavilanes F, Gonzalez-Ros JM, Peterson DL. J. Biol. Chem. 1982; 257: 7770–7777.
35. Howard CR, Burrell CJ. Prog. Med. Virol. 1976; 22: 36–103.
36. Neurath AR, Kent SB. Adv. Virus Res. 1988; 34: 65–142.
37. Heermann KH, Goldmann U, Schwartz W, Seyffarth T, Baumgarten H, Gerlich WH. J. Virol. 1984; 52: 396–402.
38. Heermann KH, Kruse F, Seifer M, Gerlich WH. Intervirology 1987; 28: 14–25.
39. Crowther RA, Kiselev NA, Bottcher B, Berriman JA, Borisova GP, Ose V, Pumpens P. Cell 1994; 77: 943–950.
40. Kenney JM, von Bonsdorff CH, Nassal M, Fuller SD. Structure 1995; 3: 1009–1019.
41. Summers J, O'Connell A, Millman I. Proc. Natl Acad. Sci. USA 1975; 72: 4597–4601.
42. Kaplan PM, Greenman RL, Gerin JL, Purcell RH, Robinson WS. J. Virol. 1973; 12: 995–1005.
43. Robinson WS, Greenman RL. J. Virol. 1974; 13: 1231–1236.
44. Bartenschlager R, Junker-Niepmann M, Schaller H. J. Virol. 1990; 64: 5324–5332.
45. Bartenschlager R, Schaller H. EMBO J. 1992; 11: 3413–3420.
46. Albin C, Robinson W. J. Virol. 1980; 34: 297–302.
47. Feitelson MA, Marion PL, Robinson WS. J. Virol. 1982; 43: 741–748.
48. Hu J, Toft DO, Seeger C. EMBO J. 1997; 16: 59–68.
49. Arbuthnot P, Capovilla A, Kew M. J. Gastroenterol. Hepatol. 2000; 15: 357–368.
50. Ou JH. J. Gastroenterol. Hepatol. 1997; 12: S178–S187.
51. Ganem D, Varmus HE. Annu. Rev. Biochem. 1987; 56: 651–693.
52. Miller RH, Kaneko S, Chung CT, Girones R, Purcell RH. Hepatology 1989; 9: 322–327.
53. Will H, Reiser W, Weimer T, Pfaff E, Buscher M, Sprengel R, Cattaneo R, Schaller H. J. Virol. 1987; 61: 904–911.
54. Ou JH, Bao H, Shih C, Tahara SM. J. Virol. 1990; 64: 4578–4581.
55. Nassal M, Junker-Niepmann M, Schaller H. Cell 1990; 63: 1357–1363.
56. Yen TSB. Semin Virol 1993; 4: 33–42.
57. Wu HL, Chen PJ, Tu SJ, Lin MH, Lai MY, Chen DS. J. Virol. 1991; 65: 1680–1686.
58. Huang J, Liang TJ. Mol. Cell. Biol. 1993; 13: 7476–7486.
59. Huang ZM, Yen TS. J. Virol. 1994; 68: 3193–3199.
60. Donello JE, Beeche AA, Smith GJ III, Lucero GR, Hope TJ. J. Virol. 1996; 70: 4345–4351.
61. Huang Y, Wimler KM, Carmichael GG. EMBO J. 1999; 18: 1642–1652.

62. Obert S, Zachmann-Brand B, Deindl E, Tucker W, Bartenschlager R, Schaller H. EMBO J. 1996; 15: 2565–2574.

63. Terre S, Petit MA, Brechot C. J. Virol. 1991; 65: 5539–5543.

64. Gunther S, Sommer G, Iwanska A, Will H. Virology 1997; 238: 363–371.

65. Sommer G, van Bommel F, Will H. Virology 2000; 271: 371–381.

66. Soussan P, Garreau F, Zylberberg H, Ferray C, Brechot C, Kremsdorf D. J. Clin. Investig. 2000; 105: 55–60.

67. Huang HL, Jeng KS, Hu CP, Tsai CH, Lo SJ, Chang C. Virology 2000; 275: 398–410.

68. zu Putlitz J, Tong S, Wands JR. Virology 1999; 254: 245–256.

69. Wagner M, Alt M, Hofschneider PH, Renner M. J. Gen. Virol. 1999; 80: 2673–2683.

70. De Meyer S, Gong ZJ, Suwandhi W, van Pelt J, Soumillion A, Yap SH. J. Viral Hepat. 1997; 4: 145–153.

71. Galle PR, Schlicht HJ, Kuhn C, Schaller H. Hepatology 1989; 10: 459–465.

72. Shih CH, Li LS, Roychoudhury S, Ho MH. Proc. Natl Acad. Sci. USA 1989; 86: 6323–6327.

73. Kuroki K, Eng F, Ishikawa T, Turck C, Harada F, Ganem D. J. Biol. Chem. 1995; 270: 15022–15028.

74. Tong S, Li J, Wands JR. J. Virol. 1995; 69: 7106–7112.

75. Tong S, Li J, Wands JR. J. Virol. 1999; 73: 8696–8702.

76. Li JS, Tong SP, Wands JR. J. Virol. 1996; 70: 6029–6035.

77. Li J, Tong S, Wands JR. J. Biol. Chem. 1999; 274: 27658–27665.

78. Breiner KM, Schaller H. J. Virol. 2000; 74: 2203–2209.

79. Thomas HC, Jacyna M, Waters J, Main J. Semin. Liver Dis. 1988; 8: 342–349.

80. Gerlich WH, Lu X, Heermann KH. J. Hepatol. 1993; 17: S10–S14.

81. Budkowska A, Bedossa P, Groh F, Louise A, Pillot J. J. Virol. 1995; 69: 840–848.

82. Franco A, Paroli M, Testa U, Benvenuto R, Peschle C, Balsano F, Barnaba V. J. Exp. Med. 1992; 175: 1195–1205.

83. Persing DH, Varmus HE, Ganem D. J. Virol. 1987; 61: 1672–1677.

84. Gripon P, Le Seyec J, Rumin S, Guguen-Guillouzo C. Virology 1995; 213: 292–299.

85. Bruss V, Hagelstein J, Gerhardt E, Galle PR. Virology 1996; 218: 396–399.

86. Macrae DR, Bruss V, Ganem D. Virology 1991; 181: 359–363.

87. Summers J, Smith PM, Huang MJ, Yu MS. J. Virol. 1991; 65: 1310–1317.

88. Pontisso P, Petit MA, Bankowski MJ, Peeples ME. J. Virol. 1989; 63: 1981–1988.

89. Pontisso P, Ruvoletto MG, Gerlich WH, Heermann KH, Bardini R, Alberti A. Virology 1989; 173: 522–530.

90. Leenders WP, Glansbeek HL, de Bruin WC, Yap SH. Hepatology 1990; 12: 141–147.

91. Neurath AR, Kent SB, Strick N, Parker K. Cell 1986; 46: 429–436.

92. Le Seyec J, Chouteau P, Cannie I, Guguen-Guillouzo C, Gripon P. J. Virol. 1999; 73: 2052–2057.

93. Petit MA, Capel F, Dubanchet S, Mabit H. Virology 1992; 187: 211–222.

94. Pontisso P, Ruvoletto MG, Tiribelli C, Gerlich WH, Ruol A, Alberti A. J. Gen. Virol. 1992; 73: 2041–2045.

95. Neurath AR, Strick N, Li YY. J. Exp. Med. 1989; 176: 1561–1569.

96. Treichel U, Meyer zum Buschenfelde KH, Stockert RJ, Poralla T, Gerken G. J. Gen. Virol. 1994; 75: 3021–3029.

97. Ryu CJ, Cho DY, Gripon P, Kim HS, Guguen-Guillouzo C, Hong HJ. J. Virol. 2000; 74: 110–116.

98. Budkowska A, Quan C, Groh F, Bedossa P, Dubreuil P, Bouvet JP, Pillot J. J. Virol. 1993; 67: 4316–4322.

99. Dash S, Rao KV, Panda SK. J. Med. Virol. 1992; 37: 116–121.
100. de Bruin WC, Hertogs K, Leenders WP, Depla E, Yap SH. J. Gen. Virol. 1995; 76: 1047–1050.
101. Mehdi H, Kaplan MJ, Anlar FY, Yang X, Bayer R, Sutherland K, Peeples ME. J. Virol. 1994; 68: 2415–2424.
102. Mehdi H, Yang X, Peeples ME. Virology 1996; 217: 58–66.
103. Hertogs K, Leenders WP, Depla E, De Bruin WC, Meheus L, Raymackers J, Moshage H, Yap SH. Virology 1993; 197: 549–557.
104. Wang SX, Sun YT, Sui SF. Biochem. J. 2000; 348: 103–106.
105. Stefas E, Rucheton M, Graafland H, Moynier M, Sompeyrac C, Bahraoui EM, Veas F. AIDS Res. Hum. Retroviruses 1997; 13: 97–104.
106. Gong ZJ, De Meyer S, van Pelt J, Hertogs K, Depla E, Soumillion A, Fevery J, Yap SH. Hepatology 1999; 29: 576–584.
107. De Meyer S, Gong ZJ, Hertogs K, Depla E, van Pelt JF, Roskams T, Maertens G, Yap SH. J. Viral Hepat. 2000; 7: 104–114.
108. De Meyer S, Depla E, Maertens G, Soumillion A, Yap SH. J. Viral Hepat. 1999; 6: 277–285.
109. De Meyer S, Gong Z, Depla E, Maertens G, Yap SH. J. Hepatol. 1999; 31: 783–790.
110. Wright JF, Kurosky A, Wasi S. Biochem. Biophys. Res. Commun. 1994; 198: 983–989.
111. Huang RT, Lichtenberg B, Rick O. FEBS Lett. 1996; 392: 59–62.
112. Bruss V, Gerhardt E, Vieluf K, Wunderlich G. Intervirology 1996; 39: 23–31.
113. Grgacic EVL, Kuhn C, Schaller H. J. Virol. 2000; 74: 2455–2458.
114. Grgacic EV, Schaller H. J. Virol. 2000; 74: 5116–5122.
115. Rodriguez-Crespo I, Gomez-Gutierrez J, Nieto M, Peterson DL, Gavilanes F. J. Gen. Virol. 1994; 75: 637–639.
116. Rodriguez-Crespo I, Nunez E, Gomez-Gutierrez J, Yelamos B, Albar JP, Peterson DL, Gavilanes F. J. Gen. Virol. 1995; 76: 301–308.
117. Rodriguez-Crespo I, Nunez E, Yelamos B, Gomez-Gutierrez J, Albar JP, Peterson DL, Gavilanes F. Virology 1999; 261: 133–142.
118. Rodriguez-Crespo I, Yelamos B, Albar JP, Peterson DL, Gavilanes F. Biochim. Biophys. Acta 2000; 1463: 419–428.
119. McGarvey MJ, Goldin RD, Karayiannis P, Thomas HC. J. Viral Hepat. 1996; 3: 67–73.
120. Kann M, Bischof A, Gerlich WH. J. Virol. 1997; 71: 1310–1316.
121. Kock J, Schlicht HJ. J. Virol. 1993; 67: 4867–4874.
122. Kramvis A, Kew MC. J. Viral Hepat. 1998; 5: 357–367.
123. Jeong JK, Yoon GS, Ryu WS. J. Virol. 2000; 74: 5502–5508.
124. Schaaf SG, Beck J, Nassal M. J. Biol. Chem. 1999; 274: 37787–37794.
125. Pollack JR, Ganem D. J. Virol. 1994; 68: 5579–5587.
126. Beck J, Nassal M. J. Virol. 1997; 71: 4971–4980.
127. Beck J, Nassal M. Mol. Cell. Biol. 1998; 18: 6265–6272.
128. Tavis JE, Massey B, Gong Y. J. Virol. 1998; 72: 5789–5796.
129. Nassal M. Intervirology 1999; 42: 100–116.
130. Hu J, Seeger C. Proc. Natl Acad. Sci. USA 1996; 93: 1060–1064.
131. Kim SS, Shin HJ, Cho YH, Rho HM. Arch. Virol. 2000; 145: 1305–1320.
132. Cho G, Park SG, Jung G. Biochem. Biophys. Res. Commun. 2000; 269: 191–196.
133. Nassal M, Schaller H. J. Viral Hepat. 1996; 3: 217–226.
134. Yao E, Gong Y, Chen N, Tavis JE. J. Virol. 2000; 74: 8648–8657.
135. Lanford RE, Notvall L, Lee H, Beames B. J. Virol. 1997; 71: 2996–3004.
136. Lanford RE, Kim YH, Lee H, Notvall L, Beames B. J. Virol. 1999; 73: 1885–1893.

124

137. Gong Y, Yao E, Stevens M, Tavis JE. J. Gen. Virol. 2000; 81: 2059–2065.
138. Jiang H, Loeb DD. J. Virol. 1997; 71: 5345–5354.
139. Ho TC, Jeng KS, Hu CP, Chang C. J. Virol. 2000; 74: 9010–9018.
140. Miller RH, Tran CT, Robinson WS. Virology 1984; 139: 53–63.
141. Havert MB, Loeb DD. J. Virol. 1997; 71: 5336–5344.
142. Loeb DD, Tian R, Gulya KJ, Qualey AE. J. Virol. 1998; 72: 6565–6573.
143. Perri S, Ganem D. J. Virol. 1997; 71: 8448–8455.
144. Kock J, Wieland S, Blum HE, von Weizsacker F. J. Virol. 1998; 72: 9116–9120.
145. Birnbaum F, Nassal M. J. Virol. 1990; 64: 3319–3330.
146. Zhou S, Standring DN. Proc. Natl Acad. Sci. USA 1992; 89: 10046–10050.
147. Chang C, Zhou S, Ganem D, Standring DN. J. Virol. 1994; 68: 5225–5231.
148. Konig S, Beterams G, Nassal M. J. Virol. 1998; 72: 4997–5005.
149. Koschel M, Thomssen R, Bruss V. J. Virol. 1999; 73: 2153–2160.
150. Preikschat P, Borisova G, Borschukova O, Dislers A, Mezule G, Grens E, Kruger DH, Pumpens P, Meise H. J. Gen. Virol. 1999; 80: 1777–1788.
151. Metzger K, Bringas R. J. Gen. Virol. 1998; 79: 587–590.
152. Bottcher B, Wynne SA, Crowther RA. Nature 1997; 386: 88–91.
153. Conway JF, Cheng N, Zlotnick A, Wingfield PT, Stahl SJ, Steven AC. Nature 1997; 386: 91–94.
154. Wynne SA, Crowther RA, Leslie AG. Mol. Cell 1999; 3: 771–780.
155. Zlotnick A, Stahl SJ, Wingfield PT, Conway JF, Cheng N, Steven AC. FEBS Lett. 1998; 431: 301–304.
156. Conway JF, Cheng N, Zlotnick A, Stahl SJ, Wingfield PT, Belnap DM, Kanngiesser U, Noah M, Steven AC. J. Mol. Biol. 1998; 279: 1111–1121.
157. Liao W, Ou J-H. J. Virol. 1995; 69: 1025–1029.
158. Kann M, Thomssen R, Kochel HG, Gerlich WH. Arch. Virol. Suppl. 1993; 8: 53–62.
159. Kann M, Gerlich WH. J. Virol. 1994; 68: 7993–8000.
160. Gazina EV, Fielding JE, Lin B, Anderson DA. J. Virol. 2000; 74: 4721–4728.
161. Lan YT, Li J, Liao W, Ou J. Virology 1999; 259: 342–348.
162. Yeh CT, Liaw YF, Ou JH. J. Virol. 1990; 64: 6141–6147.
163. Eckhardt SG, Milich DR, McLachlan A. J. Virol. 1991; 65: 575–582.
164. Kann M, Sodeik B, Vlachou A, Gerlich WH, Helenius A. J. Cell Biol. 1999; 145: 45–55.
165. Zlotnick A, Cheng N, Stahl SJ, Conway JF, Steven AC, Wingfield PT. Proc. Natl Acad. Sci. USA 1997; 94: 9556–9561.
166. Dworetzky SI, Feldherr CM. J. Cell Biol. 1988; 106: 575–584.
167. Qiao M, Macnaughton TB, Gowans EJ. Virology 1994; 201: 356–363.
168. Aiba N, McGarvey MJ, Waters J, Hadziyannis SJ, Thomas HC, Karayiannis P. Hepatology 1997; 26: 1311–1317.
169. Garcia PD, Ou JH, Rutter WJ, Walter P. J. Cell Biol. 1988; 106: 1093–1104.
170. Ou JH, Yeh CT, Yen TS. J. Virol. 1989; 63: 5238–5243.
171. Yeh CT, Hong LH, Ou JH, Chu CM, Liaw YF. Arch. Virol. 1996; 141: 425–438.
172. Duclos-Vallee JC, Capel F, Mabit H, Petit MA. J. Gen. Virol. 1998; 79: 1665–1670.
173. Jyh-Hwa K, Ling-Pai T. J. Virol. 1998; 72: 3796–3803.
174. Huovila AP, Eder AM, Fuller SD. J. Cell Biol. 1992; 118: 1305–1320.
175. Bruss V, Ganem D. Proc. Natl Acad. Sci. USA 1991; 88: 1059–1063.
176. Ueda K, Tsurimoto T, Matsubara K. J. Virol. 1991; 65: 3521–3529.
177. Lu X, Mehta A, Dadmarz M, Dwek R, Blumberg BS, Block TM. Proc. Natl Acad. Sci. USA 1997; 94: 2380–2385.

178. Loffler-Mary H, Werr M, Prange R. Virology 1997; 235: 144–152.
179. Poisson F, Severac A, Hourioux C, Goudeau A, Roingeard P. Virology 1997; 228: 115–120.
180. Tan WS, Dyson MR, Murray K. J. Mol. Biol. 1999; 286: 797–808.
181. Loffler-Mary H, Dumortier J, Klentsch-Zimmer C, Prange R. Virology 2000; 270: 358–367.
182. Bruss V, Thomssen R. J. Virol. 1994; 68: 1643–1650.
183. Bruss V. J. Virol. 1997; 71: 9350–9357.
184. Hui EK, Yi YS, Lo SJ. J. Gen. Virol. 1999; 80: 2647–2659.
185. Koschel M, Oed D, Gerelsaikhan T, Thomssen R, Bruss V. J. Virol. 2000; 74: 1–7.
186. Bottcher B, Tsuji N, Takahashi H, Dyson MR, Zhao S, Crowther RA, Murray K. EMBO J. 1998; 17: 6839–6845.
187. Gerelsaikhan T, Tavis JE, Bruss V. J. Virol. 1996; 70: 4269–4274.
188. Wei Y, Tavis JE, Ganem D. J. Virol. 1996; 70: 6455–6458.
189. Hourioux C, Touze A, Coursaget P, Roingeard P. J. Gen. Virol. 2000; 81: 1099–1101.
190. Huang J, Kwong J, Sun EC, Liang TJ. J. Virol. 1996; 70: 5582–5591.
191. Sirma H, Weil R, Rosmorduc O, Urban S, Israel A, Kremsdorf D, Brechot C. Oncogene 1998; 16: 2051–2063.
192. Hu Z, Zhang Z, Doo E, Coux O, Goldberg AL, Liang TJ. J. Virol. 1999; 73: 7231–7240.
193. Zhang Z, Torii N, Furusaka A, Malayaman N, Hu Z, Liang TJ. J. Biol. Chem. 2000; 275: 15157–15165.

Hepatitis C virus

A. Scott Muerhoff and George J. Dawson
Infectious Diseases Research and Development, Abbott Diagnostics Division, Abbott Laboratories, North Chicago, Illinois, USA

The development of diagnostic tests for hepatitis A and B viruses in the 1970s, provided exclusionary criteria for the identification of another form of viral hepatitis, called non-A, non-B (NANB) hepatitis which had spread predominantly via transfusion of blood and blood products and intravenous drug abuse. The identification of the infectious agent responsible for NANB hepatitis was made possible only through the use of molecular cloning methods which allowed the isolation and sequencing of a portion of the viral genome [1]. This agent came to be known as hepatitis C virus (HCV). It is now known that HCV infects more than 170 million people worldwide and that in more than 50% of cases the infection is chronic. In the United States there are approximately 4 million people infected and 30,000 new infections are estimated to occur annually. HCV is responsible for 8000–10,000 deaths annually in the United States. The high morbidity and mortality due to HCV infection has provided impetus for ongoing research into the biology, epidemiology, pathogenicity, treatment and prevention of HCV infection.

HCV Genome organization

HCV consists of a single-stranded RNA genome of positive polarity surrounded by a nucleocapsid protein core enclosed in a lipid membrane or envelope. The RNA genome is approximately 9400–9500 nucleotides in length. The organization of the coding regions along the linear genome (Fig. 1) resembles that of other flaviviruses [2] and the recently discovered GB viruses [3–5]. Based upon phylogenetic analysis of various nonstructural proteins [e.g. the RNA-dependent RNA polymerase (RdRp)], its genome organization, and biological properties, HCV has been recently classified into a distinct genus called the "hepciviruses" within the Flaviviridae family (Fig. 2) [6]. The HCV genome possesses a large open reading frame (ORF) encoding a polyprotein precursor of 3010–3033 amino acids depending on the particular isolate [7,8]. The individual viral proteins involved in replication and packaging are derived by proteolytic processing of the polyprotein by both cellular and viral proteases to at least 10 distinct entities [9,10]. Like other members of the Flaviviridae [e.g. yellow fever virus, bovine viral diarrhea virus (BVDV)], HCV structural genes (core and envelope) are encoded near the $5'$-end of the genome, followed by the proteases and helicase, the helicase cofactor and the replicase. Noncoding regions (NCR) thought to be important in replication are found at each end of the genome.

The $5'$ NCR of HCV is 324–341 nucleotides in length depending on the particular isolate [11,12]. Distinct areas within the $5'$ NCR are highly conserved amongst all HCV

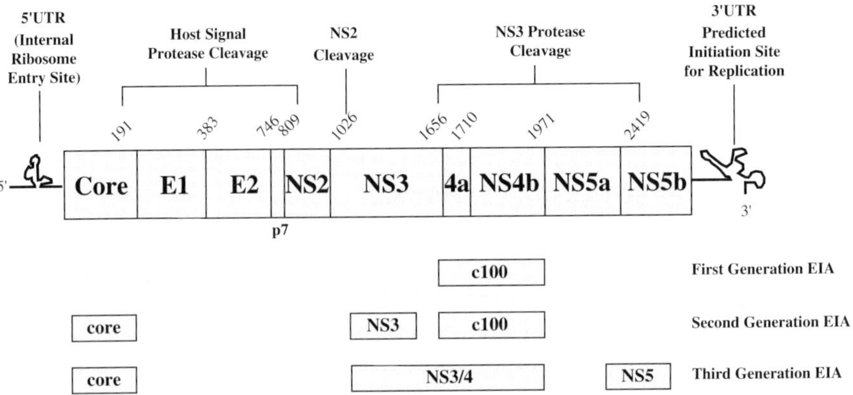

Fig. 1. HCV Genome organization and proteins used in blood screening assays. The carboxy-terminal amino acid number at which cleavage is believed to occur is shown above the polyprotein; the enzyme responsible for the cleavage is indicated. The regions of the core, NS3, NS4, and NS5 that are included in the recombinant proteins used for ELISA assays are indicated below in the genome.

genotypes, suggesting their importance in the virus life cycle. These invariant areas serve as useful targets for molecular diagnostics in determining the presence of HCV infection (see below). An internal ribosome entry site (IRES) upstream of the initiating methionine residue of the core protein has been identified [13,14]; thus, translation of the viral polyprotein is believed to proceed via a cap-independent manner [15]. The 3' NCR of HCV is composed of three regions: a segment of variable sequence and length immediately following the stop codon of the polyprotein followed by a poly(U) tract of varying length and a highly conserved 98 nucleotide sequence [16–19] that has shown to be essential for replication *in vivo* [20]. Low levels of HCV replication have been observed in HuH7 or HepG2 cells for synthetic, essentially full-genome length RNAs that have replaced the 3' NCR with a poly(A) sequence or that terminate at the naturally occurring poly(U) tract [21,22].

Until recently, there were no efficient cell culture models for the study of HCV replication. Only the chimpanzee animal model was available to study the natural history of disease. Recent advances in molecular biology have provided several new approaches to the study of HCV replication, thereby providing means to screen and identify novel therapeutic compounds that interfere with the virus life cycle. Several groups have succeeded in the production of infectious full-genome length molecular clones of HCV [23–25]. These investigators have shown that intrahepatic inoculation of chimpanzees with full-length RNA transcripts produced from the clones is infectious as shown by (i) the production of antibodies to HCV nonstructural and structural proteins, and (ii) detection of RNA in the serum for weeks after inoculation. The availability of these clones allowed investigators to demonstrate that a complete 5'-untranslated region and the poly(U-UC) and conserved region, but not the variable region, of the 3'-untranslated region of the genome are critical for *in vivo* replication [26]. More recently, Lohmann and colleagues [27] described the replication in Huh7 cells of an HCV genotype 1b subgenomic replicon expressing a selectable marker. However, only one in one million

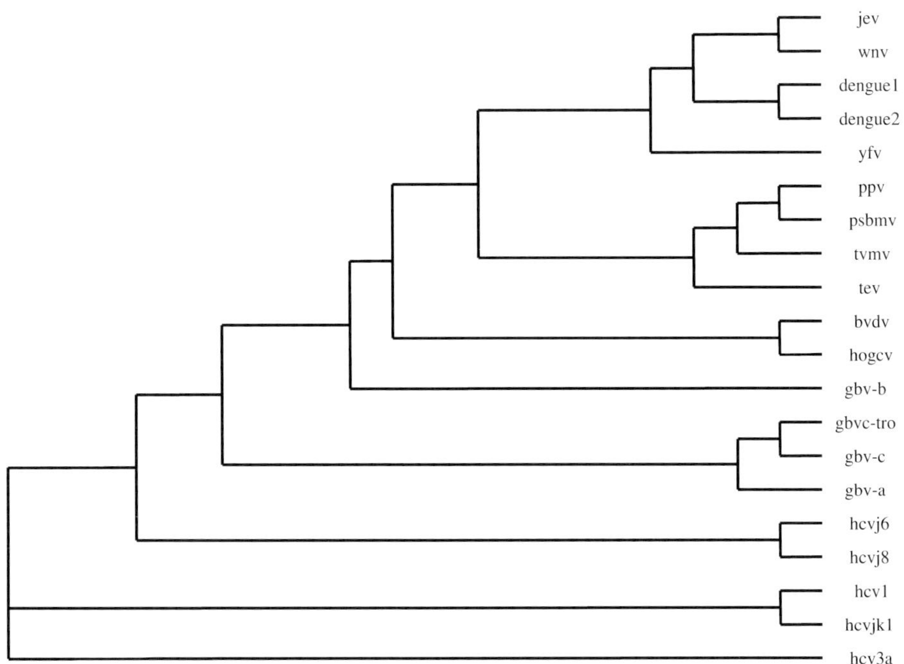

Fig. 2. Unrooted phylogenetic tree demonstrating the relationship between HCV isolates, the GB viruses and other members of the Flaviviridae. Full-length helicase sequences from the NS3 regions of each representative virus was aligned using the PILEUP program of the Wisconsin Package (version 10). The tree was then generated using the PROTDIST and NEIGHBOR programs of the PHYLIP package [362]. The distance scale is indicated at the bottom and represents distances in substitutions per position. Abbreviations: HCV, hepatitis C virus; GBV, GB virus; GBVCtro, GB V-C related virus from *Pan troglodytes*; BVDV, bovine viral diarrhea virus; HOGCV, hog cholera virus; JEV, Japanese encephalitis virus; WNV, West Nile virus; YFV, yellow fever virus.

cells were believed to support replication. Further development of this system identified mutant replicons containing amino acid substitutions within the NS5A gene that correlated with their ability to infect and replicate in 10% of the cultured Huh7 cells [28]. This model system was used to show that interferon-alpha could inhibit replication independent of the so-called interferon sensitivity determining region. Recent reviews [29,30] provide an excellent in-depth examination of models of HCV replication.

HCV structural gene products

Core protein

The proteins encoded by the HCV viral genome are translated as a single polyprotein that is co- or post-translationally processed into the individual viral proteins (Fig. 1). HCV structural proteins are encoded in the N-terminal one-third of the viral polyprotein. These proteins are thought to be cleaved from the nascent polyprotein in the lumen of the

130

endoplasmic reticulum by the host-encoded signal peptidase. The core (or nucleocapsid) protein comprises the first 191 amino acids of the polyprotein. Analogous proteins are encoded by both pesti- and flaviviruses, though the flavivirus analog is further processed into two distinct proteins. HCV core is profoundly basic in nature (calculated overall pI of 12.0 for HCV-1a from amino acids 1–150) and is thought to be involved in genomic RNA binding for encapsidation. The region from amino acids 1–120 of core contains the vast majority of the antigenic determinants for this protein [31–34].

The core protein is cleaved from the polyprotein at residue 191 resulting in a protein with molecular weight of approximately 23 kDa (p23); this cleavage also generates the amino-terminus of the E1 glycoprotein. A subsequent cleavage of core at residue 174 removes the E1 signal sequence from the carboxy-terminus of core resulting in a 21 kDa molecular weight protein (p21) [35], which is the most abundant form of the protein detected following expression of cloned core sequences in mammalian cells [36–39]. Mutational analyses upstream and downstream of residue 191 that abolish cleavage are in agreement with the requirements for protein processing by the host signal peptidase complex as proposed by von Heijne [40,41]. Production of mature p21 resulting from cleavage at residue 174 is also believed to occur via the host signal peptidase [8,42,43]. The p21 form of core has been found in the serum of infected individuals further suggesting that this is the mature form of the capsid protein [35]. *In vitro* translation studies indicate that cleavage of the protein requires microsomal membranes for processing [36,44] and that the hydrophobic domain at the carboxy-terminus acts as a signal sequence to direct the downstream E1 protein to the ER membrane [36].

HCV core is presumed to function primarily as the major component of the viral nucleocapsid. Unfortunately, the absence of a cell culture model for propagation of HCV has made visualization and biophysical characterization of HCV viral particles rather difficult. There have been reports that describe the visualization of virus particles in the liver of infected chimpanzees and virus-like particles (VLP) were detected by electron microscopy in cytoplasmic vesicles in HCV-infected Daudi cells [45]. The appearance of cytoplasmic tubular structures, originally described for chimpanzee hepatocytes in association with HCV infection, were observed as was the immunoperoxidase staining of the VLP using antibodies against core and envelope. More recently, the visualization of VLP produced in insect cells expressing recombinant core and envelope proteins was reported [46,47]. These studies support the idea that the core protein is a major structural protein of the hepatitis C virion.

HCV core protein expressed in mammalian cell culture exhibits cytoplasmic localization associated with granular structures on the endoplasmic reticulum (ER) as determined by electron microscopy and immunostaining [36,38,43,48]. Fractionation of cells expressing core protein indicate that core is isolated with the membrane fraction [49] with an apparently small fraction of the intracellular core protein (p21 form) found associated with the nuclear membrane or inside the nucleus. Interestingly, the expression of core from HCV1, the prototype isolate of HCV (genotype 1a), in rat embryo fibroblasts results in predominantly nuclear localization [50] with the detection of 16 kDa species. The "p16" species has been observed by others, but only when using the core sequence derived from HCV-1 [37,51]. Nuclear localization is observed frequently when forms of core lacking the carboxy-terminal hydrophobic domain (i.e. amino acids 153–191) are expressed [35,37–39,52–54]. In fact, the presence of three distinct nuclear localization signals within the amino-terminal domain of core has been reported [55]. However, there have been other studies where nuclear localization has not been demonstrated [56–58]. Thus, there still remains some uncertainty about the ability of core to localize to the nucleus of infected cells.

The reported effects of intracellular HCV core protein expression are numerous and include modulation of gene expression, RNA binding, association with ribosomal subunits, and interaction with various signal transduction systems [59,60]. One of the more intriguing observations about HCV core protein is its apparent ability to affect lipid metabolism. Evidence for this comes from a variety of studies including the association with lipid droplets within the cytoplasm of infected cells [49,61] and also with apolipoprotein II [54,56] but not apolipoprotein A1 [54]. Two lines of transgenic mice expressing core in their liver demonstrate hepatic steatosis [62,63] suggesting that HCV core may have a significant effect on lipid metabolism and may play a critical role in the development of fatty liver in humans [64,65], a common condition among HCV positive individuals [64,66] although there are many other clinical conditions with which steatosis is associated [67]. Since HCV core protein appears to influence, or at least be associated with, so many cellular processes, it will be critical to develop the proper tools (e.g. animal models and cell culture) with which to distinguish between the effects observed *in vitro* with the pathogenesis of HCV infection *in vivo*. Expression of high levels of core protein in mammalian cells in culture may result in subcellular localization that would not normally be seen in naturally infected cells. In addition, over-expression could have effects on cellular function (i.e. gene expression) that are artefactual. Ultimate resolution of this question will require a cell culture system for viral replication.

Envelope proteins

HCV encodes two distinct envelope proteins, E1 and E2. Both proteins are N-linked glycosylated at multiple sites [8,68,69] contributing significantly to their observed molecular mass upon SDS-PAGE. E1 is a 192 amino acid protein of 31–37 kDa with 3–5 potential N-linked glycosylation sites. Despite the high degree of amino acid sequence divergence in this protein, several regions are highly conserved, most likely being residues of functional importance. HCV E2 is a 426–430 amino acid protein with an apparent molecular weight of 58–70 kDa. E1 and E2 are analogous to the BVDV envelope glycoproteins [1,11,70]. Both possess hydrophobic C-terminal domains that may function to anchor the proteins into the ER membrane while the N-terminal domain is translocated to the lumenal side [71]. HCV E1 and E2 proteins are cleaved from the polyprotein by host signal peptidase (Fig. 1) and E2 is further processed to release another protein termed p7. The p7 region of E2 is 63 amino acids in length and is followed immediately by a hydrophobic domain that may function to direct E2 to the ER allowing subsequent cleavage of p7. Processing of the E2/p7 junction appears to be rather inefficient resulting in the production of two E2 species, each with different C-termini [72,73]. It is not currently known whether p7 functions as a structural or nonstructural protein.

Intra- and intermolecular interactions between E1 and E2 proteins have been studied by expression of the proteins in mammalian cells or insect cells in culture. Antibodies to E1 can precipitate E1 and E2, in addition, NS3 and NS4 proteins can also be found in these immunoprecipitates [74] suggesting a complex interaction among structural and nonstructural proteins during viral replication. There is no evidence to suggest covalent linkage between envelope proteins [74] although intramolecular disulfide bonds are believed to form during the maturation process of E1 [75,76]. A recent report by Baumert and colleagues [46] demonstrated the coimmunoprecipitation of E2 with core and E1 following expression in insect cells, which extends earlier reports using mammalian expression systems to examine E1/E2 interactions [77] and core/E1 interactions [78]. Immunostaining of cells expressing envelope proteins show localization in the ER and not at the plasma membrane suggesting that they are retained in the ER and not shuttled through the Golgi [77]. Additional evidence that HCV envelope proteins do not progress beyond the ER membrane comes from the observation that the complex types of glycans typically associated with proteins processed through the *trans* Golgi are not found on E1 or E2 [76]. Retention of E1 and E2 in the ER may indicate that HCV virions assemble in a manner similar to the flaviviruses which do not bud from the plasma membrane but assemble in vesicles within the ER which are then released from the cells via an exocytotic pathway [79,80]. Work by Baumert et al. [46,47,81] has demonstrated that HCV Core, E1 and E2 expressed in insect cells can form VLP. These VLPs exhibit the same immunological characteristics as the individual proteins and may therefore be useful as vaccines.

HCV E2 protein possesses a region of significant amino acid sequence variation near the amino-terminus, often referred to as the hypervariable region 1 or HVR-1 [82,83]. The existence of this hypervariability may be the result of immunological selection of variants in the infected individuals [84–86]. Direct evidence for immune selection on E2 comes from examination of E2 sequences from an HCV-infected patient with agammaglobulinemia: the E2 hypervariable region amino acid sequence was conserved in excess of two years [87]. Similarly, in an experimentally infected chimpanzee with no detectable antibody response, the E2 HVR-1 amino acid sequence was conserved for six years [88]. Interestingly, in the seventh year following infection the animal developed an anti-E2 hypervariable region antibody response, which resulted in subsequent variation within this region. The E2 HVR-1 is the only known target protein encoded by HCV for neutralizing antibodies [84,89,90]; however, in spite of this, it has been shown that there is a lack of immune protection in chimpanzees sequentially infected with different strains of HCV [91]. Adding to the conundrum of HCV immunity is the cross-reactivity between anti-E2 antibodies induced by HCV isolates with dissimilar HVR1 sequences [92,93] and the ability of ELISA assays that utilize a single E2 protein to detect anti-E2 antibodies in the vast majority (>97%) of HCV RNA-positive individuals [94]. Thus, the ability of HCV to establish persistent infection must include mechanisms other than hypervariation within the E2 gene. By comparison, GB virus C (GBV-C), which is a close relative of HCV [4], establishes persistent infections in many individuals regardless of immune status but shows little or no variability within the E2 gene or any other region of the genome [95,96]. In contrast to HCV, however, most GBV-C RNA + individuals who seroconvert to anti-GBV-C E2 demonstrate the loss of detectable GBV-C RNA in the serum shortly thereafter [97–99]. Thus, while closely related, these two viruses have evolved specific mechanisms for viral persistence.

Hepatitic C virus E2 is believed to play an important role in the early life cycle of viral replication, i.e. cell membrane attachment. Recently, CD81, a member of the tetraspanin family of membrane proteins, was identified as the putative HCV receptor [100] based on its interaction with purified E2 and virus particles *in vitro*. CD81 is organized in four highly hydrophobic transmembrane domains and two hydrophilic domains that form extracellular loops. The large extracellular loop (LEL) of CD81 was shown to be sufficient to bind HCV by specific interaction with E2 [100]. Pretreatment of HCV-containing plasma with sera from chimpanzees that had been "vaccinated" with recombinant E1 and E2 blocked binding of HCV to CD81, providing indirect evidence that CD81 was an HCV receptor. Monoclonal antibodies that recognize two conserved HCV E2 epitopes have also been reported to inhibit binding of recombinant E2 to CD81 demonstrating that at least two conserved regions are important for E2–CD81 interactions [101]. A recent report [102] concluded that CD81 attachment does not efficiently internalize bound ligands, as determined through the use of a CD81-specific monoclonal antibody that binds to the same region on CD81 as HCV E2: only 30% of CD81 was internalized following 12-h incubation with the monoclonal antibody. Unfortunately, a similar experiment using purified HCV E2 or HCV VLP was not reported. Thus, it is not known if HCV E2, or intact virus, would elicit similar internalization rates. From these experiments it is unclear whether CD81 mediates both attachment and entry or whether accessory proteins are required for entry.

It should be noted that CD81 expression is not limited to hepatocytes [103] and therefore the tropism of HCV for hepatocytes cannot be explained by binding to CD81 alone. Recent evidence indicates that the low density lipoprotein receptor (LDLr) may be the cellular receptor for HCV on hepatocytes [104,105]. Using HCV particles isolated from infectious human plasma as a source of virus, Wunschmann and colleagues [106] demonstrated LDL receptor-dependent binding of HCV to MOLT-4 cells. Purified E2 protein binding to MOLT-4 cells was inhibited by soluble CD-81 but not soluble LDLr. Experiments designed to determine the extent of internalization of HCV demonstrated that the amount of viral entry correlated with LDLr expression and was independent of CD81 expression. These results suggest that, while HCV E2 binds CD81 specifically, binding and entry of virus requires the LDL receptor. Similar results were obtained with BVDV and GB virus C/hepatitis G virus [104]. The exact mechanism by which virus is internalized remains to be elucidated.

Nonstructural gene products

Cysteinyl protease (NS2)

Downstream of the envelope proteins genes on the HCV genome lies the coding regions for the nonstructural proteins. Immediately following the E2 gene are encoded two proteases, followed by an RNA helicase, a protease cofactor and an RdRp. The viral

134

proteases are encoded within NS2-3 (24 kDa) and NS3 (68 kDa). The NS23 protease is autocatalytic and responsible for the liberation of NS2 from NS3. The NS2-3 cleavage site (KGWRLL^APIT) is highly conserved among HCV genotypes reported to date [107] suggesting that this enzyme has rather narrow substrate specificity. However, studies utilizing site-directed mutagenesis indicate that many substitutions within the cleavage site are tolerated, including many at the P1 and P1' positions [108,109]. Several mutations that drastically reduced activity were observed for the P1–P1' positions [109]. These results indicate that recognition and cleavage of substrate by the NS2-3 protease is more dependent upon structure and less upon the amino acid sequence itself. The NS2-3 protease has been localized to the carboxy-terminal 130 residues of NS2 and the first 180 amino acids of NS3, i.e. amino acids 827–1207 [107] or 898–1325 [110]. These same studies have identified two amino acids, His-952 and Cys-993, with functional importance to the NS2-3 protease, implying that NS2-3 protease is of the cysteinyl variety. However, the NS2-3 protease activity is stimulated by zinc [110] and inhibited by EDTA suggesting it is a metalloprotease. The zinc-binding domain (i.e. cysteines 1123, 1125, 1171, and His-1175) appears to be distant from the residues believed to form the NS2-3 active site. Mutational analysis of the cysteine residues that participate in tetra-coordinating zinc inhibit cleavage at the NS2/NS3 site [71,110]. Other reports indicate that zinc is required for proper folding of the NS2-3 protease and the structural integrity of the NS3 protease [111–115]. Thus, zinc may participate in NS2-3 activity by maintaining structure without being directly involved in catalysis.

HCV Serine protease (NS3)

A second protease, localized between Gly-1049 and Ser-1215, is a classic trypsin-like serine protease analogous to serine proteases in flavi- and pestiviruses [110,116–119]. This enzyme is responsible for cleavage at the junctions between the NS3/NS4A, NS4A/NS4B, NS4B/NS5A, and NS5A/NS5B proteins. Residues His-1083, Asp-1107, and Ser-1165 constitute the catalytic triad of the enzyme and are conserved in all HCV genotypes [117]. Amino acid substitutions at the His or Ser residue abolish cleavage of NS3/NS4A, NS4A/NS4B, NS4B/NS5A, and NS5A/NS5B. Substrate specificity studies of the NS3 protease involved HCV polyprotein sequence alignments where a cysteine or threonine residue was found in the P1 position of the cleavage site, with a serine or alanine requirement in the P1' position. Additionally, an acidic amino acid (aspartic or glutamic acid) was identified as important in the P6 position [117]. Fine mapping studies of the serine proteinase substrate by way of site-directed mutagenesis reveal that the P1 position is the most sensitive to amino acid substitution, while the P1' position was more tolerant. The acidic residue in the P6 position seemed to be dispensable for proper cleavage [120, 121]. Cleavage of all sites except NS3/NS4A is restored by providing NS3 in *trans* [118] and the NS3/4A site is the only site where catalysis does not occur with cysteine in the P1 position of the cleavage site instead, there is a threonine in the P1 position. Computer modeling experiments [122] substantiate the hypothesis that NS3/4A cleavage occurs intramolecularly (i.e. in *cis*).

The amino-terminal peptide of NS4A (amino acids 1658–1711) is an essential cofactor for this enzyme and is required for NS4B/NS5A and cleavage and facilitates NS5A/NS5B cleavage [119]. The NS3 serine protease and NS4A peptide form a stable complex [123–125] that requires the amino-terminal 22 residues of NS3 [126]. The determination of the crystal structure of the NS3 serine protease in the absence [113] or presence of NS4A cofactor [112] revealed that the structure of the enzyme is similar to trypsin, containing a beta-barrel adjacent to a cleft possessing the substrate binding pocket and the active site. Binding of the NS4A cofactor peptide induces a conformational change in the 30 amino-terminal residues of NS3 from a flexible polypeptide extending away from the protein to a beta-sheet that participates in the beta-barrel formation at the amino-terminus of the NS3 protein. Residue Phe-1180 appears to be the major substrate specificity determinant. The zinc binding domain is located near the surface of the protein and distant to the active site. Thus, the zinc does not participate directly in catalysis but is absolutely required for structural integrity: there is no activity in the absence of zinc.

The determination of the X-ray crystal structure of HCV NS3 protease has allowed the detailed study of the requirements for substrate binding. The S1 binding pocket is delimited by Leu-135 and Ala-157 on either side and Phe-154 on one end. The pocket is shallow and can accommodate the side chain of the cysteine residue present at the P1 position of the substrate [112,113,115,122]. Specificity of substrate binding appears to be governed by the presence of the P1 cysteine side chain and the acidic residue found in the P6 position, which may interact with the basic side chains of Arg-161 and Lys-165. Specificity determinants for S2-4 are essentially absent. Thus, active site binding is determined by side chains that are several residues apart (i.e. P1 and P6) with a minimum length requirement for efficient binding of synthetic peptide substrates of 10 amino acids (P6 through P4$'$ [127]).

The HCV NS3 protease exhibits inhibition by the N-terminal cleavage products of substrate peptides from the NS4A-NS4B, NS4B-NS5A, and NS5A-NS5B cleavage sites [128], suggesting that design of peptide inhibitors is possible. Thus, in spite of the shallowness of the substrate binding site, and its exposure to solvent, as well as its requirement for relatively large polypeptide substrate, several successful attempts at inhibitor design have been reported. These include an amidated alpha-carboxy-hexapeptide $K_i = 100$ nM [129], an RNA aptamer identified by *in vitro* systemic evolution $K_a = 10$ *nM* [130], and a dansylated hexapeptide $K_i = 200$ nM [131]. Recently short polypeptide analogs (i.e. tri-peptide alpha-ketoacids with difluoro aminobutyric acid in the P1 position) that exhibit K_i values of 27–67 nM for *E. coli* expressed NS3 plus 4A cofactor peptide have been identified [132]. However, these compounds were slow acting and with half-lives for dissociation of 11–16 h. The X-ray crystal structure of the NS3/4A protein with these inhibitors bound to the enzyme active site demonstrated little or no change in the conformation of the enzyme upon inhibitor binding [133]; hence, the substrate binding pocket remains rather nondescript and exposed to solvent. This underscores the previous prediction that the design of highly selective, tight-binding inhibitors of HCV NS3 protease will prove rather difficult [112,115]. Lai et al. [134] described the production of a chimeric virus between BVDV and HCV in which the BVDV cysteine protease is replaced by HCV sequences encoding the NS4A gene tethered to the NS3 protease domain. The chimeric virus was able to infect bovine cells *in vitro* and yield viable progeny virus; thus, this system may prove useful in the identification of HCV protease inhibitors. These recent results demonstrate the future promise, but difficult road ahead, in discovering candidate antivirals for treatment of HCV infection.

RNA Helicase (NS3)

The NS3 region of HCV encodes not only a trypsin-like serine protease, but also an RNA helicase [135,136] which is typical of many positive-stranded RNA viruses. The helicase activity is believed to participate in replication of the RNA genome by maintaining the template strand(s) or the nascent complementary strand(s) in a single-stranded conformation. Several lines of evidence indicate that the NS3 protein contains helicase activity. The NS3 or NS3-NS4A proteins expressed in either eukaryotic or prokaryotic systems demonstrate: nucleotide triphosphate hydrolysis and helicase activity on synthetic substrates [136–141], binding to single-stranded DNA or RNA, enhancement of NTPase activity upon binding to DNA or RNA [136,137,142–144], and the requirement of divalent cations for helicase activity [137,139,144].

The NS3 helicase activity is apparently localized to the carboxy-terminal two-thirds of the NS3 region, however, optimum catalytic efficiency is observed when the entire NS3 domain plus the NS4A region are expressed, where the NS4A domain is fused to the amino-terminus of NS3 (i.e. NS4A-NS3) [145]. This may be due to stabilization of the helicase or optimum folding of the polynucleotide binding site [146]. The availability of the X-ray crystal structure of full-length (single polypeptide) NS3/NS4A domain revealed that the protease and helicase domains are distinct but connected by a flexible region [147–149]. In the crystal structure the protease active site appears to be located at the C-terminus of the NS3 protein. This is interpreted as being an autoinhibitory conformation that is released upon binding of polyprotein substrate, which induces conformational change.

The advent of the X-ray crystal structure of HCV helicase provides a tool for the development of small molecule inhibitors of HCV helicase for use as therapeutic agents. There are numerous molecular mechanisms that could be targets for inhibitors including, NTP-binding site(s), catalytic site for NTP hydrolysis, polynucleotide binding site(s), and unwinding activity. One recent report described a modified ribavirin molecule (ribavirin-5′-triphosphate) that showed an IC50 of 40 vs. 500 uM for ribavirin [150]. While hydrolysis of the ribavirin-5′-phosphate may have decreased its potency, non-hydrolyzable analogs may possess even lower IC50s. High throuput screening assays will undoubtedly be useful in the rapid identification of helicase inhibitors [151,152] which remains a relatively new but rapidly progressing area of research [153].

NS4A and NS4B

As mentioned above, the NS4A protein (only 54 amino acids) functions as a cofactor for the NS3 serine protease. Its function is to stabilize the structure of the enzyme through interaction with the amino-terminus of the protein by contributing to the formation of the beta-barrel structure. In addition, NS4A may facilitate the anchoring of NS3 to the ER membrane through its amino-terminal hydrophobic domain [112,154]. NS4 also possesses some of the more diagnostically relevant epitopes [155–157]. Sequence variation within the NS4 region between HCV strains can negatively affect the sensitivity of HCV diagnostic assays [158–164]. Current immunodiagnostic assays utilize HCV recombinant proteins and/or peptides derived from core, NS3, NS4 and NS5 such that very high specificity and sensitivity is obtained (Fig. 1). The genotype-specific response to NS4A has been exploited to develop type-specific immunoassays [165–169]. The role of the NS4B protein in HCV replication is not clearly understood. It is a relatively hydrophobic protein of about 30 kDa. Recent evidence suggests it plays a role in the hyperphosphorylation of NS5A [170,171].

NS5A

HCV NS5A protein exists as two species when expressed in mammalian cells in culture, one of 56 kDa and the other of 58 kDa. Treatment of the proteins with phosphatase indicated that the 58 kDa form is hyperphosphorylated with respect to the 56 kDa form [172,173]. Phosphorylation occurs mainly on serine residues in the regions from 2200 to 2250 and 2350 to the carboxy-terminus of the protein [173]. The site around amino acid 2350 is phosphorylated in both 56 and 58 kDa forms, while the 58 kDa protein is phosphorylated at additional sites. These sites have been identified through site-directed mutagenesis as Ser-2197, Ser-2201, and Ser-2204 [173]. As mentioned above, the presence of NS4A appears to enhance p58 phosphorylation, possibly through the activation of cellular kinases [172]. The cellular kinase responsible for phosphorylation is not currently known.

The function of NS5A in viral replication is not entirely clear at this point; however, NS5A does appear to be a part of the viral replication complex [154,174,175] and also exhibits influence on the activity of the interferon-sensitive, double-stranded RNA-dependent protein kinase PKR [176]. PKR is a major enzyme in the pathways involved in host defense against viral infection; it represses translation by phosphorylation of the alpha-subunit of the eukaryotic translation initiation factor eIF2. It is believed that HCV NS5A interacts with PKR and inhibits its ability to phosphorylate eIF2, thereby allowing translation to proceed [177–179]. The region(s) of NS5A responsible for this action may include amino acids 2209–2248 [177].

The earliest observation that led to the hypothesis that NS5A might be involved in modulating the HCV response to interferon came from examination of HCV genome sequences from interferon-treated individuals. Enomoto and colleagues [180,181] observed that among HCV 1b-infected Japanese patients treated with interferon, there were mutations that clustered around the E2 HVR-1 region as well as the carboxy-terminal half of NS5A. *In vitro* studies indicate that a region of NS5A including the ISDR binds to RNA-dependent protein kinase (PKR), thus preventing PKR from suppressing viral protein translation [177]. If four or more mutations occur within this region, the mutated NS5A protein is unable to bind to PKR, thus allowing PKR to inhibit viral

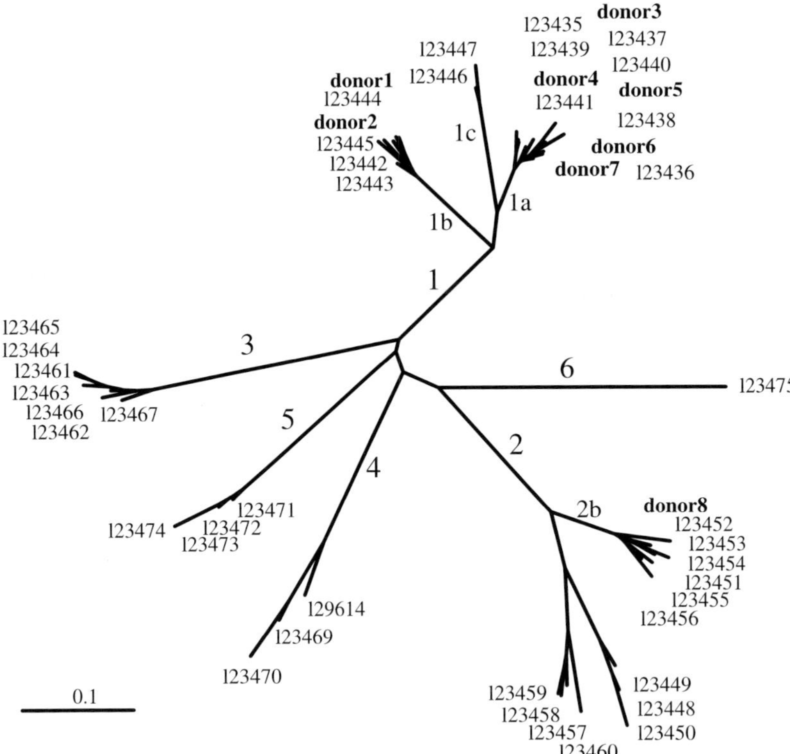

Fig. 3. Unrooted phylogenetic tree of HCV NS5 sequences showing the 6 major genotypes of HCV. The sequences analyzed were taken from Ref. [363] and are indicated by their GenBank accession numbers. NS5 sequences from HCV-positive blood donors were determined by the authors (unpublished data). The tree was produced using methods previously described [364].

Methods used to determine genotype include (i) phylogenetic analysis of various regions of the genome, for example 5′-UTR, core, E1, NS5b, (ii) restriction fragment length polymorphism and single-strand conformational polymorphism, which rely on analysis of amplicons generated from the 5′-UTR, and the (iii) line-probe assay [215, 216]. Each method has advantages and disadvantages related to time, expense, specificity and sensitivity. Analysis of sequences from more than one portion of the genome has become the gold standard for genotyping although many clinicians currently rely upon the line-probe assay known as InnoLIPA II from Innogenetics [217]. This assay relies upon the specific hybridization of 5′-UTR amplicon sequences to genotype-specific oligonucleotide probes immobilized onto a solid support. The resulting banding pattern determines the genotype and/or subtype. The need to determine HCV genotype in infected individuals is determined by the correlation between antiviral treatment outcome and genotype. Recent clinical trials of combination therapy of peginterferon and ribavirin demonstrated that individuals infected with genotype 1 had 29−41% sustained viral response compared to 73–77% sustained viral response in those infected with non-type 1 isolates [218–220]. Thus, while it is important to determine the major genotype there is currently no clinical need to determining subtype.

HCV isolates that cannot be classified into any of the original six major groups continue to be described. Isolates classified as genotype 4a, 4b, 4c, or 4d were identified based upon analysis of E1 sequences [221]. An isolate from an Indonesian patient was classified as a new subtype, designated 1d [222], and a subsequent study by Hotta et al. [223] demonstrated that HCV-1d was found in 15 of 61 (25%) HCV antibody positive Indonesian patients diagnosed with either chronic hepatitis or liver cirrhosis. Analysis of sequences derived from the E1 and NS5b genes demonstrated the existence of HCV genotypes 2d, 4e, 4f, 4g, and 4h [213]. Novel genotypes of HCV were identified in Nepal [224] and classified as genotypes 3b, 3c, 3d, 3e, and 3f based upon phylogenetic analysis of sequences obtained from the $5'$-terminal 1500 bases and the $3'$-terminal 1200 bases of the viral genome. The number of HCV types has expanded to include groups 7, 8, and 9 from Vietnam [225] and include subtypes 7a, 7b, 8a, 8b, and 9a. These investigators also found it necessary to analyze a larger portion of the viral genome (i.e. 1093 bp from NS5b) to obtain nonoverlapping distributions of phylogenetic distances between types. HCV isolates from Indonesia have been classified into novel genotypes, i.e. 2e and 2f and 10a and 11a [226].

The existence of HCV types 7–9 and 11 has been met with some controversy. It has been suggested that sequences from groups 7–9 and 11 should be included in group 6 due to the smaller intergroup sequence variability observed within groups 7–9 and 11 as compared to groups 1–6 [227–230]. The full-length sequences of HCV isolates corresponding to the provisional 7th, 8th, and 9th groups were recently reported [231]. Analysis of these and 30 other full-length HCV genome sequences via comparison of pairwise distances indicated that 11 groups could be delineated, but that there was some overlap in the distance values between isolates in groups 6–9. The investigators did not use bootstrapping (statistical analysis) to support their groupings. Earlier analysis of subgenomic regions of isolates belonging to putative groups 7–11 using bootstrap resampling found that these novel variants grouped together with the previously described genotypes (i.e. genotypes 1–6) [228]. Recently published recommendations establish methods to be used for the classification of HCV [230]. These include the use of bootstrapping to determine the robustness of the sequence groupings rather than relying on the attainment of nonoverlapping pairwise distance values to distinguish groups. It is now generally accepted that the hepciviruses consist of six major groups (genotypes) with various numbers of subtypes within each.

The prevalence of HCV genotypes demonstrate a clear correlation with geographic origin [214,232]. In the United States genotypes 1a and 1b are most prevalent [233]. In Japan genotype 1b is found in up to 73% of HCV-infected individuals [234]. In Italy, it was shown that 42% of 79 patients tested were infected with genotype 1 while 45% were infected with genotype 2 and only 4% with genotype 3 [235]. In Germany, genotype 1b was the most common (58%), although genotype 3 was found to be very common among intravenous drug users (35%) [236]. This was also found in another European study [237] in which genotype 3a was found in 63% of intravenous drug abusers but in only 10% of

142

blood donors; (50% of all donors were infected with genotype 1b). These authors concluded that there were two independent HCV epidemics occurring, one that affects blood donors and those without known route of infection and the other involving intravenous drug abusers. HCV genotype 4 is prevalent in North Africa and the Middle East [238,239]. HCV genotype 5 was first reported in South Africa [240] where it accounts for up to 80% of all infections [214]. Genotypes 7, 8 and 9 have only been found in Vietnamese patients [225] while genotypes 10 and 11 were identified originally in patients from Jakarta, Indonesia [226]. A recent report from Russia describes the existence of an intergenotypic recombinant form of HCV [241]. This recombinant, tentatively designated RF1_2k/1b, is a hybrid between the $5'$-terminal portion (through NS2) of a genotype 2k isolate and the $3'$-portion of a genotype 1b isolate. The existence of this recombinant raises important issues with respect to methods used to genotype HCV isolates and the tailoring of antiviral treatment based upon genotype (discussed below).

Clinical implications of HCV genotypes

The genetic diversity of HCV has had an impact on several areas of clinical importance. One of these is in diagnostic assays. As discussed above (see "NS4A and NS4B"), and in more detail later (see "Antibody detection"), sequence variation in the NS4 region can lead to moderately reduced sensitivity of assays using recombinant antigens derived from a single genotype of HCV (see also Ref. [242]). In addition, the ability to detect HCV RNA by reverse transcription PCR can be greatly influenced by the choice of target sequences for the PCR primers. Current methods utilize regions from the $5'$-NCR that are very highly conserved among all HCV genotypes resulting in assays that are very sensitive and specific (see "Nucleic acid testing"). The impact of HCV genomic sequence heterogeneity on diagnostic tests was significant among the first generation assays, but many of these shortcomings have since been overcome by the inclusion of recombinant proteins from other regions of the genome that exhibit lower sequence variability.

HCV sequence heterogeneity may also impact pathogenesis. Though a significant amount of research has been done in this area, it remains rather controversial since some studies indicate there is no difference in pathogenicity between genotypes while other studies indicate some genotypes are more pathogenic. For example, it has been estimated that, excluding other common causes, steatosis is present in up to 30% of chronic HCV patients [243]. Subsequently, several studies postulated an association between steatosis and HCV genotype 3, but not genotype 1, among chronically infected individuals [244, 245]. Additional studies have shown similar associations [246–248]. More convincing evidence for a direct involvement of the HCV in liver steatosis was provided in a study conducted among patients with sustained viral response to antiviral therapy: responders infected with HCV genotype 3 had a significant reduction in steatosis whereas those infected with genotype 1 there was no change in steatosis whether or not they responded to treatment [249,250].

Another study showed the rate of progression to chronic infection was 92% for HCV genotype 1b and 35–50% for all other genotypes [251]. Numerous other studies have linked HCV genotype 1b with more severe liver disease, a more aggressive course, increased propensity for development of cirrhosis, and even hepatocellular carcinoma [252]. In contrast, Zeuzem et al. [253] found that among chronically infected HCV patients there was no correlation between HCV genotype and serum RNA levels, liver function tests or liver histology, although patients infected with genotypes 1b and 2a–c tended to be older. An investigation of chronically infected patients from the Benelux countries found no significant relationship between severity of fibrosis or liver inflammation and the infecting HCV genotype [254]. Many other studies have reported similar findings regarding the lack of association between HCV genotype 1b and hepatic disease severity [252]. It has been hypothesized that the discrepancies between the findings in these studies are may be due to the fact that many individuals who are infected with HCV genotype 1b are older than those infected with any of the other genotypes [255]; thus, these patients may have been infected for a longer time. When the duration of infection was taken into account, the distribution of genotypes was equal between moderate and severe hepatitis cases. Similar results have been reported elsewhere [256,257]. Thus, the association between HCV type 1b and severe liver disease may be due to its long history of infection rather than intrinsically high pathogenicity.

Screening and diagnostic tests for HCV infection

Detection of exposure to HCV

There are two general strategies for detecting exposure to HCV. One relies on the detection of antibodies to HCV, which, when confirmed positive by supplemental antibody tests, indicates past or ongoing HCV infection. Antibody testing is utilized to prevent transmission of HCV by eliminating seropositive blood donors from the donor pool, and to determine if HCV infection is associated with cases of acute or chronic hepatitis. The second strategy relies on the detection of viral-specific molecules (e.g. HCV RNA or HCV core proteins) in serum/plasma, and is indicative of active HCV replication. These tests are most useful in detecting HCV infection during the "window period" (time between exposure and seroconversion), and in determining viral presence and titer in chronically infected individuals, especially as a means to monitor the efficacy of antiviral therapy. If viral RNA is detected, nucleotide sequencing or other techniques may be performed to determine the genotype of the sample or to analyze isolate-specific sequences.

Antibody detection

Due to the unavailability of HCV virions obtained either directly from infected individuals or from HCV infected cell lines, serologic tests have relied on recombinant antigens or synthetic peptides. The first generation anti-HCV screening tests were based on detection of antibodies directed against a recombinant protein (HCV genotype 1a) originating from sequences located in the nonstructural region 4 (NS4) protein (C100-3)

144

[1,258]. Although the implementation of these assays was a major breakthrough in identifying individuals exposed and/or infected with HCV [259], the assays failed to detect antibodies in approximately 10% of individuals having chronic HCV infection and up to 10–30% of individuals with acute HCV infection. Further, the assays frequently failed to detect HCV antibodies in blood donors capable of transmitting HCV to recipients [260,261].

The second generation anti-HCV assays have incorporated recombinant proteins originating from three different regions of the HCV genome (HCV genotype 1a), including amino acid sequences from the HCV core, NS3, and NS4 protein [262,263]. The amino acid sequences from NS3 and core regions are more immunogenic and are more highly conserved than the NS4 region allowing improved detection of antibodies in individuals infected with all HCV genotypes. The second generation assays detect antibodies in close to 100% of chronic HCV cases [264] and in nearly 100% of the acute cases by 12 weeks post infection [265]. The second generation assays showed a marked improvement over the first generation test in identifying HCV infected blood donors [260,261]. The NS4 antigen utilized in the first generation assay has variable amino acid sequences across HCV genotypes, leading to inadequate detection of antibodies in HCV samples other than genotype 1. The type specific antibody response to NS4 protein, in fact, has been utilized as a means to serotype HCV infections [165]. Although NS3 and core are highly conserved, the antibody responses to genotype-specific antigens are higher among individuals infected with a homologous genotype [242]. However, the reduced titers are not sufficient to produce false negative results with the genotype 1a antigens utilized in the major antibody tests.

The third generation test includes a recombinant protein expressing amino acid sequences from the NS5 region, as well as antigens from the core, NS3 and NS4 [266]. Some studies have indicated a slight improvement in sensitivity in comparing the third generation test to second generation tests [266,267], but this improvement is largely attributed to changes in the NS3 protein rather than the inclusion of NS5 [268]. Thus, some assays utilizing second generation type reagents, without the inclusion of NS5, appear to be as sensitive or more sensitive than the third generation test, partly due to the implementation of microparticles as the solid phase [269].

Supplemental antibody tests

In general, specimens that are reactive in the antibody test are retested in duplicate; if one of the two duplicate test results are reactive, the sample is considered as repeatably reactive. Supplemental assays are performed to "confirm" the results of the primary test. The supplemental assays use the same antigens as those in the primary test, but allow separate results to be obtained for each component of the primary test. The first supplemental tests, including the recombinant immunoblot assay (RIBA-1), was composed of an NS4 recombinant antigen and a small NS4 synthetic peptide [270]. Alternatively, antibodies to NS4 could be confirmed by a neutralization test and peptide antibody tests allowing for detection of antibodies to different regions of the NS4 protein [271]. Later, RIBA-2 [157] and the Abbott Matrix [272] were developed as the supplemental tests for second generation screening assays, and were composed of individualized recombinant antigens and/or peptides to core, NS3, and NS4. Later, the RIBA-3 assay was developed, which includes two recombinant proteins (one from NS3 and one from NS5) and peptides representing the core and NS4 antigens. The second and third generation supplemental assays have been utilized for the second and third generation antibody tests. Samples are considered positive in these assays when reactivity is noted for at least two distinct regions of HCV. Samples that are reactive for only one protein are considered indeterminate.

In the clinical setting, anti-HCV tests are utilized to assist in the diagnosis of patients with signs and symptoms of hepatitis. About 95% of the seropositive samples observed in clinical laboratories are positive for HCV RNA [273,274]. Thus, supplemental antibody testing is not required in clinical laboratories due to the high predictive value of an antibody positive result. In the blood screening setting, approximately 65% of the repeatably reactive samples obtained with second or third generation tests are positive in supplemental testing, and about 25% are considered indeterminate [275]. Approximately 50–60% of the samples that are positive with supplemental tests are HCV RNA-positive [276,277], but only very few (approximately 1%) of the indeterminate samples are HCV RNA-positive. Thus, the correlation between a repeatably reactive result and HCV RNA positivity is weaker in blood screening sites than in clinical laboratories.

Other serologic markers

The anti-HCV IgM class antibody response against HCV core protein may be detected in acutely infected individuals who are exposed to HCV by transfusion [278]. The anti-HCV IgM responses are transient, lasting for about 2–6 weeks, and are co-detected with anti-HCV IgG, but do not precede detection of IgG class antibodies against HCV. Some studies have suggested that the HCV IgM detection may not be restricted to acutely infected individuals. For example, anti-HCV IgM was detected in about one-third of anti-HCV IgG positive blood donors [279] and IgM was frequently detected in HCV seropositive patients, possibly associated with active HCV replication [280,281]. In contrast, no correlation was found between the presence of anti-HCV IgM and serum HCV RNA or serum transaminase levels [282]. In summary, there is inconsistent evidence that anti-HCV IgM detection is useful during acute infections, because the IgM response is detected transiently and in only some acutely infected subjects. In addition, the utility of anti-HCV IgM as a surrogate marker for ongoing viral replication has been supplanted in recent years by the widespread use of HCV RNA testing to determine active viral replication (see below).

Utilizing a novel expression system, the second envelope protein (E2) of HCV has been expressed in mammalian cells [94], purified, and evaluated to determine its utility as a serologic marker for HCV infection. The correlation between detection of antibodies to E2 and detection of HCV RNA was quite high at about 97%. In addition, HCV RNA-positive samples that were indeterminate in supplemental assays, were frequently antibody positive for antibodies to E2, thus increasing the confirmation rate of reactive samples, especially those obtained from screening blood donors [94]. Thus, the E2 protein is a candidate for inclusion in future generation serologic assays for HCV detection, either to be included in a primary antibody test or as an adjunct marker in the supplemental assays.

In 1990, Japanese researchers described an autoantibody (anti-GOR) that was detected in 81% of chronic NANB patients, and in about 2% of healthy blood donors [283]. Further studies suggested that anti-GOR is frequently detected in individuals with autoimmune hepatitis type 2 [characterized by the presence of antibodies against liver-kidney-microsomes (LKM-1) proteins], and may be of use in characterizing patients with chronic hepatitis [284]. Another report argued that the immune response against GOR sequence was not an autoimmune phenomenon, but simply represented a cross reaction between the antibody response to HCV core (amino acids 4–20) with the GOR47-1 clone (amino acid residues 15–29) [285].

Nucleic acid testing

In spite of the increased sensitivity of second and third generation serologic assays to detect HCV antibodies, antibody tests fail to detect exposure to HCV until, on average, approximately 70 days after infection [286]. Further, the detection of antibody provides little value in determining the progression of HCV disease, or in determining the efficacy of antiviral therapy. Over the last decade nucleic acid tests (NAT) designed to detect HCV RNA have been utilized to study disease progression, monitor antiviral therapy and screen blood donors. In addition, clinical laboratories may use qualitative HCV RNA testing as an alternative to supplemental antibody testing in establishing that a seropositive individual has been infected with HCV; in general, 95% of the seropositives in the clinical laboratory are positive for HCV RNA. In many cases, HCV -infected persons enrolling in an antiviral therapy program are required to have a positive HCV RNA result before treatment is initiated. HCV NAT has become an indispensable tool in the diagnosis and management of HCV.

HCV RNA detection assays can be either qualitative or quantitative. Qualitative tests are used for diagnosis and blood donor screening while quantitative assays are most often used to monitor patient response to drug therapy. Three major methods are currently in use: the branched chain DNA (bDNA) signal amplification assay, traditional reverse transcription PCR (RT-PCR), and transcription mediated amplification (TMA). The bDNA assay is a quantitative test and the TMA assay is a qualitative test; RT-CPR can be either qualitative or quantitative. The bDNA assay relies on capture of HCV RNA molecules on microwells coated with oligonucleotides complementary to HCV RNA [287]. A series of probes complementary to the $5'$ untranslated region of the genome are hybridized to the captured HCV RNA, mediating the capture of branched DNA molecules capable of binding to alkaline-phosphatase labeled probes. The signal generated via a chemiluminescent substrate is proportional to the level of target sequences in the original sample. The RT-PCR assay, first described in 1990 [288], requires the synthesis of cDNA followed by the amplification of the cDNA by using primers complementary to the $5'$-UTR. Amplicons generated can be detected by agarose gel electrophoresis and ethidium bromide staining which can be followed by Southern blotting for greater sensitivity.

Alternatively, the amplicons may be detected by capture on a solid phase and then hybridized to oligonucleotide probes conjugated with a reporter molecule. There are several versions of PCR-based HCV RNA tests that may be either be qualitative or quantitative; in many cases the quantitative assays utilize an internal control that competes with the HCV RNA target for amplification or detection. Viral load is computed by determining the ratio of the HCV-specific amplicon to the internal control amplicon. The third type of assay utilizes TMA which relies on the generation of cDNA by reverse transcriptase enzyme and employing a primer that contains sequences complementary to HCV RNA linked to T7 RNA polymerase promoter site. In the presence of T7 RNA polymerase, 100–1000 copies of HCV RNA transcripts are produced from each cDNA template. The RNA generated in this step, serves as the template for cDNA synthesis thereby starting the cycle again. Amplified product is detected via hybridization of acridinium-labeled DNA complementary to the HCV RNA transcript.

Nucleic acid testing for diagnosis or blood donor screening

HCV RNA detection assays currently licensed by the US FDA for diagnosis of HCV infection are the Roche Amplicor and Roche COBAS Amplicor (semi-automated version of the former). The limit of sensitivity for the manual assay is 50 IU/ml while the semi-automated COBAS test has a limit of sensitivity of 60 IU/ml. There are currently three HCV RNA detection tests approved for use in the United States for blood donor screening (Table 1). The Gen-Probe TMA assay (Procleix) detects both HCV and HIV RNA in pools of up to 16 members. It was recently shown that this assay has excellent sensitivity for detection of HCV genotypes 1–6 and for HIV-1 group M subtypes A–G [289]. The early version of the HCV TMA assay had a sensitivity of less than 50 HCV RNA copies/ml [290,291]; however, the latest version (VERSANT HCV RNA Qualitative) detects as little as 5.3 or 29 copies/ml in 95% of all clinical samples examined [292] or 5 IU/ml (41 copies/ml) using the WHO HCV RNA standard [293]. The UltraQual test from NGI has a limit of detection of 5–8 copies/ml and was approved by the US FDA for screening 512 member pools of donor plasma or serum intended for fractionation. For a recent review of these methods see Ref. [294].

Table 1

United States FDA approved HCV RNA detection assays for plasma donor screening

Tradename(s)	Format	Sample	Limit of detection	Use	Manufacturer	FDA Approval date
Procleix	HIV-1/HCV Nucleic acid test (TMA)	Plasma	100 copies/ml	Donor screen	Gen-Probe	2/8/2002
UltraQual HCV RT-PCR assay	RT-PCR	Plasma or serum	5–8 copies/ml	Donor screen	NGI	9/18/2001
COBAS AmpliScreen HCV test	RT-PCR	Plasma or serum	60 IU/ml	Donor screen	Roche	12/2/2002

HCV RNA testing of blood donors has been implemented in the US, Germany and several other countries. Preliminary results indicate that HCV RNA can be detected in about 1 in 333,000 to about 1 in 500,000 donations, for seronegative volunteer blood donors in the United States and Germany, respectively [295,296]. The detection of HCV RNA occurs about three times more frequently than for HIV [295,296], but not as frequently as predicted (approximately 1 in 125,000) in the model proposed prior to the implementation of testing [286].

Currently, HCV RNA testing is performed on serum or plasma. However, several recent studies [297–300] have shown that whole blood from individuals chronically infected with HCV contain significantly more RNA than plasma. Using an "in-house" RT-PCR test, 50 chronic HCV patients were tested for HCV RNA: 35 were RNA-positive when using whole blood as the RNA source compared to 28 when using plasma tested. Only 26 were positive when plasma was tested using the Roche Amplicor test [299]. These results warrant investigation of the use of whole blood assay for HCV RNA in the blood screening setting since it may be possible to identify HCV-infected donors with undetectable serum RNA levels but who may have cell-associated viral RNA that is readily detectable. The whole blood assay may also be beneficial for monitoring the outcome of antiviral therapy.

Quantitative nucleic acid tests

HCV RNA quantitative assays currently report results in International Units per ml, however, this was not always the case. Previously, each manufacturer or laboratory reported results in RNA copies per ml or genome equivalents per ml. Given the variations in assay methods and precision it was difficult to equate RNA levels determined by one assay with that of another. The World Health Organization (WHO) therefore, established an HCV standard to which manufacturers can calibrate their assays [301,302]. A recent review provides a table of conversion factors for various HCV RNA quantitative assays to convert from IU/ml to the nonstandardized copies/ml [303]. The specificity of these assays is generally greater than 98% and quantification is independent of genotype. For these assays, differences between measurements taken with the same method of less than 0.5 log fall within the proven variability of the assay and should therefore not be considered significant.

Quantitative determination of HCV RNA levels in patient serum or plasma is an indirect measure of the extent of viremia and has been used as a measure of response to antiviral drug therapy. Quantitative assays currently in use, but not yet FDA approved, are listed in Table 2. The Roche Amplicor Monitor and semi-automated COBAS Amplicor Monitor assays utilize RT-PCR technology and were recently calibrated to report results in International Units [304,305]. Each assay has a linear range of quantification from 600 to 500,000 IU/ml. The second generation bDNA test is highly reproducible and precise, but cannot be utilized for samples having low viral loads since the sensitivity limit of the assay is about 100,000 HCV RNA copies/ml. This assay has been utilized for several years in monitoring antiviral therapy [306]. The third generation bDNA assay shows significant enhancement in genotype detectability and sensitivity

Table 2

Assays for quantification of HCV RNA in human plasma or serum

Tradename(s)	Format	Quantification range	Manufacturer
Amplicor HCV Monitor v2.0	RT-PCR	600– < 500,000 IU/ml	Roche
COBAS Amplicor Monitor	Semi-automated RT-PCR	600– < 500,000 IU/ml	Roche
Versant HCV RNA 2.0	bDNA	200,000–120 million genome equivalents per ml	Bayer
Versant HCV RNA 3.0	bDNA	615–7,692,000 IU/ml	Bayer
SuperQuant	RT-PCR, competitive	30–1,470,000 IU/ml	National Genetics Institute
LCx HCV RNA	RT-PCR, competitive	23–2,325,000 IU/ml	Abbott Laboratories

over the bDNA 2.0 assay [307,308] with a dynamic range of 615–7.7 million IU/ml. The SuperQuant assay from NGI utilizes competitive RT-PCR and exhibits greater sensitivity than the tests mentioned above (lowest level quantified: 30 IU/ml) but has a smaller dynamic range than the bDNA 3.0 assay. Finally, the LCx HCV RNA Quantitative assay developed by Abbott Laboratories exhibits a dynamic range of 23–2,300,000 IU/ml making this assay the most sensitive quantitative assay available. This assay has not been FDA-approved but is expected to become available for commercial use in the European community sometime in 2003.

The HCV RNA quantitative assays listed in Table 2 can be used to monitor patients' response to antiviral therapy. According to the EASL International Consensus Conference on Hepatitis C [309] the clinically relevant threshold for tailoring treatment among HCV-infected patients is 800,000 IU/ml. A recent study by Pawlotsky and colleagues [302] demonstrated that there is no perfect correlation between assays for samples at or near the 800,000 IU/ml level. Samples with RNA values below 800,000 IU/ml in the COBAS Amplicor v2.0 assay contained more than 2,000,000 copies/ml RNA based on the NGI Superquant assay and, for several samples, the inverse was observed. Thus, given the inherent bias between assays, and the imprecision of most assays of nearly 0.5 log IU/mL, one must be careful when interpreting test results that are close to the so-called decision threshold of 800,000 IU/ml. Other factors such as HCV genotype age, fibrosis stage, alcohol use, etc. must be weighed before treatment regime is established.

HCV antigen testing

Two different types of serologic assays have been developed which permit detection of HCV core antigen in serum. One assay format detects HCV core antigens in seronegative or seropositive hepatitis C patients [310,311], while the other assay format is designed to detect core antigens only in volunteer blood donors prior to seroconversion [312,313]. The EIA assays developed by Ortho-Clinical Diagnostics (New Jersey, USA) are quantitative assays. The prototype chemiluminescent assay developed by Abbott Laboratories is qualitative, but signals correlated well with HCV RNA levels in the serum [314].

150

HCV core antigen is detected in the serum of infected individuals who are HCV RNA-positive but anti-HCV-negative, thus providing a potential alternative method to nucleic acid testing for the identification of newly infected individuals [313,315,316]. In seroconversion panels, the HCV antigen is detected much earlier than antibody [312–314,317]. As shown in Table 3, Ortho's third generation antibody screening test and the Abbott PRISM antibody screening test detect HCV antibodies on day 62 of the seroconversion panel, while the HCV antigen test detects exposure 38 days earlier on day 24, the same day on which HCV RNA is first detected. The first bleed in the panel was negative for all HCV markers. These data clearly indicate that antigenemia is detectable several weeks prior to seroconversion. A prototype combination antibody/antigen assay for Abbott's PRISM screening instrument is shown to detect those samples that are antibody negative/antigen positive, as well as those that are positive for antibody and antigen (Table 3) (Shah DO, Chang CD, Jiang LX, Cheng KY, Muerhoff AS, Gutierrez RA, Leary TP, Desai SM, Batac-Herman IV, Salbilla VA, Haller AS, Stewart JL, Dawson GJ (2003) Transfusion, vol 43: 1067–1074). The HCV "Combo" assay format allows for greater sensitivity over existing licensed antibody screening assays by allowing detection of antibody and antigen in a single test thereby closing the seroconversion window to within a few days of NAT.

The HCV antigen test has comparable sensitivity to the HCV bDNA version 2.0 test in detecting exposure to HCV in patients infected with different HCV genotypes [318], and in monitoring antiviral therapy [311,319]. Individuals who respond to interferon are more likely to have lower levels of HCV core antigen than nonresponders [311]. A recent study of 657 samples from patients on antiviral therapy utilizing the Ortho-Clinical Diagnostics Total HCV core Ag Assay demonstrated that the assay could be used for patient viral load monitoring and for evaluation of baseline viral load prior to commencement of treatment [320]. It was shown that 1 pg/ml of total core Ag is

Table 3

Seroconversion profile for HCV plasma donor, indicating reactivity with various HCV tests

Day	S/CO Values for various serological assays[a]				
	Prism HCV Ab	Ortho HCV 3.0 Ab	Abbott HCV Ag	Abbott Combo[b]	HCV RNA[c]
0	0.09	0.16	0.46	0.31	<100
24	0.11	0.13	**5.9**	**2.98**	7,413,100
27	0.10	0.16	**19.6**	**13.1**	>10,000,000
31	0.11	0.11	**50.0**	**27.1**	>10,000,000
62	**5.19**	**4.1**	**26.2**	**15.5**	>10,000,000
64	**5.22**	**4.1**	**30.6**	**24.8**	>10,000,000
69	**5.91**	**4.1**	**9.2**	**7.3**	>10,000,000
71	**6.29**	**4.1**	**4.6**	**5.08**	>10,000,000

[a]S/CO values >1.0 considered reactive.
[b]Abbott Combo assay for PRISM measures antibody and core antigen simultaneously.
[c]HCV RNA levels in copies/ml determined Abbott LCx HCV RNA quantitative assay (limit of detection 100 copies/ml, quantification limit of 10 million copies/ml).

equivalent to approximately 8000 HCV RNA IU/ml. However, because total core antigen did not correlate well with HCV RNA values below 20,000 IU/mL, this level was established as the lower limit of viral load measurement for patients on therapy.

Natural history

Acute infection

HCV infection occurs primarily through parenteral exposure although occasionally infection may occur via sexual or perinatal transmission [321]. Following exposure, the virus enters a susceptible hepatocyte and viral replication occurs. There is an eclipse phase of approximately 10 days during which time there is no evidence of viral presence, where viral RNA is undetectable, serum transaminase levels are within normal limits, with no evidence of a specific immune response to HCV [286]. About 10 days following exposure HCV RNA is detectable, often with viral loads between 100,000 and 120,000,000 genome equivalents per ml of serum. Several weeks later, there is typically an increase in ALT levels (Fig. 4) indicating inflamation of the liver and destruction of hepatocytes, mediated by the killing of HCV-infected cells by CTL activity [322]. Often CD4 + T-cells, macrophages and CD8 + T-cells are present adjacent to dead or dying hepatocytes [323], indicating a role in killing HCV-infected hepatocytes. At about the same time that ALT elevations occur, the appearance of specific antibodies directed to HCV proteins is detected, usually within about 70 days post-infection [286].

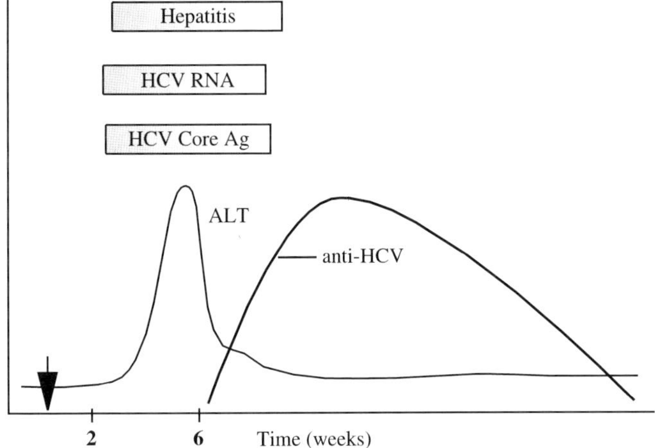

Fig. 4. Natural history of acute, resolving hepatitis C virus infection in humans. Time of infection is indicated by the arrow. HCV RNA and core antigen are detected within 1–2 weeks post-infection with elevations in serum ALT levels occurring within 5–10 weeks following exposure. Seroconversion typically occurs within 6–12 weeks post-infection (using third generation EIAs). In this profile, acute disease is followed by clearance of viral RNA from the serum and a normalization of serum ALT values.

152

These specific antibodies may interfere with viral entry into host cells, and may, through oponsonization, lead to the ingestion of viruses via macrophages [324]. However, the humoral immune response is believed to be incapable of eliminating the virus from infected cells. About 75% of infected individuals are asymptomatic during the acute phase of infection. However, some individuals experience symptoms such as malaise, weakness, upper right quadrant pain, and in some cases jaundice [325]. About 85% of HCV-infected individuals are unable to clear the virus, thus, HCV RNA and antibodies, and, in some cases, elevated ALT values may be observed continually or intermittently for many years [325].

Viral clearance

About 15% of HCV-infected individuals clear their viral infection in the weeks and months following exposure [326–328]. Viral clearance has recently been characterized as the continued absence of HCV RNA in blood samples from a person previously infected with HCV, and the normalization of ALT levels. In one study, the duration of viremia after seroconversion had a mean time of 19 months prior to viral clearance [327]. Individuals who have lower viral RNA titers and lower ALT peaks during the acute phase of infection tend to clear the virus [327]. There is accumulating evidence that HCV viral clearance is closely linked to a strong T-cell response against viral proteins, allowing selective destructive of infected cells displaying HCV antigens. As the virus replicates and HCV-specific gene products egress from infected cells, antigen presenting cells (e.g. dendritic cells and macrophages) display HCV antigens to CD4 + T-helper cells which activate CD8 + cytotoxic killer cells and natural killer cells. The T-cell-specific response targets NS3 and NS4 proteins with one highly conserved sequence in NS3 (aa 1248-1261) being a prime target of CD8 + cytotoxic T-cells [329,330]. Virus-infected cells are killed via the induction of apoptosis by CD8 positive CTLs. Some individuals who mount a strong T-cell response show evidence of long-term viral clearance [330–332], characterized by continued absence of HCV RNA in serum. Other infected individuals may mount a strong T-cell response, resulting in transient viral clearance, but HCV RNA may reappear as the T-cell response declines in intensity [330]. The cellular immune system is believed to play a significant role in limiting viral replication in individuals who remain chronically infected, so that in most individuals the disease progression is slow. Thus, viral persistence may be viewed as the failure of the host immune system to eliminate the virus, due potentially to inadequate stimulation of the CTL response.

Chronic infection

A typical profile of individuals with chronic HCV infection includes continued detection of HCV RNA, presence of antibodies to HCV and intermittent elevations in serum ALT levels (Fig. 5). About one-third of individuals with chronic hepatitis have persistently normal ALT levels, yet these individuals may have histological evidence of liver disease [333]. In general, most patients with chronic hepatitis are asymptomatic, and are unaware

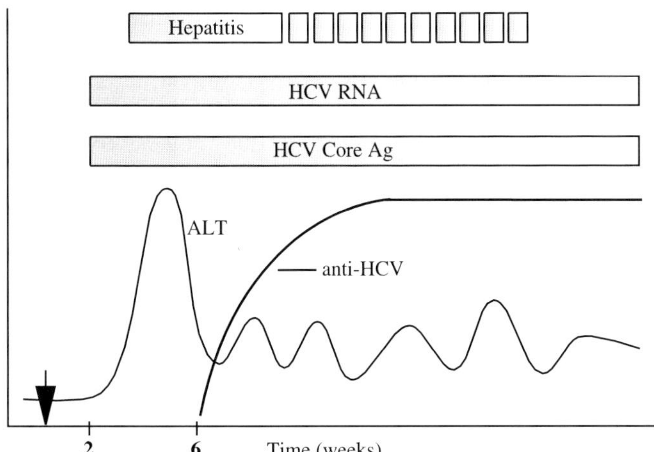

Fig. 5. Natural history of chronic hepatitis C virus infection in humans. Time of infection is indicated by the arrow. HCV RNA and core antigen are detected within 1–2 weeks post-infection with elevations in serum ALT levels occurring within 5–10 weeks. Seroconversion typically occurs within 6–12 weeks post-infection (using third generation EIAs). Chronic disease is characterized by constant or intermittent viremia and fluctuating serum ALT levels.

of their disease. Among those with symptoms, fatigue is the most commonly recognized problem. There are many possible clinical outcomes for patients with chronic HCV infection ranging from chronic hepatitis characterized by low level hepatic inflammation, to serious liver disease including cirrhosis and/or hepatocellular carcinoma. However, the likelihood of a given individual progressing to serious liver disease appears to be partly dependent upon how long the individuals has been infected, since in general, if untreated, HCV appears to be a slowly progressive disease. As noted above, there is not a high degree of correlation between HCV RNA titers and disease progression.

One of the major factors contributing to more rapidly advancing liver disease is the ingestion of alcohol. Chronic alcoholism in HCV patients increases the likelihood that cirrhosis and hepatocellular carcinoma may develop in HCV patients [334,335]. In a recent survey of chronically infected patients it was found that an increased risk for end-stage liver disease (expressed as ascites, esophageal varices, or encephalopathy) was associated with intake of relatively large amounts of alcohol (260 g or more per week) but not with moderate intake [328]. Thus, it is apparent that there are other as yet unidentified factors that mitigate the affects of alcohol consumption among HCV-infected individuals.

One confounding problem in predicting clinical outcome is that the date of first infection with HCV is often not clear. Individuals presenting with symptomatic HCV infection for the first time, may actually have been infected with HCV for many years. Most of the HCV natural history studies have been performed on individuals who became infected with HCV via transfusion, wherein the date of first exposure to HCV is documented. In a recent review compiling separate studies on patients infected with HCV via blood transfusion, the risk of progression to cirrhosis has been estimated to be

100% for patients 50 years of age or older at the onset of infection and after 15 years duration of infection. The rate of progression to liver disease is much slower for younger patients with the same duration of infection [336]. However, the risk of progression to serious liver disease may be more likely in transfused patients than in non-transfused patients, due to exposure to a higher viral load, and due to the observation that transfused patients are older than non-transfused patients at the time of their respective exposures [337]. A recent study examining the rate of HCV genome sequence evolution among Japanese patients infected with genotype 1b vs. US subjects infected with genotype 1a provided an estimate of the approximate time of introduction of the virus and its spread throughout the two populations. The data suggested that the epidemic in the United States is still in its early stages and the incidence of hepatocellular carcinoma among HCV-infected patients will increase significantly for the next 20 years. This model assumes no significant contribution of nonviral factors to the incidence of hepatocellular carcinoma in Japan.

Antiviral therapy

Even before the discovery of HCV, studies had suggested that interferon therapy was beneficial to some patients with NANB hepatitis [338], with these cases later recognized as being due to HCV. Until recently, most of the HCV therapeutic regimens consisted of monotherapy with 3 million units (MU) of interferon-alpha (IFN-a) administered three times a week for 24–48 weeks. In general, about 50% of the treated patients showed a biochemical response (reduction of ALT levels to within normal limits), with about half of these responders suffering a relapse at the end of treatment, characterized by elevated ALT levels. This regimen and associated results continued over the next 10 years, with only mild improvements being realized with longer treatment schedules or higher doses.

With the availability of HCV RNA testing, one could determine not only biochemical response (changes in ALT values), but also the virological response which is based upon the change in HCV RNA titer during and after treatment. The new goal of antiviral treatment is the normalization of ALT values and the elimination of HCV RNA from the infected person. Patients that respond to treatment by reducing HCV RNA to undetecable levels in their blood stream for 6 months or more at the end of treatment are referred to as complete or sustained responders. Some patients have normalization of the serum transaminase levels but are unable to clear viral RNA from their blood streams. These individuals are more likely to have a relapse [339] and are referred to as incomplete responders. Other treated patients do not normalize their transaminase levels and do not clear HCV RNA from their blood steams and are classified as nonresponders.

The requirements for patient enrollment in an HCV antiviral therapy program includes a positive HCV antibody test, elevated serum aminotransferase levels (ALT or AST) over a period of 6 months or more, the detection of HCV RNA in serum, and chronic hepatitis as diagnosed by liver biopsy [340]. In some clinics, a biopsy is required prior to the initiation of treatment. In general, patients with cirrhosis are not encouraged to undergo treatment since very few of these patients show either a sustained biochemical or a virological response to interferon. Other factors that are associated with an unfavorable outcome of interferon treatment include high serum titers, infection with HCV genotype 1, and high degree of diversity in the viral quasi-species during the baseline period [264,341].

One of the major factors associated with a sustained response to interferon treatment is a dramatic reduction in HCV RNA in the first few weeks after treatment has been initiated [342]. In general, sustained viral responders cleared HCV RNA (less than 100 HCV RNA copies/ml) by 2–6 weeks after treatment, while relapsers had slower decreases in HCV RNA during the same period. Thus, it may be possible to utilize shorter treatment periods to separate responders from nonresponders, reducing the cost and side effects associated with continuing therapy among nonresponders. Other treatment options may then be followed. However, there is some evidence that continued antiviral therapy may have benefits other than viral clearance. For example, some patients with cirrhosis, who have a transient or sustained biochemical response to interferon, have a reduced risk for developing hepatocellular carcinoma [343]. In one study of over 1600 patients, it was noted that the rate of progression to hepatocellular carcinoma in the interferon-treated group was 7.6% over a 10-year period, while the progression rate was 12.4% among the untreated group [344].

Combination therapy with interferon and ribavarin are more effective than interferon alone in treating HCV-infected patients. In one study, the overall sustained response rates for the 48-week treatment period was 43% for individuals receiving dual therapy and 19% for individuals receiving interferon alone [345]. For the combination therapy, the sustained rate was 64% for those infected with genotypes 2 or 3, but only 31% for individuals infected with genotypes 1, 2, 5 or 6 (most patients were genotype 1). For the patients treated with interferon alone, the sustained rates were 33% for individuals infected with genotypes 2 or 3 and 11% for individuals infected with HCV genotypes 1, 4, 5, or 6. Similar results were obtained in a second study, demonstrating about a 2-fold increase in the sustained response rates for individuals treated with the combination therapy regimen [346]. In another study, one-third of those individuals who were relapsers with interferon monotherapy had a sustained response with the combination therapy [347]. Until very recently, the dual therapy using interferon and ribavirin was being recommended as the current regimen of choice for treating chronic HCV infection [340]. However, substantially better results have been achieved by combination therapy consisting of pegylated interferon-alpha and ribivarin. Two randomized controlled trials of pegylated-interferon and ribavirin that included over 2600 patients [218,220], clearly demonstrated 29–56% sustained viral response compared to other studies using peginterferon alone (7–30% sustained viral response) [219]. When the efficacy of the dual treatment was examined by genotype, it was shown that those infected with genotype 1 had sustained viral response rate of 29–51% compared to 73–77% for those infected with other genotypes. Patients treated for 48 weeks showed a higher rate of response than those treated for only 24 weeks, although a 24-week course was adequate for those infected with genotype 2 or 3. Further studies are need to examine the efficacy of these treatment regimes among HCV patients with other mitigating factors such as HIV-coinfection, alcohol abuse, renal disease, etc.

Future options for treatment

One of the newer treatment options being explored is a new generation of interferon, called pegylated interferon (PEG interferon) used in combination with ribivarin to enhance the antiviral response rate. The PEG interferon increases the biological half-life of interferon. Published data indicate that a higher percentage of patients had a sustained viral response to PEG interferon plus ribivarin than to PEG interferon alone [348]. Additional studies are underway comparing the antiviral effect of PEG-interferon plus ribavarin vs. the current dual therapy regimen. Although dual therapy (interferon plus ribavarin) has been effective in increasing the response rate in previously untreated patients or in relapsers, only about 10% of the patients who were nonresponders to interferon monotherapy showed a response to interferon plus ribavarin [349]. In a recent study, nonresponders to interferon treatment were treated with triple antiviral therapy (interferon, ribavirin and amantadine) or with dual therapy (interferon plus ribavarin). Patients undergoing triple therapy had higher sustained biochemical and virological responses (57 and 48%, respectively) when compared to patients undergoing dual therapy, where the rates were 10 and 2%, respectively [350]. Additional studies are needed to determine the effect of triple therapy on previously untreated patients, relapsers, and individuals with cirrhosis.

Protective immunity

There are several lines of evidence that indicate humoral antibodies may be an important component of the immune system that need to be stimulated if a vaccine is to be successful. Conceptually, it is apparent that the HVR1 region of HCV E2 may be important in viral persistence since there are linear epitopes that change over time resulting in variants that are nonreactive with previous immune responses [351]. Antibodies to HVR1 of HCV E2 can prevent infection *in vitro* [352,353]. In addition, antibodies to a second domain on E2 outside of HVR1, are also capable of preventing binding of HCV E2 to cells [353]. Data indicate that there are at least two or three cross-genotype conserved epitopes on E2, which are targets of antibody-mediated prevention of E2 binding to the CD81 [101], the putative cellular receptor for HCV. Rabbit hyperimmune serum containing antibodies to HCV E2 was able to limit, and in some cases prevent, HCV infection in chimpanzees [354]. However, viral escape mutants (HCV variants differing significantly from the predominant viral strain utilized in the immunization) were present in the viral challenge dose, resulting in the establishment of a chronic HCV infection in one of the two treated chimpanzees. In another study, five of seven chimpanzees vaccinated with expressed recombinant envelope glycoproteins were protected from a low dose challenge with a homologous strain of HCV [355]. Subsequent studies indicate that the serum from protected chimpanzees were positive in an *in vitro* assay measuring the neutralization of virus binding to host cells [353].

Since the initial study indicating that protein expression readily occurs following intramuscular injection of naked DNA [356], it became apparent that this methodology may be quite useful for vaccine development [357]. For HCV, this strategy may be needed, since viral clearance has been linked to strong CD8 + T-cell response. A major advantage of DNA-based vaccines over protein-based vaccines is that nucleic acid vaccines will present HCV antigens produced intracellularly that are required for strong MHC class I restricted CD8 + T-cell response.

In a recent study, chimpanzees were treated with plasmid DNA containing HCV E2 protein and challenged with the homologous, monoclonal strain of HCV [358]. Both vaccinated chimpanzees were productively infected, and developed hepatitis shortly after viral challenge. However, both chimpanzees cleared the virus by about 10 weeks post-infection, and viral RNA was no longer detected in serum for the next several months. Both antibody responses and cytotoxic T-cell responses were noted in the challenged animals. These data support the concept that a vaccine utilizing E2 can prevent chronic infection with HCV, even though it is unclear as to whether it is the antibody response or the cytotoxic T-cell response that played the more important role in viral clearance. One clear observation was that antibodies to HCV E2 were detected prior to challenge, and even though the challenge dose was relatively small (100 chimpanzee infectious doses), HCV replicated in hepatocytes and established an acute phase infection and disease in the vaccinated animals. Thus, the generation of antibodies to HCV E2 appears not to be sufficient to provide protection against experimental infection with HCV, even when the challenge inoculum is a low-titered, homologous, monoclonal strain of HCV. In another recent study, mice immunized with a recombinant plasmid encoding HCV core protein was able to induce both humoral and cellular (CTL) responses to HCV [359,360]. While these results are interesting, the biological significance of a DNA vaccine encoding HCV core, as a stand-alone vaccine or in conjunction with other constructs, has not been evaluated. Precautions must be observed in implementing DNA vaccines since under experimental conditions both HCV core and NS3 play a role in transforming cells *in vitro* [361]. In summary, there have been many advances in the last few years in determining how to induce humoral and cellular immune responses in experimentally vaccinated hosts. As our understanding of the life cycle of HCV and the hosts immune response are enhanced, there are likely to be associated improvements in vaccine design.

Conclusion

Since the initial report in 1989 on the isolation of the HCV genome, our understanding of the virus genetic organization, the proteolytic processing of the precursor polyprotein and the function of the individual gene products have been resolved in the absence of having isolated the virus in mammalian cell culture. Additionally, the introduction of effective blood screening assays has resulted in a remarkable decrease in the incidence of post-transfusion HCV infection in developed countries. With the availability of enhanced tests for detecting HCV nucleic acids or proteins directly, there is greater accuracy both in preventing transmission of HCV through the use of blood products and in determining the activity of HCV in untreated patients and in patients being treated with antiviral therapy. More recent studies on genotype analysis and drug therapies have been instrumental in understanding disease progression following initial infection and will continue to aid in further treatment regimes. The recent advances in antiviral therapy suggest very clearly

158

that HCV is a curable disease, and that as new therapies and combinations of therapies are utilized, there may be further improvements in the rate of viral clearance. Finally, future studies concerning the immunology of the disease may provide a basis on which to immunize populations, thus preventing initial infection much in the case of hepatitis B virus.

References

1. Choo Q-L, Kuo G, Weiner AJ, Overby LR, Bradley DW, Houghton M. Science 1989; 244: 359–362.
2. Major ME, Feinstone SM. Hepatology 1997; 25: 1527–1538.
3. Birkenmeyer LG, Desai SM, Muerhoff AS, Leary TP, Simons JN, Montes CC, Mushahwar IK. J Med Virol 1998; 56: 44–51.
4. Leary TP, Muerhoff AS, Simons JN, Pilot-Matias TJ, Erker JC, Chalmers MC, Schlauder GG, Dawson GJ, Desai SM, Mushahwar IK. J Med Virol 1996; 48: 60–67.
5. Muerhoff AS, Leary TP, Simons JN, Pilot-Matias TJ, Dawson GJ, Erker JC, Chalmers ML, Schlauder GG, Desai SM, Mushahwar IK. J Virol 1995; 69: 5621–5630.
6. Murphy F, Fauquet C, Bishop D, Ghabrial S, Jarvis A, Martelli G, Mayo M, Summers M (editors). Virus Taxonomy. Sixth Report of the International Committee on Taxonomy of Viruses Vienna: Springer-Verlag; 1995.
7. Choo Q-L, Richman KH, Han JH, Berger K, Lee C, Dong C, Gallegos C, Coit D, Medina-Selby A, Barr PJ, Weiner AJ, Bradley DW, Houghton M. Proc Natl Acad Sci USA 1991; 88: 2451–2455.
8. Grakoui A, Wychowski C, Lin C, Feinstone SM, Rice CM. J Virol 1993; 67: 1385–1395.
9. Bartenschlager R. J Viral Hepat 1999; 6: 165–181.
10. Reed KE, Rice CM. Curr Stud Hematol Blood Transfus 1998: 1–37.
11. Han JH, Shyamala V, Richman KH, Brauer MJ, Irvine B, Urdea MS, Tekamp-Olson P, Kuo G, Choo Q-L, Houghton M. Proc Natl Acad Sci USA 1991; 88: 1711–1715.
12. Trowbridge R, Gowans EJ. J Viral Hepat 1998; 5: 95–98.
13. Tsukiyama-Kohara K, Iizuka N, Kohara M, Nomoto A. J Virol 1992; 66: 1476–1483.
14. Wang C, Sarnow P, Siddiqui A. J Virol 1993; 67: 3338–3344.
15. Hagedorn C, Rice C (editors). The Hepatitis C Viruses Berlin: Springer-Verlag; 2000.
16. Kolykhalov AA, Feinstone SM, Rice CM. J Virol 1996; 70: 3363–3371.
17. Tanaka T, Kato N, Cho MJ, Shimotohno K. Biochem Biophys Res Commun 1995; 215: 744–749.
18. Tanaka T, Kato N, Cho MJ, Sugiyama K, Shimotohno K. J Virol 1996; 70: 3307–3312.
19. Yamada N, Tanihara K, Takada A, Yorihuzi T, Tsutsumi M, Shimomura H, Tsuji T, Date T. Virology 1996; 223: 255–261.
20. Kolykhalov AA, Mihalik K, Feinstone SM, Rice CM. J Virol 2000; 74: 2046–2051.
21. Dash S, Halim AB, Tsuji H, Hiramatsu N, Gerber MA. Am J Pathol 1997; 151: 363–373.
22. Yoo BJ, Selby MJ, Choe J, Suh BS, Choi SH, Joh JS, Nuovo GJ, Lee HS, Houghton M, Han JH. J Virol 1995; 69: 32–38.
23. Hong Z, Beaudet-Miller M, Lanford RE, Guerra B, Wright-Minogue J, Skelton A, Baroudy BM, Reyes GR, Lau JY. Virology 1999; 256: 36–44.
24. Kolykhalov AA, Agapov EV, Blight KJ, Mihalik K, Feinstone SM, Rice CM. Science 1997; 277: 570–574.

25. Yanagi M, Purcell RH, Emerson SU, Bukh J. Proc Natl Acad Sci USA 1997; 94: 8738–8743.
26. Yanagi M, St Claire M, Emerson SU, Purcell RH, Bukh J. Proc Natl Acad Sci USA 1999; 96: 2291–2295.
27. Lohmann V, Korner F, Koch J, Herian U, Theilmann L, Bartenschlager R. Science 1999; 285: 110–113.
28. Blight KJ, Kolykhalov AA, Rice CM. Science 2000; 290: 1972–1974.
29. Bartenschlager R, Lohmann V. J Gen Virol 2000; 81: 1631–1648.
30. Grakoui A, Hanson HL, Rice CM. Hepatology 2001; 33: 489–495.
31. Goeser T, Muller HM, Ye J, Pfaff E, Theilmann L. Virology 1994; 205: 462–469.
32. Jolivet-Reynaud C, Dalbon P, Viola F, Yvon S, Paranhos-Baccala G, Piga N, Bridon L, Trabaud MA, Battail N, Sibai G, Jolivet M. J Med Virol 1998; 56: 300–309.
33. Sallberg M, Pumpen P, Zhang ZX, Lundholm P, Gusars I, Ruden U, Wahren B, Magnius LO. J Med Virol 1994; 43: 62–68.
34. Sallberg M, Ruden U, Wahren B, Magnius LO. J Clin Microbiol 1992; 30: 1989–1994.
35. Yasui K, Wakita T, Tsukiyama-Kohara K, Funahashi SI, Ichikawa M, Kajita T, Moradpour D, Wands JR, Kohara M. J Virol 1998; 72: 6048–6055.
36. Santolini E, Migliaccio G, La Monica N. J Virol 1994; 68: 3631–3641.
37. Lo SY, Masiarz F, Hwang SB, Lai MM, Ou JH. Virology 1995; 213: 455–461.
38. Suzuki R, Matsuura Y, Suzuki T, Ando A, Chiba J, Harada S, Saito I, Miyamura T. J Gen Virol 1995; 76: 53–61.
39. Liu Q, Tackney C, Bhat RA, Prince AM, Zhang P. J Virol 1997; 71: 657–662.
40. von Heijne G. Nucleic Acids Res 1986; 14: 4683–4690.
41. Nielsen H, Engelbrecht J, Brunak S, Heijne GV. Protein Engng 1997; 10: 1–6.
42. Hijikata M, Kato N, Ootsuyama Y, Nakagawa M, Shimotohno K. Proc Natl Acad Sci USA 1991; 88: 5547–5551.
43. Selby MJ, Choo QL, Berger K, Kuo G, Glazer E, Eckart M, Lee C, Chien D, Kuo C, Houghton M. J Gen Virol 1993; 74: 1103–1113.
44. Hussy P, Langen H, Mous J, Jacobsen H. Virology 1996; 224: 93–104.
45. Shimizu YK, Feinstone SM, Kohara M, Purcell RH, Yoshikura H. Hepatology 1996; 23: 205–209.
46. Baumert TF, Ito S, Wong DT, Liang TJ. J Virol 1998; 72: 3827–3836.
47. Baumert TF, Wellnitz S, Aono S, Satoi J, Herion D, Tilman Gerlach J, Pape GR, Lau JY, Hoofnagle JH, Blum HE, Jake Liang T. Hepatology 2000; 32: 610–617.
48. Harada S, Watanabe Y, Takeuchi K, Suzuki T, Katayama T, Takebe Y, Saito I, Miyamura T. J Virol 1991; 65: 3015–3021.
49. Moradpour D, Englert C, Wakita T, Wands JR. Virology 1996; 222: 51–63.
50. Ray RB, Lagging LM, Meyer K, Ray R. J Virol 1996; 70: 4438–4443.
51. Lo SY, Selby M, Tong M, Ou JH. Virology 1994; 199: 124–131.
52. Chang SC, Yen JH, Kang HY, Jang MH, Chang MF. Biochem Biophys Res Commun 1994; 205: 1284–1290.
53. Marusawa H, Hijikata M, Chiba T, Shimotohno K. J Virol 1999; 73: 4713–4720.
54. Sabile A, Perlemuter G, Bono F, Kohara K, Demaugre F, Kohara M, Matsuura Y, Miyamura T, Brechot C, Barba G. Hepatology 1999; 30: 1064–1076.
55. Shih CM, Chen CM, Chen SY, Lee YH. J Virol 1995; 69: 1160–1171.
56. Barba G, Harper F, Harada T, Kohara M, Goulinet S, Matsuura Y, Eder G, Schaff Z, Chapman MJ, Miyamura T, Brechot C. Proc Natl Acad Sci USA 1997; 94: 1200–1205.
57. Buratti E, Baralle FE, Tisminetzky SG. Cell Mol Biol (Noisy-le-grand) 1998; 44: 505–512.

58. Moradpour D, Wakita T, Wands JR, Blum HE. Biochem Biophys Res Commun 1998; 246: 920–924.

59. McLauchlan J. J Viral Hepat 2000; 7: 2–14.

60. Naganuma A, Nozaki A, Tanaka T, Sugiyama K, Takagi H, Mori M, Shimotohno K, Kato N. J Virol 2000; 74: 8744–8750.

61. Hope RG, McLauchlan J. J Gen Virol 2000; 81: 1913–1925.

62. Moriya K, Fujie H, Yotsuyanagi H, Shintani Y, Tsutsumi T, Matsuura Y, Miyamura T, Kimura S, Koike K. Jpn J Med Sci Biol 1997; 50: 169–177.

63. Moriya K, Yotsuyanagi H, Shintani Y, Fujie H, Ishibashi K, Matsuura Y, Miyamura T, Koike K. J Gen Virol 1997; 78: 1527–1531.

64. Baiocchi L, Tisone G, Palmieri G, Rapicetta M, Pisani F, Orlando G, Casciani CU, Angelico M. Liver Transpl Surg 1998; 4: 441–447.

65. Czaja AJ, Carpenter HA, Santrach PJ, Moore SB. J Hepatol 1998; 29: 198–206.

66. Hourigan LF, Macdonald GA, Purdie D, Whitehall VH, Shorthouse C, Clouston A, Powell EE. Hepatology 1999; 29: 1215–1219.

67. Fong DG, Nehra V, Lindor KD, Buchman AL. Hepatology 2000; 32: 3–10.

68. Matsuura Y, Harada S, Suzuki R, Watanabe Y, Inoue Y, Saito I, Miyamura T. J Virol 1992; 66: 1425–1431.

69. Lanford RE, Notvall L, Chavez D, White R, Frenzel G, Simonsen C, Kim J. Virology 1993; 197: 225–235.

70. Miller RH, Purcell RH. Proc Natl Acad Sci USA 1990; 87: 2057–2061.

71. Reed KE, Rice CM. Curr Top Microbiol Immunol 2000; 242: 55–84.

72. Lin C, Lindenbach BD, Pragai BM, McCourt DW, Rice CM. J Virol 1994; 68: 5063–5073.

73. Mizushima H, Hijikata M, Asabe S, Hirota M, Kimura K, Shimotohno K. J Virol 1994; 68: 6215–6222.

74. Matsuura Y, Suzuki T, Suzuki R, Sato M, Aizaki H, Saito I, Miyamura T. Virology 1994; 205: 141–150.

75. Dubuisson J, Rice CM. J Virol 1996; 70: 778–786.

76. Deleersnyder V, Pillez A, Wychowski C, Blight K, Xu J, Hahn YS, Rice CM, Dubuisson J. J Virol 1997; 71: 697–704.

77. Dubuisson J, Hsu HH, Cheung RC, Greenberg HB, Russell DG, Rice CM. J Virol 1994; 68: 6147–6160.

78. Lo SY, Selby MJ, Ou JH. J Virol 1996; 70: 5177–5182.

79. Leary K, Blair CD. J Ultrastruct Res 1980; 72: 123–129.

80. Ishak R, Tovey DG, Howard CR. J Gen Virol 1988; 69: 325–335.

81. Baumert TF, Vergalla J, Satoi J, Thomson M, Lechmann M, Herion D, Greenberg HB, Ito S, Liang TJ. Gastroenterology 1999; 117: 1397–1407.

82. Ogata N, Alter HJ, Miller RH, Purcell RH. Proc Natl Acad Sci USA 1991; 88: 3392–3396.

83. Kurosaki M, Enomoto N, Marumo F, Sato C. Hepatology 1993; 18: 1293–1299.

84. Weiner AJ, Geysen HM, Christopherson C, Hall JE, Mason TJ, Saracco G, Bonino F, Crawford K, Marion CD, Crawford KA, et al. Proc Natl Acad Sci USA 1992; 89: 3468–3472.

85. Farci P, Alter HJ, Wong DC, Miller RH, Govindarajan S, Engle R, Shapiro M, Purcell RH. Proc Natl Acad Sci USA 1994; 91: 7792–7796.

86. Shimizu YK, Igarashi H, Kiyohara T, Cabezon T, Farci P, Purcell RH, Yoshikura H. Virology 1996; 223: 409–412.

87. Kumar U, Monjardino J, Thomas HC. Gastroenterology 1994; 106: 1072–1075.

88. van Doorn LJ, Capriles I, Maertens G, DeLeys R, Murray K, Kos T, Schellekens H, Quint W. J Virol 1995; 69: 773–778.

89. Hijikata M, Kato N, Ootsuyama Y, Nakagawa M, Ohkoshi S, Shimotohno K. Biochem Biophys Res Commun 1991; 175: 220–228.

90. Kato N, Ootsuyama Y, Ohkoshi S, Nakazawa T, Sekiya H, Hijikata M, Shimotohno K. Biochem Biophys Res Commun 1992; 189: 119–127.

91. Farci P, Alter HJ, Govindarajan S, Wong DC, Engle R, Lesniewski RR, Mushahwar IK, Desai SM, Miller RH, Ogata N, et al. Science 1992; 258: 135–140.

92. Scarselli E, Cerino A, Esposito G, Silini E, Mondelli MU, Traboni C. J Virol 1995; 69: 4407–4412.

93. Zibert A, Kraas W, Meisel H, Jung G, Roggendorf M. J Virol 1997; 71: 4123–4127.

94. Lesniewski R, Okasinski G, Carrick R, Van Sant C, Desai S, Johnson R, Scheffel J, Moore B, Mushahwar I. J Med Virol 1995; 45: 415–422.

95. Muerhoff AS, Smith DB, Leary TP, Erker JC, Desai SM, Mushahwar IK. J Virol 1997; 71: 6501–6508.

96. Simons JN, Desai SM, Mushahwar IK. In: The Hepatitis C Viruses. (Hagedorn CH, Rice CM, editors), vol. 242, Berlin: Springer-Verlag; 1999; pp. 341–375.

97. Pilot-Matias TJ, Carrick RJ, Coleman PF, Leary TP, Surowy TK, Simons JN, Muerhoff AS, Buijk SL, Chalmers ML, Dawson GJ, Desai SM, Mushahwar IK. Virology 1996; 225: 282–292.

98. Dille BJ, Surowy TK, Gutierrez RA, Coleman PF, Knigge MF, Carrick RJ, Aach RD, Hollinger FB, Stevens CE, Barbosa LH, Nemo GJ, Mosley JW, Dawson GJ, Mushahwar IK. J Infect Dis 1997; 175: 458–461.

99. Gutierrez RA, Dawson GJ, Knigge MF, Melvin SL, Heynen CA, Kyrk CR, Young CE, Carrick RJ, Schlauder GG, Surowy TK, Dille BJ, Coleman PF, Thiele DL, Lentino JR, Pachucki C, Mushahwar IK. J Med Virol 1997; 53: 167–173.

100. Pileri P, Uematsu Y, Campagnoli S, Galli G, Falugi F, Petracca R, Weiner AJ, Houghton M, Rosa D, Grandi G, Abrignani S. Science 1998; 282: 938–941.

101. Allander T, Drakenberg K, Beyene A, Rosa D, Abrignani S, Houghton M, Widell A, Grillner L, Persson MA. J Gen Virol 2000; 81: 2451–2459.

102. Petracca R, Falugi F, Galli G, Norais N, Rosa D, Campagnoli S, Burgio V, Di Stasio E, Giardina B, Houghton M, Abrignani S, Grandi G. J Virol 2000; 74: 4824–4830.

103. Levy S, Todd SC, Maecker HT. Annu Rev Immunol 1998; 16: 89–109.

104. Agnello V, Abel G, Elfahal M, Knight GB, Zhang QX. Proc Natl Acad Sci USA 1999; 96: 12766–12771.

105. Monazahian M, Bohme I, Bonk S, Koch A, Scholz C, Grethe S, Thomssen R. J Med Virol 1999; 57: 223–229.

106. Wunschmann S, Medh JD, Klinzmann D, Schmidt WN, Stapleton JT. J Virol 2000; 74: 10055–10062.

107. Grakoui A, McCourt DW, Wychowski C, Feinstone SM, Rice CM. Proc Natl Acad Sci USA 1993; 90: 10583–10587.

108. Hirowatari Y, Hijikata M, Tanji Y, Nyunoya H, Mizushima H, Kimura K, Tanaka T, Kato N, Shimotohno K. Arch Virol 1993; 133: 349–356.

109. Reed KE, Grakoui A, Rice CM. J Virol 1995; 69: 4127–4136.

110. Hijikata M, Mizushima H, Akagi T, Mori S, Kakiuchi N, Kato N, Tanaka T, Kimura K, Shimotohno K. J Virol 1993; 67: 4665–4675.

111. De Francesco R, Urbani A, Nardi MC, Tomei L, Steinkuhler C, Tramontano A. Biochemistry 1996; 35: 13282–13287.

112. Kim JL, Morgenstern KA, Lin C, Fox T, Dwyer MD, Landro JA, Chambers SP, Markland W, Lepre CA, O'Malley ET, Harbeson SL, Rice CM, Murcko MA, Caron PR, Thomson JA. Cell 1996; 87: 343–355.

113. Love RA, Parge HE, Wickersham JA, Hostomsky Z, Habuka N, Moomaw EW, Adachi T, Hostomska Z. Cell 1996; 87: 331–342.

114. Stempniak M, Hostomska Z, Nodes BR, Hostomsky Z. J Virol 1997; 71: 2881–2886.

115. Yan Y, Li Y, Munshi S, Sardana V, Cole JL, Sardana M, Steinkuehler C, Tomei L, De Francesco R, Kuo LC, Chen Z. Protein Sci 1998; 7: 837–847.

116. Eckart MR, Selby M, Masiarz F, Lee C, Berger K, Crawford K, Kuo C, Kuo G, Houghton M, Choo QL. Biochem Biophys Res Commun 1993; 192: 399–406.

117. Grakoui A, McCourt DW, Wychowski C, Feinstone SM, Rice CM. J Virol 1993; 67: 2832–2843.

118. Tomei L, Failla C, Santolini E, De Francesco R, La Monica N. J Virol 1993; 67: 4017–4026.

119. Failla C, Tomei L, De Francesco R. J Virol 1994; 68: 3753–3760.

120. Kolykhalov AA, Agapov EV, Rice CM. J Virol 1994; 68: 7525–7533.

121. Bartenschlager R, Ahlborn-Laake L, Yasargil K, Mous J, Jacobsen H. J Virol 1995; 69: 198–205.

122. Pizzi E, Tramontano A, Tomei L, La Monica N, Failla C, Sardana M, Wood T, De Francesco R. Proc Natl Acad Sci USA 1994; 91: 888–892.

123. Bartenschlager R, Lohmann V, Wilkinson T, Koch JO. J Virol 1995; 69: 7519–7528.

124. Failla C, Tomei L, De Francesco R. J Virol 1995; 69: 1769–1777.

125. Lin C, Thomson JA, Rice CM. J Virol 1995; 69: 4373–4380.

126. Satoh S, Tanji Y, Hijikata M, Kimura K, Shimotohno K. J Virol 1995; 69: 4255–4260.

127. Steinkuhler C, Urbani A, Tomei L, Biasiol G, Sardana M, Bianchi E, Pessi A, De Francesco R. J Virol 1996; 70: 6694–6700.

128. Steinkuhler C, Biasiol G, Brunetti M, Urbani A, Koch U, Cortese R, Pessi A, De Francesco R. Biochemistry 1998; 37: 8899–8905.

129. Ingallinella P, Altamura S, Bianchi E, Taliani M, Ingenito R, Cortese R, De Francesco R, Steinkuhler C, Pessi A. Biochemistry 1998; 37: 8906–8914.

130. Fukuda K, Vishnuvardhan D, Sekiya S, Hwang J, Kakiuchi N, Taira K, Shimotohno K, Kumar PK, Nishikawa S. Eur J Biochem 2000; 267: 3685–3694.

131. Fattori D, Urbani A, Brunetti M, Ingenito R, Pessi A, Prendergast K, Narjes F, Matassa VG, De Francesco R, Steinkuhler C. J Biol Chem 2000; 275: 15106–15113.

132. Narjes F, Brunetti M, Colarusso S, Gerlach B, Koch U, Biasiol G, Fattori D, De Francesco R, Matassa VG, Steinkuhler C. Biochemistry 2000; 39: 1849–1861.

133. Di Marco S, Rizzi M, Volpari C, Walsh MA, Narjes F, Colarusso S, De Francesco R, Matassa VG, Sollazzo M. J Biol Chem 2000; 275: 7152–7157.

134. Lai VC, Zhong W, Skelton A, Ingravallo P, Vassilev V, Donis RO, Hong Z, Lau JY. J Virol 2000; 74: 6339–6347.

135. Kato N, Hijikata M, Nakagawa M, Ootsuyama Y, Muraiso K, Ohkoshi S, Shimotohno K. FEBS Lett 1991; 280: 325–328.

136. Suzich JA, Tamura JK, Palmer-Hill F, Warrener P, Grakoui A, Rice CM, Feinstone SM, Collett MS. J Virol 1993; 67: 6152–6158.

137. Jin L, Peterson DL. Arch Biochem Biophys 1995; 323: 47–53.

138. Kim DW, Gwack Y, Han JH, Choe J. Biochem Biophys Res Commun 1995; 215: 160–166.

139. Preugschat F, Averett DR, Clarke BE, Porter DJT. J Biol Chem 1996; 271: 24449–24457.

140. Tai CL, Chi WK, Chen DS, Hwang LH. J Virol 1996; 70: 8477–8484.

141. Heilek GM, Peterson MG. J Virol 1997; 71: 6264–6266.

142. Gwack Y, Kim DW, Han JH, Choe J. Biochem Biophys Res Commun 1996; 225: 654–659.

143. Hong Z, Ferrari E, Wright-Minogue J, Chase R, Risano C, Seelig G, Lee CG, Kwong AD. J Virol 1996; 70: 4261–4268.

144. Morgenstern KA, Landro JA, Hsiao K, Lin C, Gu Y, Su MS, Thomson JA. J Virol 1997; 71: 3767–3775.

145. Howe AY, Chase R, Taremi SS, Risano C, Beyer B, Malcolm B, Lau JY. Protein Sci 1999; 8: 1332–1341.

146. Gallinari P, Brennan D, Nardi C, Brunetti M, Tomei L, Steinkuhler C, De Francesco R. J Virol 1998; 72: 6758–6769.

147. Cho HS, Ha NC, Kang LW, Chung KM, Back SH, Jang SK, Oh BH. J Biol Chem 1998; 273: 15045–15052.

148. Kim JL, Morgenstern KA, Griffith JP, Dwyer MD, Thomson JA, Murcko MA, Lin C, Caron PR. Structure 1998; 6: 89–100.

149. Yao N, Reichert P, Taremi SS, Prosise WW, Weber PC. Struct Fold Des 1999; 7: 1353–1363.

150. Borowski P, Mueller O, Niebuhr A, Kalitzky M, Hwang LH, Schmitz H, Siwecka MA, Kulikowsk T. Acta Biochim Pol 2000; 47: 173–180.

151. Hsu CC, Hwang LH, Huang YW, Chi WK, Chu YD, Chen DS. Biochem Biophys Res Commun 1998; 253: 594–599.

152. Kyono K, Miyashiro M, Taguchi I. Anal Biochem 1998; 257: 120–126.

153. Yao N, Weber PC. Antiviral Ther 1998; 3: 93–97.

154. Hijikata M, Mizushima H, Tanji Y, Komoda Y, Hirowatari Y, Akagi T, Kato N, Kimura K, Shimotohno K. Proc Natl Acad Sci USA 1993; 90: 10773–10777.

155. Chien DY, Choo QL, Tabrizi A, Kuo C, McFarland J, Berger K, Lee C, Shuster JR, Nguyen T, Moyer DL. Proc Natl Acad Sci USA 1992; 89: 10011–10015.

156. Mori S, Ohkoshi S, Hijikata M, Kato N, Shimotohno K. Jpn J Cancer Res 1992; 83: 264–268.

157. Van der Poel CL, Cuypers HT, Reesink HW, Weiner AJ, Quan S, Di Nello R, Van Boven JJ, Winkel I, Mulder-Folkerts D, Exel-Oehlers PJ, et al. Lancet 1991; 337: 317–319.

158. McOmish F, Chan SW, Dow BC, Gillon J, Frame WD, Crawford RJ, Yap PL, Follett EA, Simmonds P. Transfusion 1993; 33: 7–13.

159. Nagayama R, Tsuda F, Okamoto H, Wang Y, Mitsui T, Tanaka T, Miyakawa Y, Mayumi M. J Clin Investig 1993; 92: 1529–1533.

160. Alonso C, Qu D, Lamelin JP, de Sanjose S, Vitvitski L, Li J, Berby F, Lambert V, Cortey ML, Trepo C. J Clin Microbiol 1994; 32: 211–212.

161. McOmish F, Yap PL, Dow BC, Follett EA, Seed C, Keller AJ, Cobain TJ, Krusius T, Kolho E, Naukkarinen R, et al. J Clin Microbiol 1994; 32: 884–892.

162. Zein NN, Rakela J, Persing DH. Mayo Clin Proc 1995; 70: 449–452.

163. Dhaliwal SK, Prescott LE, Dow BC, Davidson F, Brown H, Yap PL, Follett EA, Simmonds P. J Med Virol 1996; 48: 184–190.

164. Zein NN, Germer JJ, Wendt NK, Schimek CM, Thorvilson JN, Mitchell PS, Persing DH. J Clin Microbiol 1997; 35: 311–312.

165. Simmonds P, Rose KA, Graham S, Chan SW, McOmish F, Dow BC, Follett EA, Yap PL, Marsden H. J Clin Microbiol 1993; 31: 1493–1503.

166. Chang JC, Seidel C, Ofenloch B, Jue DL, Fields HA, Khudyakov YE. Virology 1999; 257: 177–190.

167. Kobayashi M, Chayama K, Arase Y, Tsubota A, Saitoh S, Suzuki Y, Ikeda K, Matsuda M, Koike H, Hashimoto M, Kumada H. J Gastroenterol 1999; 34: 505–509.

168. Schroter M, Feucht HH, Schafer P, Zollner B, Laufs R. J Clin Microbiol 1999; 37: 2576–2580.
169. Fabrizi F, Martin P, Quan S, Dixit V, Brezina M, Conrad A, Polito A, Gitnick G. Am J Kidney Dis 2000; 35: 832–838.
170. Koch JO, Bartenschlager R. J Virol 1999; 73: 7138–7146.
171. Neddermann P, Clementi A, De Francesco R. J Virol 1999; 73: 9984–9991.
172. Kaneko T, Tanji Y, Satoh S, Hijikata M, Asabe S, Kimura K, Shimotohno K. Biochem Biophys Res Commun 1994; 205: 320–326.
173. Tanji Y, Kaneko T, Satoh S, Shimotohno K. J Virol 1995; 69: 3980–3986.
174. Lin C, Wu JW, Hsiao K, Su MS. J Virol 1997; 71: 6465–6471.
175. Ishido S, Fujita T, Hotta H. Biochem Biophys Res Commun 1998; 244: 35–40.
176. Gale MJ Jr., Korth MJ, Tang NM, Tan SL, Hopkins DA, Dever TE, Polyak SJ, Gretch DR, Katze MG. Virology 1997; 230: 217–227.
177. Gale M Jr., Blakely CM, Kwieciszewski B, Tan SL, Dossett M, Tang NM, Korth MJ, Polyak SJ, Gretch DR, Katze MG. Mol Cell Biol 1998; 18: 5208–5218.
178. Gale M Jr., Katze MG. Pharmacol Ther 1998; 78: 29–46.
179. Gale MJ Jr., Korth MJ, Katze MG. Clin Diagn Virol 1998; 10: 157–162.
180. Enomoto N, Sakuma I, Asahina Y, Kurosaki M, Murakami T, Yamamoto C, Izumi N, Marumo F, Sato C. J Clin Investig 1995; 96: 224–230.
181. Enomoto N, Sakuma I, Asahina Y, Kurosaki M, Murakami T, Yamamoto C, Ogura Y, Izumi N, Marumo F, Sato C. N Engl J Med 1996; 334: 77–81.
182. Maekawa S, Enomoto N, Kurosaki M, Nagayama K, Marumo F, Sato C. J Med Virol 2000; 61: 303–310.
183. Chayama K, Tsubota A, Kobayashi M, Okamoto K, Hashimoto M, Miyano Y, Koike H, Koida I, Arase Y, Saitoh S, Suzuki Y, Murashima N, Ikeda K, Kumada H. Hepatology 1997; 25: 745–749.
184. Kurosaki M, Enomoto N, Murakami T, Sakuma I, Asahina Y, Yamamoto C, Ikeda T, Tozuka S, Izumi N, Marumo F, Sato C. Hepatology 1997; 25: 750–753.
185. Casato M, Agnello V, Pucillo LP, Knight GB, Leoni M, Del Vecchio S, Mazzilli C, Antonelli G, Bonomo L. Blood 1997; 90: 3865–3873.
186. Hofgartner WT, Polyak SJ, Sullivan DG, Carithers RL Jr., Gretch DR. J Med Virol 1997; 53: 118–126.
187. Pawlotsky JM, Germanidis G, Neumann AU, Pellerin M, Frainais PO, Dhumeaux D. J Virol 1998; 72: 2795–2805.
188. Saiz JC, Lopez-Labrador FX, Ampurdanes S, Dopazo J, Forns X, Sanchez-Tapias JM, Rodes J. J Infect Dis 1998; 177: 839–847.
189. Frangeul L, Cresta P, Perrin M, Lunel F, Opolon P, Agut H, Huraux JM. Hepatology 1998; 28: 1674–1679.
190. Sarrazin C, Kornetzky I, Ruster B, Lee JH, Kronenberger B, Bruch K, Roth WK, Zeuzem S. Hepatology 2000; 31: 1360–1370.
191. Polyak SJ, McArdle S, Liu SL, Sullivan DG, Chung M, Hofgartner WT, Carithers RL Jr., McMahon BJ, Mullins JI, Corey L, Gretch DR. J Virol 1998; 72: 4288–4296.
192. Taylor DR, Shi ST, Romano PR, Barber GN, Lai MM. Science 1999; 285: 107–110.
193. Koonin EV, Dolja VV. Crit Rev Biochem Mol Biol 1993; 28: 375–430.
194. Behrens SE, Tomei L, De Francesco R. EMBO J 1996; 15: 12–22.
195. Lohmann V, Korner F, Herian U, Bartenschlager R. J Virol 1997; 71: 8416–8428.
196. Yuan ZH, Kumar U, Thomas HC, Wen YM, Monjardino J. Biochem Biophys Res Commun 1997; 232: 231–235.

197. Al RH, Xie Y, Wang Y, Hagedorn CH. Virus Res 1998; 53: 141–149.

198. Yamashita T, Kaneko S, Shirota Y, Qin W, Nomura T, Kobayashi K, Murakami S. J Biol Chem 1998; 273: 15479–15486.

199. Blight KJ, Rice CM. J Virol 1997; 71: 7345–7352.

200. Lohmann V, Roos A, Korner F, Koch JO, Bartenschlager R. Virology 1998; 249: 108–118.

201. Carroll SS, Sardana V, Yang Z, Jacobs AR, Mizenko C, Hall D, Hill L, Zugay-Murphy J, Kuo LC. Biochemistry 2000; 39: 8243–8249.

202. Ago H, Adachi T, Yoshida A, Yamamoto M, Habuka N, Yatsunami K, Miyano M. Struct Fold Des 1999; 7: 1417–1426.

203. Bressanelli S, Tomei L, Roussel A, Incitti I, Vitale RL, Mathieu M, De Francesco R, Rey FA. Proc Natl Acad Sci USA 1999; 96: 13034–13039.

204. Lesburg CA, Cable MB, Ferrari E, Hong Z, Mannarino AF, Weber PC. Nat Struct Biol 1999; 6: 937–943.

205. Jager J, Pata JD. Curr Opin Struct Biol 1999; 9: 21–28.

206. Georgiadis MM, Jessen SM, Ogata CM, Telesnitsky A, Goff SP, Hendrickson WA. Structure 1995; 3: 879–892.

207. Hansen JL, Long AM, Schultz SC. Structure 1997; 5: 1109–1122.

208. Patterson JL, Fernandez-Larsson R. Rev Infect Dis 1990; 12: 1139–1146.

209. Lohmann V, Roos A, Korner F, Koch JO, Bartenschlager R. J Viral Hepat 2000; 7: 167–174.

210. Simmonds P, Alberti A, Alter HJ, Bonino F, Bradley DW, Brechot C, Brouwer JT, Chan SW, Chayama K, Chen DS, et al. Hepatology 1994; 19: 1321–1324.

211. Simmonds P. Hepatology 1995; 21: 570–583.

212. Simmonds P, Smith DB, McOmish F, Yap PL, Kolberg J, Urdea MS, Holmes EC. J Gen Virol 1994; 75: 1053–1061.

213. Stuyver L, van Arnhem W, Wyseur A, Hernandez F, Delaporte E, Maertens G. Proc Natl Acad Sci USA 1994; 91: 10134–10138.

214. Bukh J, Miller RH, Purcell RH. Semin Liver Dis 1995; 15: 41–63.

215. Arens M. J Clin Virol 2001; 22: 11–29.

216. Nolte FS. Mol Diagn 2001; 6: 265–277.

217. Stuyver L, Wyseur A, van Arnhem W, Hernandez F, Maertens G. J Clin Microbiol 1996; 34: 2259–2266.

218. Manns MP, McHutchison JG, Gordon SC, Rustgi VK, Shiffman M, Reindollar R, Goodman ZD, Koury K, Ling M, Albrecht JK. Lancet 2001; 358: 958–965.

219. Di Bisceglie AM, Hoofnagle JH. Hepatology 2002; 36: S121–S127.

220. Fried MW, Shiffman ML, Reddy KR, Smith C, Marinos G, Goncales FL Jr., Haussinger D, Diago M, Carosi G, Dhumeaux D, Craxi A, Lin A, Hoffman J, Yu J. N Engl J Med 2002; 347: 975–982.

221. Bukh J, Purcell RH, Miller RH. Proc Natl Acad Sci USA 1993; 90: 8234–8238.

222. Okamoto H, Kojima M, Sakamoto M, Iizuka H, Hadiwandowo S, Suwignyo S, Miyakawa Y, Mayumi M. J Gen Virol 1994; 75: 629–635.

223. Hotta H, Handajani R, Lusida MI, Soemarto W, Doi H, Miyajima H, Homma M. J Clin Microbiol 1994; 32: 3049–3051.

224. Tokita H, Shrestha SM, Okamoto H, Sakamoto M, Horikita M, Iizuka H, Shrestha S, Miyakawa Y, Mayumi M. J Gen Virol 1994; 75: 931–936.

225. Tokita H, Okamoto H, Tsuda F, Song P, Nakata S, Chosa T, Iizuka H, Mishiro S, Miyakawa Y, Mayumi M. Proc Natl Acad Sci USA 1994; 91: 11022–11026.

277. Lok AS, Gunaratnam NT. Hepatology 1997; 26: 48S–56S.
278. Chau KH, Dawson GJ, Mushahwar IK, Gutierrez RA, Johnson RG, Lesniewski RR, Mattsson L, Weiland O. J Virol Methods 1991; 35: 343–352.
279. Kapprell HP, Majnak A, Medgyesi GA, Szabo JE, Michel G, Albrecht M, Heyermann H, Hampl H, Troonen H. Vox Sang. 1995; 68: 57–58.
280. Chen PJ, Wang JT, Hwang LH, Yang YH, Hsieh CL, Kao JH, Sheu JC, Lai MY, Wang TH, Chen DS. Proc Natl Acad Sci USA 1992; 89: 5971–5975.
281. Quiroga JA, Herrero M, Castillo I, Navas S, Pardo M, Carreno V. J Infect Dis 1994; 170: 669–673.
282. Lau JY, Davis GL, Kniffen J, Qian KP, Urdea MS, Chan CS, Mizokami M, Neuwald PD, Wilber JC. Lancet 1993; 341: 1501–1504.
283. Mishiro S, Hoshi Y, Takeda K, Yoshikawa A, Gotanda T, Takahashi K, Akahane Y, Yoshizawa H, Okamoto H, Tsuda F, et al. Lancet 1990; 336: 1400–1403.
284. Michel G, Ritter A, Gerken G, Meyer zum Buschenfelde KH, Decker R, Manns MP. Lancet 1992; 339: 267–269.
285. Hosein B, Fang X, Wang CY. Lancet 1992; 339: 871.
286. Busch MP, Kleinman SH, Jackson B, Stramer SL, Hewlett I, Preston S. Transfusion 2000; 40: 143–159.
287. Urdea MS, Horn T, Fultz TJ, Anderson M, Running JA, Hamren S, Ahle D, Chang CA. Nucleic Acids Symp Ser 1991: 197–200.
288. Weiner AJ, Kuo G, Bradley DW, Bonino F, Saracco G, Lee C, Rosenblatt J, Choo QL, Houghton M. Lancet 1990; 335: 1–3.
289. Linnen JM, Gilker JM, Menez A, Vaughn A, Broulik A, Dockter J, Gillotte-Taylor K, Greenbaum K, Kolk DP, Mimms LT, Giachetti C. J Virol Methods 2002; 102: 139–155.
290. Schiff ER, de Medina M, Kahn RS. Semin Liver Dis 1999; 19(Suppl 1): 3–15.
291. Mimms L, Linnen L, Kolk D, Burrell T, Hotaling S, Klemm M, Nelson P, Greenbaum K, McDonough S, Giachetti C. Vox Sang. 2000; 78(Suppl 1): 0076.
292. Gorrin G, Friesenhahn M, Lin P, Sanders M, Pollner R, Eguchi B, Pham J, Roma G, Spidle J, Nicol S, Wong C, Bhade S, Comanor L. J Clin Microbiol 2003; 41: 310–317.
293. Ross RS, Viazov SO, Hoffmann S, Roggendorf M. J Clin Lab Anal 2001; 15: 308–313.
294. Grant PR, Busch MP. Transfus Med 2002; 12: 229–242.
295. Roth WK, Buhr S, Drosten C, Seifried E. Vox Sang. 2000; 78(Suppl 2): 257–259.
296. Stramer S, Waldman K, Davis J, Brodsky J, Dodd R. Vox Sang. 2000; 78(Suppl 1): 70.
297. Schmidt WN, Wu P, Han JQ, Perino MJ, LaBrecque DR, Stapleton JT. J Infect Dis 1997; 176: 20–26.
298. Schmidt WN, Wu P, Brashear D, Klinzman D, Phillips MJ, LaBrecque DR, Stapleton JT. Hepatology 1998; 28: 1110–1116.
299. Stapleton JT, Klinzman D, Schmidt WN, Pfaller MA, Wu P, LaBrecque DR, Han J, Phillips MJ, Woolson R, Alden B. J Clin Microbiol 1999; 37: 484–489.
300. Schmidt WN, Stapleton JT, LaBrecque DR, Mitros FA, Kirby PA, Phillips MJ, Brashear DL. Hepatology 2000; 31: 737–744.
301. Saldanha J, Lelie N, Heath A. Vox Sang. 1999; 76: 149–158.
302. Pawlotsky JM, Bouvier-Alias M, Hezode C, Darthuy F, Remire J, Dhumeaux D. Hepatology 2000; 32: 654–659.
303. Pawlotsky JM. Hepatology 2002; 36: S65–S73.
304. Lee SC, Antony A, Lee N, Leibow J, Yang JQ, Soviero S, Gutekunst K, Rosenstraus M. J Clin Microbiol 2000; 38: 4171–4179.

305. Konnick EQ, Erali M, Ashwood ER, Hillyard DR. J Clin Microbiol 2002; 40: 768–773.

306. Gretch DR. Hepatology 1997; 26: 43S–47S.

307. Germer JJ, Heimgartner PJ, Ilstrup DM, Harmsen WS, Jenkins GD, Patel R. J Clin Microbiol 2002; 40: 495–500.

308. Trimoulet P, Halfon P, Pohier E, Khiri H, Chene G, Fleury H. J Clin Microbiol 2002; 40: 2031–2036.

309. Benahmou J-P, Rodes J, Alter H, Bismuth H, Desmet V, Guardia J, Heathcoate J, Lok A, Maddrey WC, K. H, M. zb, Pagliaro L, Paumgartner G, Sherlock S. J Hepatol 1999; 30: 959–961.

310. Kobayashi M, Chayama K, Arase Y, Tsubota A, Saitoh S, Suzuki Y, Ikeda K, Matsuda M, Koike H, Hashimoto M, Kumada H. J Gastroenterol 1999; 34: 94–99.

311. Tanaka E, Ohue C, Aoyagi K, Yamaguchi K, Yagi S, Kiyosawa K, Alter HJ. Hepatology 2000; 32: 388–393.

312. Aoyagi K, Ohue C, Iida K, Kimura T, Tanaka E, Kiyosawa K, Yagi S. J Clin Microbiol 1999; 37: 1802–1808.

313. Peterson J, Green G, Iida K, Caldwell B, Kerrison P, Bernich S, Aoyagi K, Lee SR. Vox Sang. 2000; 78: 80–85.

314. Muerhoff AS, Jiang L, Shah DO, Gutierrez RA, Patel J, Garolis C, Kyrk CR, Leckie G, Frank A, Stewart JL, Dawson GJ. Transfusion 2002; 42: 349–356.

315. Grant PR, Sims CM, Tedder RS. Transfusion 2002; 42: 1032–1036.

316. Nubling CM, Unger G, Chudy M, Raia S, Lower J. Transfusion 2002; 42: 1037–1045.

317. Icardi G, Ansaldi F, Bruzzone BM, Durando P, Lee S, de Luigi C, Crovari P. J Clin Microbiol 2001; 39: 3110–3114.

318. Dickson RC, Mizokami M, Orito E, Qian KP, Lau JY. Transplantation 1999; 68: 1512–1516.

319. Tanaka T, Lau JY, Mizokami M, Orito E, Tanaka E, Kiyosawa K, Yasui K, Ohta Y, Hasegawa A, Tanaka S, et al. J Hepatol 1995; 23: 742–745.

320. Bouvier-Alias M, Patel K, Dahari H, Beaucourt S, Larderie P, Blatt L, Hezode C, Picchio G, Dhumeaux D, Neumann AU, McHutchison JG, Pawlotsky JM. Hepatology 2002; 36: 211–218.

321. Alter MJ. Hepatology 1997; 26: 62S–65S.

322. Cerny A, Chisari FV. Hepatology 1999; 30: 595–601.

323. Lefkowitch JH, Schiff ER, Davis GL, Perrillo RP, Lindsay K, Bodenheimer HC Jr., Balart LA, Ortego TJ, Payne J, Dienstag JL, et al. Gastroenterology 1993; 104: 595–603.

324. Liang TJ, Rehermann B, Seeff LB, Hoofnagle JH. Ann Intern Med 2000; 132: 296–305.

325. Hoofnagle JH. Hepatology 1997; 26: 15S–20S.

326. Alter MJ, Margolis HS, Krawczynski K, Judson FN, Mares A, Alexander WJ, Hu PY, Miller JK, Gerber MA, Sampliner RE. N Engl J Med 1992; 327: 1899–1905.

327. Villano SA, Vlahov D, Nelson KE, Cohn S, Thomas DL. Hepatology 1999; 29: 908–914.

328. Thomas DL, Astemborski J, Rai RM, Anania FA, Schaeffer M, Galai N, Nolt K, Nelson KE, Strathdee SA, Johnson L, Laeyendecker O, Boitnott J, Wilson LE, Vlahov D. J Am Med Assoc 2000; 284: 450–456.

329. Diepolder HM, Zachoval R, Hoffmann RM, Wierenga EA, Santantonio T, Jung MC, Eichenlaub D, Pape GR. Lancet 1995; 346: 1006–1007.

330. Pape GR, Gerlach TJ, Diepolder HM, Gruner N, Jung M, Santantonio T. J Viral Hepat 1999; 6: 36–40.

331. Missale G, Bertoni R, Lamonaca V, Valli A, Massari M, Mori C, Rumi MG, Houghton M, Fiaccadori F, Ferrari C. J Clin Investig 1996; 98: 706–714.

332. Rehermann B. J Viral Hepat 1999; 6(Suppl 1): 31–35.
333. Jamal MM, Soni A, Quinn PG, Wheeler DE, Arora S, Johnston DE. Hepatology 1999; 30: 1307–1311.
334. Schiff ER. Hepatology 1997; 26: 39S–42S.
335. Yoshihara H, Noda K, Kamada T. Recent Dev Alcohol 1998; 14: 457–469.
336. Marcellin P, Asselah T, Boyer N. Hepatology 2002; 36: S47–S56.
337. Rodger AJ, Roberts S, Lanigan A, Bowden S, Brown T, Crofts N. Hepatology 2000; 32: 582–587.
338. Hoofnagle JH, Mullen KD, Jones DB, Rustgi V, Di Bisceglie A, Peters M, Waggoner JG, Park Y, Jones EA. N Engl J Med 1986; 315: 1575–1578.
339. Tong MJ, Blatt LM, McHutchison JG, Co RL, Conrad A. Hepatology 1997; 26: 1640–1645.
340. McHutchison J, Hoofnagle J. In: Hepatitis C (Liang T, Hoofnagle J, editors). New York: Academic Press; 2000; pp. 201–239.
341. Martinot-Peignoux M, Marcellin P, Pouteau M, Castelnau C, Boyer N, Poliquin M, Degott C, Descombes I, Le Breton V, Milotova V, et al. Hepatology 1995; 22: 1050–1056.
342. Lee WM, Reddy KR, Tong MJ, Black M, van Leeuwen DJ, Hollinger FB, Mullen KD, Pimstone N, Albert D, Gardner S. Hepatology 1998; 28: 1411–1415.
343. Zeuzem S. J Viral Hepat 2000; 7: 327–334.
344. Ikeda K, Saitoh S, Arase Y, Chayama K, Suzuki Y, Kobayashi M, Tsubota A, Nakamura I, Murashima N, Kumada H, Kawanishi M. Hepatology 1999; 29: 1124–1130.
345. Poynard T, Marcellin P, Lee SS, Niederau C, Minuk GS, Ideo G, Bain V, Heathcote J, Zeuzem S, Trepo C, Albrecht J. Lancet 1998; 352: 1426–1432.
346. McHutchison JG, Gordon SC, Schiff ER, Shiffman ML, Lee WM, Rustgi VK, Goodman ZD, Ling MH, Cort S, Albrecht JK. N Engl J Med 1998; 339: 1485–1492.
347. Davis GL, Esteban-Mur R, Rustgi V, Hoefs J, Gordon SC, Trepo C, Shiffman ML, Zeuzem S, Craxi A, Ling MH, Albrecht J. N Engl J Med 1998; 339: 1493–1499.
348. Glue P, Rouzier-Panis R, Raffanel C, Sabo R, Gupta SK, Salfi M, Jacobs S, Clement RP. Hepatology 2000; 32: 647–653.
349. Schalm SW, Brouwer JT, Bekkering FC, van Rossum TG. J Hepatol 1999; 31(Suppl 1): 184–188.
350. Brillanti S, Levantesi F, Masi L, Foli M, Bolondi L. Hepatology 2000; 32: 630–634.
351. Kato N, Sekiya H, Ootsuyama Y, Nakazawa T, Hijikata M, Ohkoshi S, Shimotohno K. J Virol 1993; 67: 3923–3930.
352. Zibert A, Schreier E, Roggendorf M. Virology 1995; 208: 653–661.
353. Rosa D, Campagnoli S, Moretto C, Guenzi E, Cousens L, Chin M, Dong C, Weiner AJ, Lau JY, Choo QL, Chien D, Pileri P, Houghton M, Abrignani S. Proc Natl Acad Sci USA 1996; 93: 1759–1763.
354. Farci P, Shimoda A, Wong D, Cabezon T, De Gioannis D, Strazzera A, Shimizu Y, Shapiro M, Alter HJ, Purcell RH. Proc Natl Acad Sci USA 1996; 93: 15394–15399.
355. Choo QL, Kuo G, Ralston R, Weiner A, Chien D, Van Nest G, Han J, Berger K, Thudium K, Kuo C. Proc Natl Acad Sci USA 1994; 91: 1294–1298.
356. Wolff JA, Malone RW, Williams P, Chong W, Acsadi G, Jani A, Felgner PL. Science 1990; 247: 1465–1468.
357. Tang DC, DeVit M, Johnston SA. Nature 1992; 356: 152–154.

358. Forns X, Payette PJ, Ma X, Satterfield W, Eder G, Mushahwar IK, Govindarajan S, Davis HL, Emerson SU, Purcell RH, Bukh J. Hepatology 2000; 32: 618–625.
359. Tokushige K, Wakita T, Pachuk C, Moradpour D, Weiner DB, Zurawski VR Jr., Wands JR. Hepatology 1996; 24: 14–20.
360. Hu GJ, Wang RY, Han DS, Alter HJ, Shih JW. Vaccine 1999; 17: 3160–3170.
361. Houghton M. Curr Top Microbiol Immunol 2000; 242: 327–339.
362. Felsenstein J. Distributed by the author, Dept. of Genetics, University of Washington, Seattle 1993.
363. Simmonds P, Holmes EC, Cha TA, Chan SW, McOmish F, Irvine B, Beall E, Yap PL, Kolberg J, Urdea MS. J Gen Virol 1993; 74: 2391–2399.
364. Muerhoff AS, Simons JN, Leary TP, Erker JC, Chalmers ML, Pilot-Matias TJ, Dawson GJ, Desai SM, Mushahwar IK. J Hepatol 1996; 25: 379–384.

174

Who is at risk for infection?

Chronic HBV carriers are at risk for infection with HDV. Individuals who are not infected with HBV, and have not been immunized against HBV, are at risk of infection with HBV with simultaneous or subsequent infection with HDV.

Since HDV absolutely requires the support of a hepadnavirus for its own replication, inoculation with HDV in the absence of HBV will not cause hepatitis D. Alone, the viral genome indeed replicates in a helper-independent manner, but virus particles are not released.

Where is HDV a problem globally?

The HDV is present worldwide and in all age groups [14,21]. Its distribution parallels that of HBV infection, although with different prevalence rates (highest in parts of Russia, Romania, Southern Italy and the Mediterranean countries, Africa and South America). In some HBV-prevalent countries such as China, HDV infection is disproportionately low [14].

The natural reservoir is man, but HDV can be experimentally transmitted to chimpanzees and woodchucks that are infected with HBV and woodchuck hepatitis virus (WHV), respectively [19,21,24].

When is a HDV infection life-threatening?

HDV infection of chronically infected HBV-carriers may lead to fulminant acute hepatitis or severe chronic active hepatitis, often progressing to cirrhosis. Chronic hepatitis D may also lead to the development of hepatocellular carcinoma (HCC) [11].

Why is there no treatment for the disease?

Hepatitis D is a viral disease, and as such, antibiotics are of no value in the treatment of the infection.

There is no hyperimmune D globulin available for pre- or post-exposure prophylaxis.

Disease conditions may occasionally improve with administration of α-interferon [15,21,25].

Since no effective antiviral therapy is currently available for treatment of type D hepatitis, liver transplantation may be considered for cases of fulminant acute and end-stage chronic hepatitis D.

The hepatitis delta virus (HDV)

The genome of HDV is unrelated to the genomes of hepadnaviruses, of which HBV is a member. HDV is therefore not a defective-interfering particle of HBV, and should be considered as a satellite virus, a natural subviral satellite of HBV [10,11, 14,25].

Important parallels can be drawn between HDV and certain subviral agents of plants, especially the viroids, with respect to genome structure and replication mechanisms. Because of the many differences, however, HDV has been classified into the separate genus Deltavirus [13,18,25].

The genome of HDV was cloned and sequenced in 1986 [27]. HDV is a replication defective, helper (HBV) dependent ssRNA virus that requires the surface antigen of HBV (HBsAg) for the encapsidation of its own genome. The envelope proteins on the outer surface of HDV are entirely provided by HBV [16,18,25].

The outer envelope of HDV particles actually contains lipid and all three forms (S, M, and L) of HBV surface antigen (HBsAg), but predominantly the major form of HBsAg with very few middle (pre S1) and large (pre S2) proteins. This proportion (95:5:1 of S:M:L) is different from that found in HBV particles [13,14,25].

There is no evidence that the HBV-derived envelope proteins are additionally modified when they become the envelope of HDV [21].

The internal, nucleocapsid structure of HDV is composed of the viral single-stranded RNA genome and about 60 copies of delta antigen, the only HDV-encoded protein, in its large and small forms [13,25].

Synthesis of HDV results in temporary suppression of synthesis of HBV components [13].

HDV does not infect established tissue culture cell lines. Complete viral replication cycles *in vitro* are limited to primary hepatocytes, generally of woodchucks or chimpanzees, that are coinfected with a hepadnavirus or cotransfected with hepadnavirus cDNA. When experimental conditions meet these requirements, infectious HDV particles are produced [13,14,21].

In nature, HDV has only been found in humans infected with HBV. Experimentally, it can be transmitted to chimpanzees and woodchucks in the presence of HBV or WHV, respectively [13,18,19,21,24,25].

HDV replication cycle

To replicate efficiently, a virus requires the cooperation of the host cell at all stages of the replicative cycle: attachment, penetration, uncoating, provision of appropriate metabolic conditions for the synthesis of viral macromolecules, the final assembly of viral subunits and the release of new virions. HDV also requires the presence of a helper hepadnavirus to provide the protein components for its own envelope.

How HDV enters hepatocytes is still not known, but it may involve the interaction between HBsAg-L and a cellular receptor. The incoming HDV RNA is then transported

178

polyadenylation signal due to the presence of HDAg and directly synthesize full-length antigenomic HDV RNA (arrow 4). (B) Proposed new model for HDV RNA transcription and replication. The syntheses of 0.8-kb mRNA (a) and 1.7-kb monomer RNA (b) are independent and occur in parallel [17,25].

A model for RNA editing of HDV

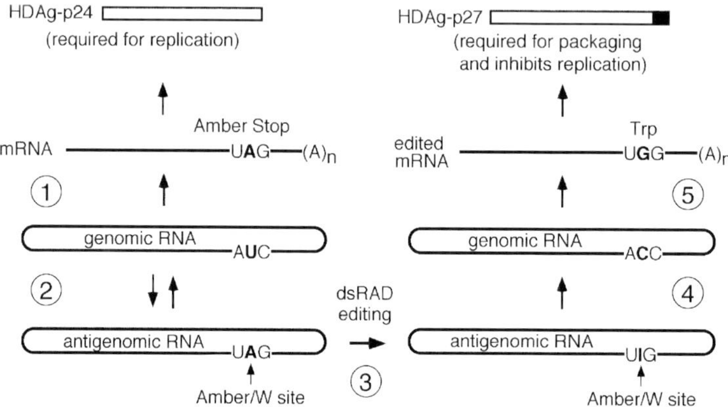

Reprinted by permission from Ref. [20], copyright (1996) Macmillan Magazines Ltd. (http:// www.nature.com).

A model for RNA editing of HDV. (1) Replication-competent genomes are transcribed to produce an mRNA encoding HDAg-p24. (2) HDAg-24 enables replication of the genome by RNA polymerase II, generating antigenomic RNA. (3) dsRAD (double-stranded-RNA-adenosine deaminase) acts on antigenomic RNA to convert the adenosine at the amber/W site to an inosine. (4) Like G, inosine prefers to pair with C; thus, after replication, the genome has a C at the amber/W site instead of a U. (5) The edited genome is transcribed to yield an mRNA encoding HDAg-p27. HDAg-p27, which contains a 19-amino-acid extension (shaded), inhibits replication and helps packaging of the HDV genome by HBV surface antigen [20].

Morphology and physicochemical properties

HDV virions are 36–43 nm, roughly spherical, enveloped particles with no distinct nucleocapsid structure. They do not have distinct spikes on their outer surface and are possibly icosahedral [13,14,25].

When the virus particle is disrupted with nonionic detergents, an internal nucleocapsid is released and HDAg becomes detectable [13,25].

The 19-nm nucleocapsid contains about 60 copies of HDAg in its two forms (24 and 27 kDa) and HDV genomic RNA [25].

The buoyant density of HDV particles is 1.25 g/cm^3 in CsCl gradients [13,14,18].

Schematic representation of viral particles found in serum of HBV–HDV-infected people

Infectious HBV particle:

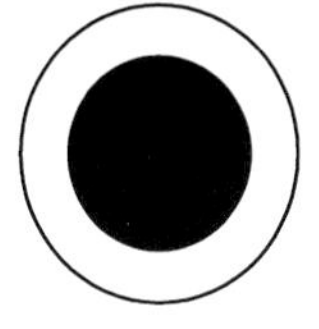

- 42 nm outer envelope containing lipid and three forms of HBsAg

- 27 nm nucleocapsid containing 180 copies of core protein and reverse transcriptase and HBV DNA

Infectious HDV particle:

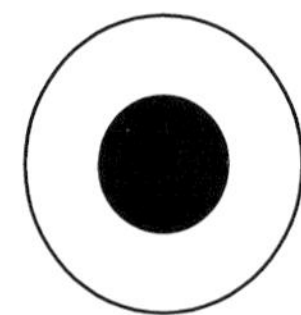

- 36 - 43 nm outer envelope containing lipid and three forms of HBsAg

- 19 nm nucleocapsid containing 60 copies of delta antigen and HDV genomic RNA

Empty noninfectious particles:

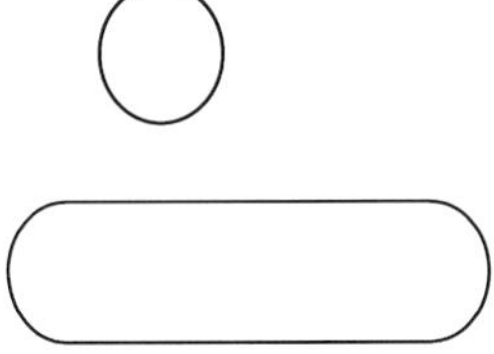

- 22 nm filaments and spheres made of lipid and mainly one form of HBsAg

From Ref. [25], with permission (http:///ww.com).

Genome and proteins

The HDV genome is a single, negative stranded, circular RNA molecule nearly 1.7 kb in length containing about 60% C + G [13,14,18,25].

HDV RNA is the only animal virus known to have a circular RNA genome [13].

A high degree of intramolecular complementarity allows about 70% of the nucleotides to be basepaired to each other to form an unbranched, double-stranded, stable, rod-shaped structure [10,13,14,18].

So far, about 14 different HDV isolates from different parts of the world have been sequenced, and all range from 1670 to 1685 nucleotides in length. Based on sequence similarities, HDV isolates can be classified into three genotypes [13].

Genotype I is the most predominant one in most areas of the world, and is associated with a broad spectrum of chronic HDV disease. Originally found in a Japanese isolate, genotype II has been found recently to predominate in Taiwan. Disease associated with genotype II might be less severe than genotype I. Genotype III is associated with outbreaks in Venezuela and Peru. It is responsible for more severe disease in the northern South American regions [5,10,11,13].

The genome contains several sense- and antisense open reading frames (ORFs), only one of which is functional and conserved. The RNA genome is replicated through an RNA intermediate, the antigenome [13,14,18].

The genomic RNA and its complement, the antigenome, can function as ribozymes to carry out self-cleavage and self-ligation reactions [13,18,25].

A third RNA present in the infected cell, also complementary to the genome, but 800-b long and polyadenylated, is the mRNA for the synthesis of the delta antigen (HDAg) [14,18,25].

The one and only protein expressed by HDV, the hepatitis delta antigen HDAg, is not exposed on the virion outer surface, but is present in the internal nucleocapsid [13,14,18].

The protein is seen as two species, of 24 and 27 kDa. The two species are identical, but the 27-kDa protein has a 19 aa longer C-terminus. The short form (195 amino acids, HDAg-S), synthesized first, is required for RNA replication; the long form (214 amino acids, HDAg-L), becoming detectable after prolonged replication, suppresses viral RNA replication and is required for packaging of the HDV genome by HbsAg [13,14,18,21,23].

The relative ratios of these two species vary from patient to patient. Two separate ORFs on different RNAs encode HDAg-S and HDAg-L. A single nucleotide at the termination codon for HDAg-S is altered by a specific post-transcriptional RNA editing event in some RNAs, so that the ORF extends for 19 additional amino acids [13].

HDAg is a non-glycosylated phosphoprotein [13,21,25]. It has an RNA-binding activity and appears to bind specifically to HDV RNA in the virus particle [28]. In infected cells, HDAg is localized in the nuclei [13,14,25].

Functional domains present in HDAg include the nuclear localization signal located within the N-terminal one-third of the protein, the RNA-binding motif present in the middle one-third of the protein and a third domain, consisting of the C-terminal 19 amino acids, possibly involved in interactions with the HBsAg during virion assembly, and in the inhibition of HDV RNA assembly [13,14,25].

The other protein present in HDV particles is HBsAg. This protein is derived from the coinfection with HBV and is essential for HDV virion assembly and virus transmission.

The three RNAs of HDV present in the infected cell

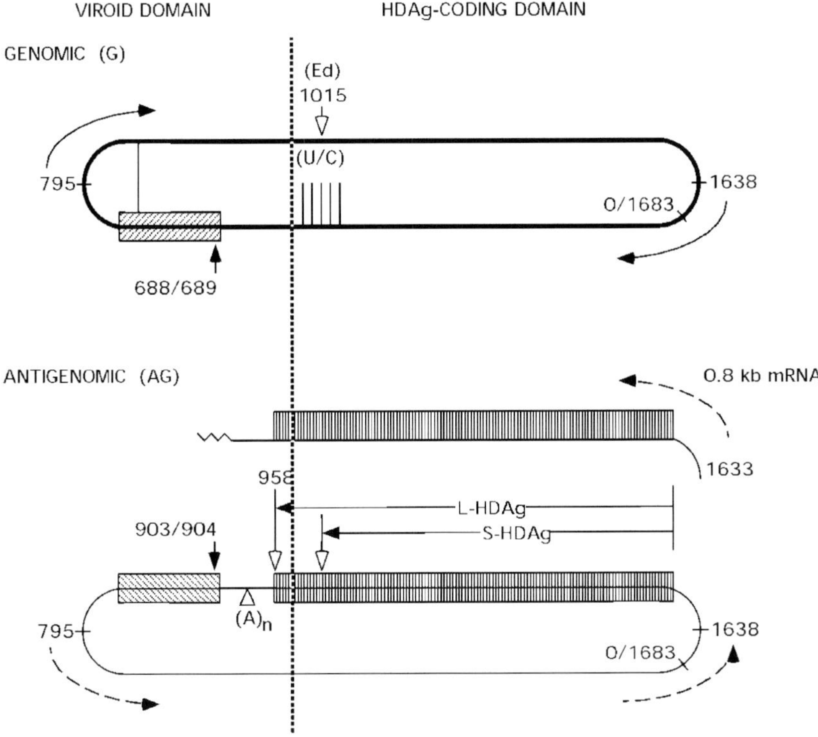

From Ref. [13], "With permission, from the Annual Review of Biochemistry, Volume 64 ©1995 by Annual Reviews http://www.AnnualReviews.org".

Schematic diagrams of the structure of HDV RNA. The antigenomic RNA and mRNA are detected only in the cells. The nucleotide numbers are according to Ref. [16] and represented in genomic orientation even on the antigenomic strand. The genomic RNA is represented in clockwise orientation, while antigenomic RNA is counterclockwise. Nucleotides 688/689 and 903/904 are ribozyme cleavage sites for genomic and antigenomic RNAs, respectively. The hatched boxes represent the ribozyme domain. Nucleotide 1015 (Ed) denotes RNA editing site. $(A)_n$ represents polyadenylation signal. The UV cross-linking site is indicated by a vertical line in the viroid domain [13].

182

Comparison of HDV genotypes

<table>
<tr><th colspan="6">Hepatitis Delta Virus Genotypes</th></tr>
<tr><th></th><th colspan="3">Genetic distance (%) between isolates of genotype:</th><th>Associated Disease?</th><th>Locations</th></tr>
<tr><th></th><th>I</th><th>II</th><th>III</th><th></th><th></th></tr>
<tr><td>I</td><td>8
(1 - 15)</td><td>23
(21 - 26)</td><td>31
(29 - 33)</td><td>moderate to severe</td><td>North America, Asia Middle East, Europe</td></tr>
<tr><td>II</td><td></td><td>7
(5 - 8)</td><td>33
(32 - 34)</td><td>moderate</td><td>East Asia</td></tr>
<tr><td>III</td><td></td><td></td><td>5
(1 - 7)</td><td>severe</td><td>South America</td></tr>
</table>

From Ref. [5], with permission.

1. Genetic variability, tentative disease associations, and geographic distributions of HDV genotypes I, II, and III [5].

<table>
<tr><th colspan="3">Comparison of HDV Genotypes I and III</th></tr>
<tr><th></th><th>Genotype I</th><th>Genotype III</th></tr>
<tr><td>Virion Formation</td><td></td><td></td></tr>
<tr><td>Transfected cells (day 10)</td><td>$\leq$.5 %</td><td>10 %</td></tr>
<tr><td>Transfected cells (day 13)</td><td>2 %</td><td>20 %</td></tr>
<tr><td>Secretion of HDAg-p27 with HBsAg (ayw)</td><td>+++</td><td>+</td></tr>
<tr><td>RNA Editing</td><td></td><td></td></tr>
<tr><td>Serum virion RNA</td><td>30%</td><td>15%</td></tr>
<tr><td>Virion RNA from transfected cell medium</td><td>18%</td><td>10%</td></tr>
<tr><td>Modified by purified dsRAD</td><td>yes</td><td>?</td></tr>
<tr><td>Target structure</td><td>base-paired</td><td>uncertain</td></tr>
<tr><td>Support of HDAg(-) RNA Replication</td><td></td><td></td></tr>
<tr><td>HDAg-p24 (I)</td><td>+++</td><td>+/-</td></tr>
<tr><td>HDAg-p24 (III)</td><td>-</td><td>+++</td></tr>
</table>

From Ref. [5], with permission.

2. Comparison of virion formation, RNA editing and RNA replication for HDV genotypes I and III [5].

Antigenicity

The intact virus particle is reactive with anti-HBsAg antibody, but not with anti-HDAg antibody.

Despite the sequence heterogeneity observed in HDV isolates from different geographical regions, there appear to be no serological differences among these isolates [14].

All HDV are antigenically related, and antibodies to HDAg do not neutralize HDV [21].

Surface epitopes unique to HDV have not been detected.

Under experimental conditions, HDV can use different hepadnaviruses as helpers. In each case, the envelope of HDV has both the physical and antigenic characteristic of the helper virus.

Stability

Because of its double-strandedness, the HDV RNA is relatively stable.

The HDV survives dry heat at 60°C for 30 h [14].

The disease

An HDV infection absolutely requires an associated HBV infection. The outcome of the disease largely depends on whether the two viruses infect simultaneously (coinfection), or whether the newly HDV-infected person is a chronically infected HBV carrier (superinfection).

Coinfection of HBV and HDV (simultaneous infection with the two viruses) results in both acute type B and acute type D hepatitis. The incubation period depends on the HBV titer of the infecting inoculum. Depending on the relative titers of HBV and HDV, a single bout or two bouts of hepatitis may be seen. Coinfections of HBV and HDV are usually acute, self-limited infections. The chronic form of hepatitis D is seen in less than 5% of HBV–HDV coinfected patients [10,21].

Acute hepatitis D occurs after an incubation period of 3–7 weeks, and a pre-icteric phase begins with symptoms of fatigue, lethargy, anorexia and nausea, lasting usually 3–7 days. During this phase, ALT and AST activities become abnormal. The appearance of jaundice is typical at the onset of the icteric phase. Fatigue and nausea persist, clay-colored stools and dark urine appear, and serum bilirubin levels become abnormal. In patients with acute, self-limiting infection, convalescence begins with the disappearance of clinical symptoms. Fatigue may persist for longer periods of time [14,21].

Superinfection of HBV and HDV (HDV infection of a chronically infected HBV carrier) causes a generally severe acute hepatitis with short incubation time that leads to chronic type D hepatitis in up to 80% of cases. Superinfection is associated with fulminant acute hepatitis and severe chronic active hepatitis, often progressive to cirrhosis [14,21].

During the acute phase of HDV infection, synthesis of both HBsAg and HBV DNA are inhibited until the HDV infection is cleared [14].

Fulminant viral hepatitis is rare, but still about 10 times more common in hepatitis D than in other types of viral hepatitis. It is characterized by hepatic encephalopathy showing changes in personality, disturbances in sleep, confusion and difficulty in concentrating, abnormal behavior, somnolence and coma. The mortality rate of fulminant hepatitis D reaches 80%. Liver transplantation is indicated [14,21].

Chronic viral hepatitis D is usually initiated by a clinically apparent acute infection. Symptoms are less severe than in acute hepatitis, and while serum ALT and AST levels are elevated, bilirubin and albumin levels and prothrombin time may be normal. In chronic hepatitis D, the HBV markers are usually suppressed [13,14,21].

Progression to cirrhosis usually takes 5–10 years, but it can appear 2 years after onset of infection. About 60–70% of patients with chronic hepatitis D develop cirrhosis. A high proportion of these patients die of hepatic failure [21].

HCC occurs in chronically infected HDV patients with advanced liver disease with the same frequency as in patients with ordinary HB. HCC may actually be more a secondary effect of the associated cirrhosis than a direct carcinogenic effect of the virus.

Taken together, three phases of chronic hepatitis D have been proposed: (i) an early active phase with active HDV replication and suppression of HBV, (ii) a second moderately active one with decreasing HDV and reactivating HBV, and (iii) a third late one with development of cirrhosis and HCC caused by replication of either virus or with remission resulting from marked reduction of both viruses [10].

The mortality rate for HDV infections lies between 2 and 20%, values that are 10 times higher than for HB [21].

Scheme of infection and clinical features

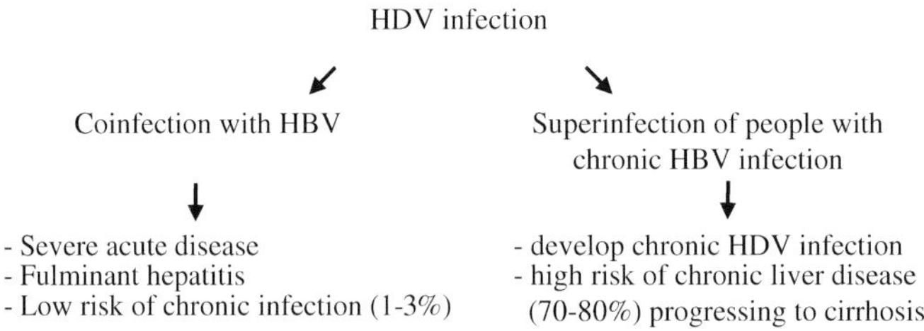

Diagnosis

Hepatitis D should be considered in any individual who is HBsAg positive or has evidence of recent HBV infection [21].

The diagnosis of acute hepatitis D is made after evaluation of serologic tests for the virus. Total anti-HDV are detected by commercially available radioimmunoassay (RIA) or enzyme immunoassay (EIA) kits [10,21].

The method of choice for the diagnosis of ongoing HDV infection should be RT-PCR, which can detect 10–100 copies of the HDV genome in infected serum [10,11,18,21].

Acute HBV–HDV coinfection:

- Appearance of HBsAg, HBeAg and HBV DNA in serum during incubation.

- Appearance of anti-HBc at the onset of clinical disease.

- Appearance of IgM anti-HD, HDV RNA, HDAg in serum.

- Anti-HDV antibodies develop late in acute phase and usually decline after infection to subdetectable levels.

- If HDAg is detectable early during infection, it disappears as anti-HDV appears.

- All markers of viral replication disappear in early convalescence, and both IgM and IgG anti-HD disappear within months to years after recovery.

HBV–HDV superinfection:

- Usually results in persistent HDV infection.

- HDV viremia appears in serum during pre-acute phase.

- High titers of IgM and IgG anti-HDV are detectable in acute phase, persisting indefinitely.

- Titer of HBsAg declines when HDAg appears in serum.

- Progression to chronicity is associated with persisting high levels of IgM anti-HD and IgG anti-HD.

- HDAg and HDV RNA remain detectable in serum and liver.

- Viremia is associated with active liver disease.

Each of the markers of HDV infection, including IgM and IgG antibodies, disappears within months after recovery. In contrast, in chronic hepatitis D, HDV RNA, HDAg, and IgM and IgG anti-HD antibodies persist [14,18].

Host immune response

Both humoral and cellular immunity are induced in patients infected with HDV [14,21].

These immune responses may provide protection from HDV re-infection, or simply modulate clinical symptoms. However, second cases of hepatitis D have not been reported [14,21].

186

Anti-HD antibodies do not always persist after acute infection is cleared. The serological evidence of past HDV infection is therefore not easy to demonstrate [14].

Typical serologic course

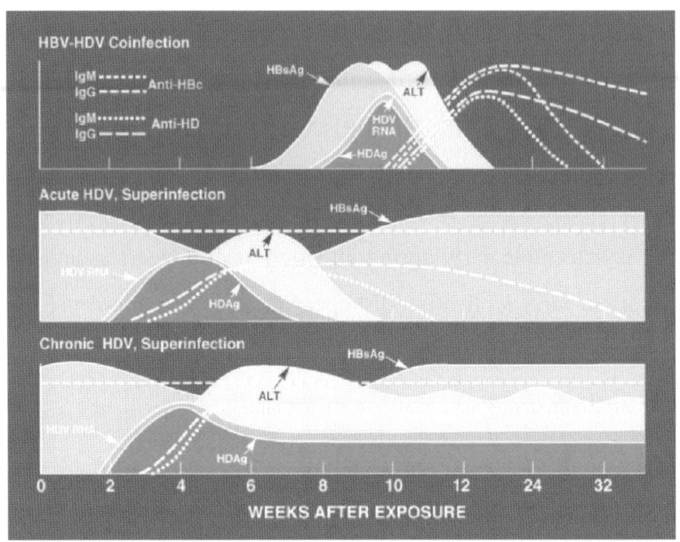

From: Ref [21], with permission (http://\ww.com).

The serological patterns of type D hepatitis: coinfection and superinfection. **Top**: coexistent acute HB and hepatitis D. **Middle**: acute hepatitis D superimposed on a chronic HBV infection. **Bottom**: acute hepatitis D progressing to chronic hepatitis, superimposed on a chronic HBV infection [21].

Prevalence

Areas of high prevalence include the Mediterranean Basin, the Middle East, Central Asia, West Africa, the Amazon Basin of South America and certain South Pacific islands [13, 14,21].

Severe, often fatal, acute and chronic type D hepatitis occurs among indigenous people of Venezuela, Colombia, Brazil, and Peru—all regions with high chronic HDV infection rates [21].

Hepatitis D is less common in Eastern Asia, but is present in Taiwan, China and India [21].

Worldwide distribution of HDV infection

The illustration below has been taken from Ref. [6]

From Ref. [6], http://www.cdc.gov/ncidod/diseases/hepatitis/slideset/hep d/slide 6.htm

Pathogenesis

Infection with both HBV and HDV is associated with more severe liver injury than HBV infection alone [13].

Pathologic changes in hepatitis D are limited to the liver, the only organ in which HDV has been shown to replicate. The histologic changes consist of hepatocellular necrosis and inflammation [21].

HDV genome replication is not acutely cytopathic, and both humoral and cellular immune mechanisms may be involved in the pathology of hepatitis D. More experimental data are needed to unravel the underlying mechanisms of HDV-induced disease [10,14, 21,25].

HBV is an essential cofactor in the evolution of hepatocellular damage [7,10].

Transmission

Transmission is similar to that of HBV:

- Bloodborne and sexual.

- Percutaneous (injecting drug use, hemophiliacs).

- Permucosal (sexual).

- Rare perinatal.

Superinfections increase the chance of HDV spread, and at the peak of an acute infection, the amount of HDV in the serum can exceed 10^{12} RNA-containing particles per ml [25].

During an HDV superinfection, the titer of HDV reaches a peak between 2 and 5 weeks post-inoculation, after which it declines in 1–2 weeks [25].

The probability of being productively coinfected, with the coinfection resulting in clinical disease, depends on both the relative and absolute amounts of the two inoculated viruses [25]. The main route of transmission is infected blood and blood products.

Risk groups

Here is a list of groups of people who are at risk of contracting HDV (see Ref. [21]):

- Intravenous drug users using HDV-contaminated injection needles.

- Promiscuous homosexual and heterosexual groups (although HDV infections are less frequent than HBV or HIV infections).

- People exposed to unscreened blood or blood products.

- Hemophiliacs (the risk has decreased in recent years due to better control of blood sources).

- Persons with clotting factor disorders (the risk has decreased in recent years due to better control of blood sources).

Incidence/epidemiology

Seroprevalence studies of anti-HD in HBsAg-positive patients has shown a worldwide but not uniform distribution [21].

Epidemics of HDV infections have been described in the Amazon Basin, the Mediterranean Basin and Central Africa.

Two epidemiologic patterns of hepatitis D infections exist: in Mediterranean countries infection is endemic among HBV carriers, and the virus is transmitted by close personal contact. In Western Europe and North America, HDV is confined to persons exposed to blood or blood products, like, e.g. intravenous drug addicts sharing unsterilized injection needles.

Worldwide, more than 10 million people are infected with HDV [10,11].

Trends

New foci of high HDV prevalence continue to be identified as in the case of the island of Okinawa in Japan, of areas of China, Northern India and Albania [10].

There is a decreasing prevalence of both acute and chronic hepatitis D in the Mediterranean area and in many other parts of the world, which has been attributed to a decline in the prevalence of chronic HBsAg carriers in the general population [10].

Immune prophylaxis

Immune prophylaxis against HDV is achieved by vaccination against HBV because HDV uses the envelope proteins of HBV. This mode of prevention is possible only for coinfections in HBV susceptible individuals [10,21].

Immunoglobulin (IG), HB specific IG and HB vaccine do not protect HBV carriers from infection with HDV.

Vaccines

No vaccines exist against HDV; however, vaccination against HBV of patients who are not chronic HBV carriers, provides protection against HDV infection.

Prevention and treatment

Since antivirals have never been as successful for the treatment of viral infections as antibiotics have been for the treatment of bacterial infections, prevention of viral diseases remains the most important weapon for their control.

Prevention

Since HDV is dependent on HBV for replication, control of HDV infection is achieved by targeting HBV infections. All measures aimed at preventing the transmission of HBV will prevent the transmission of hepatitis D. HBV vaccination is therefore recommended to avoid HBV–HDV coinfection [14].

However, there is no effective measure to prevent HDV infection of chronic HBV carriers, and prevention of HBV–HDV superinfection can only be achieved through education to reduce risk behaviors [14,21].

Promising research results indicate that in some woodchucks immunized with recombinant purified HDAg-S complete protection is possible.

HB IG and HB vaccine do not protect HBV carriers from infection by HDV.

Treatment

Currently, there is no effective antiviral therapy available for treatment of acute or chronic type D hepatitis [21].

For infected patients, massive doses of α-interferon (9 million units 3 times a week for 12 months or 5 million units daily for up to 12 months) have yielded remissions, but most patients remained positive for HDV RNA despite the improved disease conditions [21].

The effect of interferon is considered to be most likely an indirect one, possibly via an effect on the helper hepadnavirus and/or on the immune response to the infections [25].

Acyclovir, ribavirin, lamivudine and synthetic analogues of thymosin have proved ineffective [10].

Immunosuppressive agents do not have any effect on hepatitis D [14,21].

Liver transplantation has been helpful for treating fulminant acute and end-stage chronic hepatitis [11,21]. In one study, the 5-year survival rate of transplant patients for terminal delta cirrhosis was 88% with reappearance of HBsAg only in 9% under long-term anti-HBs prophylaxis [10].

Guidelines for epidemic measures

(1)   When two or more cases occur in association with some common exposure, a search for additional cases should be conducted.

(2)   Introduction of strict aseptic techniques. If a plasma derivative like antihemophilic factor, fibrinogen, pooled plasma or thrombin is implicated, the lot should be withdrawn from use.

(3)   Tracing of all recipients of the same lot in search for additional cases.

Future considerations

Whether or not immunization with HDAg can confer protection against superinfection or slow the progression of liver disease in over 350 million HBV carriers who are at risk of contracting type D hepatitis, needs to be determined [21].

Glossary

Albumin a water soluble protein. Serum albumin is found in blood plasma and is important for maintaining plasma volume and osmotic pressure of circulating blood. Albumin is synthesized in the liver. The inability to synthesize albumin is a predominant feature of chronic liver disease.

ALT alanine aminotransferase an enzyme that interconverts L-alanine and D-alanine. It is a highly sensitive indicator of hepatocellular damage. When such damage occurs, ALT is released from the liver cells into the bloodstream, resulting in abnormally high serum levels. Normal ALT levels range from 10 to 32 U/l; in women, from 9 to 24 U/l. The normal range for infants is twice that of adults.

Amino acids the basic units of proteins, each amino acid has a NH—C(R)—COOH structure, with a variable R group. There are altogether 20 types of naturally occurring amino acids.

Antibody a protein molecule formed by the immune system which reacts specifically with the antigen that induced its synthesis. All antibodies are immunoglobulins [1].

Antigen any substance which can elicit in a vertebrate host the formation of specific antibodies or the generation of a specific population of lymphocytes reactive with the substance. Antigens may be protein or carbohydrate, lipid or nucleic acid, or contain elements of all or any of these as well as organic or inorganic chemical groups attached to protein or other macromolecule. Whether a material is an antigen in a particular host depends on whether the material is foreign to the host and also on the genetic makeup of the host, as well as on the dose and physical state of the antigen [1].

Antigenome RNA molecule complementary to the viral single-stranded RNA genome.

AST aspartate aminotransferase the enzyme that catalyzes the reaction of aspartate with 2-oxoglutarate to give glutamate and oxaloacetate. Its concentration in blood may be raised in liver and heart diseases that are associated with damage to those tissues. Normal AST levels range from 8 to 20 U/l. AST levels fluctuate in response to the extent of cellular necrosis [1].

Bilirubin is the chief pigment of bile, formed mainly from the breakdown of hemoglobin. After formation it is transported in the plasma to the liver to be then excreted in the bile. Elevation of bile in the blood causes jaundice [26].

Capsid the protein coat of a virion, composed of large multimeric proteins, which closely surrounds the nucleic acid [1].

Carcinoma a malignant epithelial tumour. This is the most frequent form of cancer.

cDNA complementary DNA. DNA synthesized by RNA-directed DNA polymerase as a copy of RNA, usually isolated mRNA or viral genomic RNA. It differs in sequence from eukaryotic chromosomal DNA by the absence of introns.

Cirrhosis a chronic disease of the liver characterized by nodular regeneration of hepatocytes and diffuse fibrosis. It is caused by parenchymal necrosis followed by nodular proliferation of the surviving hepatocytes. The regenerating nodules and accompanying fibrosis interfere with blood flow through the liver and result in portal hypertension, hepatic insufficiency, jaundice and ascites.

Codon the smallest unit of genetic material that can specify an amino acid residue in the synthesis of a polypeptide chain. The codon consists of three adjacent nucleotides.

Cytopathic effects include morphological changes in the cell appearance (rounding up of cells), agglutination of red blood cells (hemagglutination assay with influenza-virus),

192

zones of cell lysis on monolayers of tissue culture or finally immortalization of animal cell lines (foci formation).

Cytoplasm the protoplasm of the cell which is outside the nucleus. It consists of a continuous aqueous solution and the organelles and inclusions suspended in it. It is the site of most of the chemical activities of the cell.

Encephalopathy an acute reaction of the brain to a variety of toxic or infective agents, without any actual inflammation such as occurs in encephalitis [1].

Endemic continuously prevalent in some degree in a community or region [26].

Endoplasmic reticulum a network or system of folded membranes and interconnecting tubules distributed within the cytoplasm of eukaryotic cells. The membranes form enclosed or semi-enclosed spaces. The endoplasmic reticulum functions in storage and transport, and as a point of attachment of ribosomes during protein synthesis.

Enzyme any protein catalyst, i.e. substance which accelerates chemical reactions without itself being used up in the process. Many enzymes are specific to the substance on which they can act, called substrate. Enzymes are present in all living matters and are involved in all the metabolic processes upon which life depends [1,26].

Epidemic an outbreak of disease such that for a limited period a significantly greater number of persons in a community or region suffer from it than is normally the case. Thus an epidemic is a temporary increase in incidence. Its extent and duration are determined by the interaction of such variables as the nature and infectivity of the casual agent, its mode of transmission and the degree of pre-existing and newly acquired immunity [26].

Epitope or antigenic determinant. The small portion of an antigen that combines with a specific antibody. A single antigen molecule may carry several different epitopes [1].

Fulminant describes pathological conditions that develop suddenly and are of great severity [1].

Genome the total genetic information present in a cell. In diploid cells, the genetic information contained in one chromosome set [1].

Golgi apparatus a cytoplasmic organelle which is composed of flattened sacs resembling smooth endoplasmic reticulum. The sacs are often cup-shaped and located near the nucleus, the open side of the cup generally facing toward the cell surface. The function of the Golgi apparatus is to accept vesicles from the endoplasmic reticulum, to modify the contents, and to distribute the products to other parts of the cell or to the cellular environment.

Hepadnavirus family of single-stranded DNA viruses of which hepatitis B virus (HBV) and woodchuck hepatitis virus (WHV) are members.

Hepatocytes liver cells [1].

Humoral pertaining to the humors, or certain fluids, of the body [1].

Icterus jaundice.

IgA antibodies IgA has antiviral properties. Its production is stimulated by aerosol immunizations and oral vaccines.

IgG antibodies IgG is the most abundant of the circulating antibodies. It readily crosses the walls of blood vessels and enters tissue fluids. IgG also crosses the placenta and confers passive immunity from the mother to the fetus. IgG protects against bacteria, viruses, and toxins circulating in the blood and lymph.

IgM antibodies IgMs are the first circulating antibodies to appear in response to an antigen. However, their concentration in the blood declines rapidly. This is diagnostically useful, because the presence of IgM usually indicates a current infection by the pathogen causing its formation. IgM consists of five Y-shaped monomers arranged in a pentamer structure. The numerous antigen-binding sites make it very effective in agglutinating antigens. IgM is too large to cross the placenta and hence does not confer maternal immunity.

Immunoglobulin (IG) is a sterile preparation of concentrated antibodies (immunoglobulins) recovered from pooled human plasma processed by cold ethanol fractionation. Only plasma that has tested negative for (i) HBsAg, (ii) antibody to human immunodeficiency virus (HIV), and (iii) antibody to hepatitis C virus (HCV) is used to manufacture IG. IG is administered to protect against certain diseases through passive transfer of antibody. The IGs are broadly classified into five types on the basis of physical, antigenic and functional variations, and labeled, respectively IgM, IgG, IgA, IgE and IgD.

Incidence the number of cases of a disease, abnormality, accident, etc. arising in a defined population during a stated period, expressed as a proportion, such as x cases per 1000 persons per year [1].

Interferon a class of proteins processing antiviral and anti-tumour activity produced by lymphocytes, fibroblasts and other tissues. They are released by cells invaded by virus and are able to inhibit virus multiplication in noninfected cells. Interferon preparations

194

have been shown to have some clinical effect as antiviral agents. The preparations so far available have produced side effects, such as fever, lassitude, and prostration, not dissimilar from those accompanying acute virus infection itself [1].

Jaundice is a yellow discoloration of the skin and mucous membranes due to excess of bilirubin in the blood, also known as icterus [26].

Lymphocyte a leukocyte of blood, bone marrow and lymphatic tissue. Lymphocytes play a major role in both cellular and humoral immunity. Several different functional and morphologic types must be recognized, i.e. the small, large, B-, and T-lymphocytes, with further morphologic distinction being made among the B-lymphocytes and functional distinction among T-lymphocytes [1].

Necrosis death of tissue [1].

Nucleotide a molecule formed from the combination of one nitrogenous base (purine or pyrimidine), a sugar (ribose or deoxyribose) and a phosphate group. It is a hydrolysis product of nucleic acid [1].

Nucleus a membrane-bounded compartment in an eukaryotic cell which contains the genetic material and the nucleoli. The nucleus represents the control center of the cell. Nuclei divide by mitosis or meiosis.

Peptide a compound of two or more amino acids linked together by peptide bonds [26].

Pleomorphic distinguished by having more than one form during a life cycle [1].

Prenylation the enzymic addition of prenyl moieties to proteins as a post-translational modification.

Prevalence is the number of instances of infections or of persons ill, or of any other event such as accidents, in a specified population, without any distinction between new and old cases [26].

Prophylaxis is the prevention of disease, or the preventive treatment of a recurrent disorder [26].

Protein large molecule made up of many amino acids chemically linked together by amide linkages. Biologically important as enzymes, structural protein and connective tissue.

Prothrombin time a test used to measure the activity of clotting factors I, II, V, VII, and X. Deficiency of any of these factors leads to a prolongation of the prothrombin time.

The test is basic to any study of the coagulation process, and it helps in establishing and maintaining anticoagulant therapy.

Reverse transcriptase RNA-directed DNA polymerase. Enzyme that synthesizes DNA according to instructions given by an RNA template.

Ribozyme an RNA molecule with catalytic activity.

RT-PCR reverse transcriptase-polymerase chain reaction. A technique commonly employed in molecular genetics through which it is possible to produce copies of DNA sequences rapidly.

Self-limited denoting a disease that tends to cease after a definite period, e.g. pneumonia [3].

Sense and antisense strands of the two strands that comprise the double helix of a DNA molecule, only sense strand contains a sequence of nucleotides that can be read out to form a protein. The complementary strand, termed the antisense strand, has a sequence of nucleotides that, if read out, would give either a garbled or a totally lacking messenger RNA [2]. An artificial, antisense, single-stranded RNA molecule of messenger RNA or of some other specific RNA transcript of a gene can hybridize with the specific RNA and thus interfere with the latter's actions or reactions.

Serum is the clear, slightly yellow fluid which separates from blood when it clots. In composition it resembles blood plasma, but with fibrinogen removed. Sera containing antibodies and antitoxins against infections and toxins of various kinds (antisera) have been used extensively in prevention or treatment of various diseases [26].

Titer a measure of the concentration or activity of an active substance.

Translation the process of forming a specific protein having its amino acid sequence determined by the codons of messenger RNA. Ribosomes and transfer RNA are necessary for translation [1].

Vaccine an antigenic preparation used to produce active immunity to a disease to prevent or ameliorate the effects of infection with the natural or "wild" organism. Vaccines may be living, attenuated strains of viruses or bacteria, which give rise to inapparent to trivial infections. Vaccines may also be killed or inactivated organisms or purified products derived from them. Formalin-inactivated toxins are used as vaccines against diphtheria and tetanus. Synthetically or genetically engineered antigens are currently being developed for use as vaccines. Some vaccines are effective by mouth, but most have to be given parenterally [1,26].

Viremia the presence of viruses in the blood, usually characterized by malaise, fever, and aching of the back and extremities [4].

Virion a structurally complete virus, a viral particle [1].

Viroid any of a class of infectious agents consisting of a single-stranded closed circular RNA lacking a capsid. The RNA does not code for proteins and is not translated; it is replicated by host cell enzymes. Viroids are known to cause several plant diseases.

Virus any of a number of small, obligatory intracellular parasites with a single type of nucleic acid, either DNA or RNA and no cell wall. The nucleic acid is enclosed in a structure called a capsid, which is composed of repeating protein subunits called capsomeres, with or without a lipid envelope. The complete infectious virus particle, called a virion, must rely on the metabolism of the cell it infects. Viruses are morphologically heterogeneous, occurring as spherical, filamentous, polyhedral, or pleomorphic particles. They are classified by the host infected, the type of nucleic acid, the symmetry of the capsid, and the presence or absence of an envelope [1].

References

1. Churchill's Illustrated Medical Dictionary. New York: Churchill Livingstone, 1989.
2. Dulbecco R. Encyclopedia of Human Biology. San Diego: Academic Press, Inc., 1991.
3. Stedman's Medical Dictionary, 26th edn. Baltimore: Williams & Wilkins, 1995.
4. Dorland's Illustrated Medical Dictionary, 29th edn. Philadelphia: WB Saunders Co., 2000.
5. Casey JL, et al. Molecular biology of HDV: analysis of RNA editing and genotype variations. In: Viral Hepatitis and Liver Disease (Rizzetto M, Purcell RH, Gerin JL, Verme G, et al. editors). Turin: Edizioni Minerva Medica; 1997; pp. 290–294.
6. Centers for Disease Control and Prevention, Epidemiology and Prevention of Viral Hepatitis A to E: An Overview; 2000; http://www.cdc.gov/ncidod/diseases/hepatitis/slideset/index.htm.
7. Davies SE, et al. Evidence that hepatitis D virus needs hepatitis B virus to cause hepatocellular damage. Am J Clin Pathol 1992; 98(6): 554–558.
8. Dingle K, et al. Initiation of hepatitis Delta virus genome replication. J Virol 1998; 72(6): 4783–4788.
9. Glenn JS, et al. Identification of a prenylation site in Delta virus large antigen. Science 1992; 256: 1331–1333.
10. Hadziyannis SJ. Hepatitis delta: an overview. In: Viral Hepatitis and Liver Disease (Rizzetto M, Purcell RH, Gerin JL, Verme G, editors). Turin: Edizioni Minerva Medica; 1997; pp. 283–289.
11. Hadziyannis SJ. Review: hepatitis delta. J Gastroenterol Hepatol 1997; 12(4): 289–298.
12. Hwang SB, Lai MMC. Isoprenylation mediates direct protein–protein interactions between hepatitis large Delta antigen and hepatitis B virus surface antigen. J Virol 1993; 67(12): 7659–7662.

13. Lai MCC. The molecular biology of hepatitis Delta virus. Annu Rev Biochem 1995; 64: 259–286.
14. Lai MMC. Hepatitis Delta virus. In: Encyclopedia of Virology (Webster RG, Granoff A, editors). London: Academic Press Ltd; 1994; pp. 574–580.
15. Lau DT, et al. Resolutions of chronic Delta hepatitis after 12 years of interferon alfa therapy. Gastroenterology 1999; 117(5): 1229–1233.
16. Makino S, et al. Molecular cloning and sequencing of a human hepatitis delta (δ) virus RNA. Nature 1987; 329: 343–346.
17. Modahl LE, Lai MMC. Transcription of hepatitis Delta antigen mRNA continues throughout hepatitis Delta virus (HDV) replication: a new model of HDV RNA transcription and replication. J Virol 1998; 72(7): 5449–5456.
18. Monjardino JP, Saldanha JA. Delta hepatitis. The disease and the virus. Br Med Bull 1990; 46(2): 399–407.
19. Negro F. Animal models of hepatitis Delta virus infection. Viral Hepat Rev 1996; 2(3): 175–185.
20. Polson AG, Bass BL, Casey JL. RNA Editing of hepatitis delta virus antigenome by dsRNA-adenosine deaminase. Nature 1996; 380: 454–456.
21. Purcell RH, Gerin JL. Hepatitis Delta virus. In: Fields Virology (Fields BN, Knipe DM, Howley PM, editors), 3rd edn. Philadelphia: Lippincott-Raven; 1996; pp. 2819–2829.
22. Rizzetto M, et al. Immunofluorescence detection of a new antigen–antibody system (d/anti-d) associated to hepatitis B virus in liver and serum of HBsAg carriers. Gut 1977; 18: 997–1003.
23. Ryu W-S, Bayer M, Taylor J. Assembly of hepatitis Delta virus particles. J Virol 1992; 66(4): 2310–2315.
24. Sureau C, et al. Cloned hepatitis Delta virus cDNA is infectious in the chimpanzee. J Virol 1989; 63(10): 4292–4297.
25. Taylor JM. Hepatitis Delta virus and its replication. In: Fields Virology (Fields BN, Knipe DM, Howley PM, editors), 3rd edn. Philadelphia: Lippincott-Raven; 1996; pp. 2809–2818.
26. Walton J, Barondess JA, Lock S. The Oxford Medical Companion. Oxford: Oxford University Press, 1994.
27. Wang K-S, et al. Structure, sequence and expression of the hepatitis delta (delta) viral genome. Nature 1986; 323: 508–514.
28. Zuccola HJ, et al. Structural basis of the oligomerization of hepatitis delta antigen. Structure 1998; 6(7): 821–830.

Viral Hepatitis
I.K. Mushahwar (editor)
© 2004 Elsevier B.V. All rights reserved.

Hepatitis E virus: current status

George G. Schlauder
Core Research and Development, Abbott Diagnostics Division, Abbott Laboratories, Abbott Park, Illinois, USA

Hepatitis E virus (HEV) is the causative agent of what previously has been referred to as enterically transmitted non-A, non-B hepatitis or "waterborne hepatitis." HEV is the major cause of epidemic hepatitis and of acute, sporadic hepatitis in developing nations [1–4]. In general, symptoms include epigastric pain, nausea and vomiting. Jaundice, discoloration of the urine and a clay-colored appearance of the stools can also occur. The severity of the illness increases with age. The overall case fatality rate has been estimated to be between 1 and 3%. However, there is a dramatic increase in fatality associated with pregnancy where the rate has been reported to range between 15 and 25% during the third trimester. Although the exact role played by this virus in these deaths has not been determined, the cause of death is generally associated with a hemorrhagic episode. There are no known chronic sequelae for this disease and the majority of patients resolve infection without any deleterious effects. The perception that HEV is a disease of concern only to regions endemic for HEV or to travelers to such areas has been recently questioned. Novel isolates of the virus have been identified in a number of patients from regions considered non-endemic for HEV [5–18]. In addition, HEV-related sequences have been identified in swine from countries both non-endemic and endemic for HEV [19–29]. These recent findings indicate that HEV may be more widely distributed than originally believed and point to the potential of a zoonotic link.

Clinical and epidemiologic characteristics

The clinical characteristics of hepatitis E infection have been established in non-human primates, as well as in humans. A general clinical profile is shown in Fig. 1. In an experimental infection in cynomolgus macaques, the incubation period was short, ranging between 21 and 45 days, averaging 24 days. HEV RNA was first detected in the bile at day 6, in the feces at day 7, and in serum at day 9. RNA remained detectable up to 45 days in the bile and up to 23 days in the serum. The peak of alanine amino transferase (ALT) levels occurred at 27 days after infection. Biochemical evidence of infection was most pronounced when the titer of the inoculum was high. However, viral replication has been observed in the absence of ALT elevations when the titer of the inoculum was low [30].

The clinical course of infection in humans has been studied by infection of a human volunteer, who ingested a clarified stool suspension from an infected individual of an outbreak in India [31]. Anorexia, epigastric pain and discoloration of the urine,

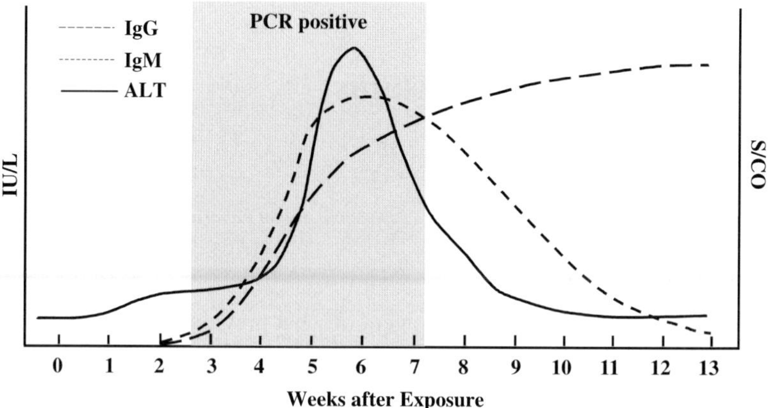

Fig. 1. Serial profile of an HEV infection. The graph shows a typical time course of markers of infection for HEV. The value of the ALT level, international units per liter (IU/l), is indicated by the solid line. The levels of the antibody responses, sample to cutoff (S/CO), to HEV are indicated by the short (IgM) and long (IgG) dashed lines. The shaded area indicates the period during which viral RNA can be detected by PCR in either the serum or stools.

characterized the ecteric phase of the infection. Elevations in ALT occurred from between 38 and 120 days with the maximum elevation occurring at 46 days after infection. HEV was detected in the stools between 38 and 46 days and in the serum between 22 and 46 days.

Documented outbreaks have occurred in Mexico and number of countries in Asia and Africa. Large outbreaks are usually associated with warm weather and poor sanitation leading to fecal contamination of drinking water. In areas where HEV is endemic, sporadic cases can occur, usually during periods between major outbreaks. Sporadic cases have also been reported in areas not considered to be endemic for HEV. These infections had been shown to be exclusively associated with travelers returning from visits to areas where HEV is endemic [32–34]. However, a few isolated cases of HEV infection had been reported in individuals with no history of travel to regions endemic for HEV, from Italy [35], Greece [36], and the Netherlands [37]. It was generally believed that such occurrences were rare in developed nations. However, seroepidemiologic studies indicate that IgG class antibodies to HEV proteins are detected in 1–2% of over 10,000 blood donors in the United States (US) and Western Europe [38]. Similar results have also been reported by other investigators [39,40]. These data indicate that HEV may be more widely distributed than originally reported. The observation of a significant seroprevalence in industrialized nations suggested that HEV may also be circulating in these regions. The identification of novel HEV isolates from patients with acute hepatitis who were from areas that were not considered endemic for HEV and who had not traveled to an HEV endemic region (described below in "Identification of novel isolates"), support the idea that the virus may be native to developed countries.

Characterization of the prototype virus

Prior to the discovery of hepatitis C virus (HCV) and HEV, diseases caused by these agents were classified as non-A, non-B hepatitis. Cases that were thought to have been transmitted enterically were sub-classified as enterically transmitted non-A, non-B hepatitis or ET-NANBH. Early studies involving electron microscopy identified virus-like particles in feces from patients with acute hepatitis.

Transmissibility of the agent was demonstrated in cynomolgus macaques and provided knowledge about viremia and the biophysical nature of the agent. Electron microscopic studies showed the virus to be non-enveloped and spherical in shape with visible surface indentations. The particle is approximately 27–34 nm in diameter. The mature complete virion bands at 1.29 g/cm^3 in potassium tartrate–glycerol density gradients. The sedimentation coefficient of the virus in sucrose is 183 s [41]. The agent was originally identified in bile from cynomolgus macaques experimentally infected with human feces obtained from patients infected during an outbreak in Burma [42]. The genome of that agent was completely characterized and designated "hepatitis E virus" [43]. It consists of a positive sense, single-stranded RNA molecule, which is polyadenylated. The genome is approximately 7.5 kb in length with three discontinuous open reading frames (ORFs) (Fig. 2). ORF 1 presumably encodes for a number of non-structural proteins. ORF 2 initiates downstream from the termination codon of ORF 1

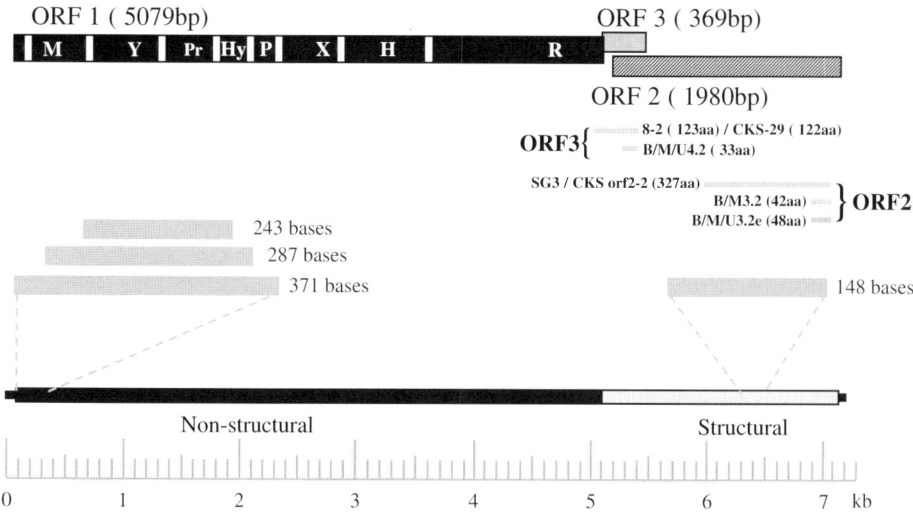

Fig. 2. Schematic representation of the HEV genome showing the relative positions of the ORF 1, ORF 2, and ORF 3, and the positions of the putative non-structural proteins (M, methyl transferase; Y, Y domain; Pr, protease; Hy, hypervariable region; P, proline-rich hinge domain; X, X domain; H, RNA helicase; R, RNA polymerase). The lines indicate the portions of the genome from which antigens have been generated for antibody assays. The bars at the bottom of the figure indicate the regions used in some genomic analyses.

and encodes for a putative capsid protein. Frequently, HEV-infected individuals produce antibodies that react with peptides or recombinant proteins derived from ORF 2. The third ORF partly overlaps both ORF 1 and ORF 2 and is also antigenic. The function of the protein encoded by ORF 3 is unknown. It has been reported that ORF 3 encodes a small phosphoprotein that associates with the cytoskeleton. Based on the overall morphology of the virus and the size and organization of the genome, the virus was tentatively classified as a member of the Caliciviridae [43–45]. However, its classification has recently been approved as an unassigned genus of "hepatitis E-like viruses" [46].

A molecular comparison of full-length HEV genomes

After the initial discovery and characterization of the original Burmese isolate of HEV, a second isolate was identified from an outbreak that had occurred in Mexico [44]. These first two prototype isolates of HEV were found to be similar in overall genomic organization but still distinct from each other having, 76% identity over the entire genome. Many of the nucleotide differences occur at the third position of the codon, such that the deduced amino acid similarities between the Burmese and Mexican strains of HEV are 83, 93 and 87%, for ORF 1, ORF 2, and ORF 3, respectively [42–44]. In the years immediately following the identification of the Burmese and Mexican prototype viruses, a number of isolates were identified from Pakistan, India and China, areas that are considered to be traditionally endemic for HEV. Interestingly, all of these isolates were found to be very closely related to the original isolate from Burma having greater than 93% nucleotide identity across the genomes [30,47]. However, the complete genome of a new isolate from China has recently been sequenced and is distinct from other Burmese-like isolates from China. This isolate, designated here as Ct1, is 74.8– 75.5% identical to the Burmese-like isolates and 74.5% identical to the Mexican isolate [48].

The first indication that acute hepatitis E infections in non-endemic areas may not be "imported," but result from an endogenous virus, was the identification of a patient from the US with acute hepatitis, who had no recent travel history to endemic regions [5,6]. Additional sequencing of the genome of this virus (US1) and the identification of a second isolate (US2) confirmed that a new genotype of HEV did exist which was distinct from previously identified isolates of HEV [7]. The two US isolates are 73.7–74.4% identical to the Burmese-like isolates, 74.5–74.7% identical to the Mexican isolate and 75.3–76.3% identical to the Chinese isolate, Ct1. Thus, four major genotypes have been proposed based on analysis of full-length sequences. Table 1 shows a comparison of some of the genomic characteristics of a number of isolates for which full-length sequence is available. The genomes range from 7170 to 7251 nucleotides in length (excluding the poly-A tail) with the US2 isolate being the longest of the reported HEV genomes [49].

Untranslated regions

The lengths of the poly-A tail vary from 3 to 26 nucleotides and no poly-A tail has been isolated for the Chinese isolate, Ct1. The US2 isolate has the longest detected poly-A tail

Table 1

Comparison of human HEV genomes

| Genotype | I | II | III | III | III | IV | IV |
Region	Burmese	Mexican	US1	US2	Japan, JRA1	Chinese, Ct1	Japan, JI4
Genome size[a]	7194 nt[b]	7170 nt	7186 nt	7251 nt	7227 nt	7232 nt	7171 nt
Poly-A tail	13	10	16	26	3	–	15
5$'$-utr[c]	27 nt	3 nt	–	35 nt	26 nt	25 nt	25 nt
ORF 1	5079 nt	5073 nt	5094 nt	5124 nt	5109 nt	5121 nt	5052 nt
	1693 aa[d]	1691 aa	1698 aa	1708 aa	1703 nt	1707 aa	1684 aa
ORF2	1980 nt	1977 nt	1980 nt	1980 nt	1980 nt	2018 nt	2013 nt
	660 aa	659 aa	660 aa	660 aa	660 aa	672 aa	671 aa
ORF3	372 nt	372 nt	366 nt	366 nt	366 nt	336 nt	342 nt
	123 aa	123 aa	122 aa	122 aa	122 aa	112 aa	114 aa
3$'$-utr	65 nt	74 nt	72 nt	72 nt	72 nt	68 nt	70 nt

[a]Excludes poly-A tail.
[b]Nucleotides.
[c]Untranslated region.
[d]Amino acids.

and the longest 5$'$-untranslated region (5$'$-utr). Lengths of identified 5$'$-utr vary from 3 to 35 nucleotides. The 5$'$-utr has not been isolated from the US1 isolate. The overlapping region of the 5$'$-utr is highly conserved with only three variable nucleotide positions between US2, B1 and Ct1. The 3 nucleotides prior to the start codon of the Mexican isolate are also conserved with these sequences. The initiating methionine for ORF 1 of the US1 strain has not been isolated, however, similarity to the US2 isolate suggests the absence of 9 amino acids at the amino terminus [49].

The 3$'$-utr sequences are less conserved than the 5$'$-utr with a number of insertions and deletions occurring about 20 nucleotides downstream of the termination codon of ORF 2. There is an insertion of 6 nucleotides within the 3$'$-utr of both of the US isolates when aligned to the Burmese isolate. The Mexican isolate has an additional 9 nucleotides and the Chinese Ct1 isolate, an additional 3 nucleotides, when compared to B1 at the same relative position. The identity between Burmese-like isolates is greater than 90% over this region. Between the US isolates there is an 87% nucleotide identity. The identities between the Burmese-like Mexican and Chinese isolates range between 75 and 77%. The largest difference in 3$'$-utr sequence is between the Ct1 isolate and the US isolates with identities of 72 and 69%.

Open reading frames

The length of the complete HEV ORF 1 polyprotein from different isolates ranges between 1691 and 1708 amino acids. The variability between isolates is due to nucleotide insertions within the hypervariable/proline-rich hinge region located between nucleotide positions 2146 and 2331 of the Burmese B1 isolate. Within this region, the US isolates contain a unique poly-C tract not found in HEV isolates from other regions.

The nucleotide or amino acid identity between geographically distributed isolates is very low within the hypervariable region. Across this region, the US isolates are 85% identical to each other at the nucleotide level and only 31–42% identical to the Burmese-like, M1 and Ct1 isolates. Similarly, most Burmese-like isolates possess 88–99% identity to each other and only 33–38% identity to M1. Within the ORF 1 polyprotein of the Burmese isolate, motifs had been identified suggesting protein function similar to many other positive-strand RNA viruses [50–52]. Within the ORF 1 sequence from all of these isolates, conserved methyltransferase, helicase, RNA-dependent RNA polymerase, Y domain, X domain, and a putative papain-like protease have been identified. Thus all isolates of HEV retain similar domain organization to rubella virus (RubV) and beet necrotic yellow vein virus (BNYVV). Based on these observations it has been suggested that these HEV RubV and BNYVV constitute a distinct monophyletic group within the alpha-like supergroup of positive-strand viruses [52].

The most highly conserved ORF protein is ORF 2, the putative capsid protein. The Burmese-like, US, and Japanese Jra1 isolates' derived proteins are 660 amino acids in length, while the Mexican protein is 659 amino acids. A deletion in the Mexican isolate occurs at amino acid position 331 of the Burmese, US, and Japanese Jra1 isolates. The Chinese, Ct1 isolate is reported to have additional 14 codons in the $5'$-end of ORF 2, which overlaps ORF 1 by 1 base [48]. An additional 11 codons also found in the Japanese isolate, JI4, in this region [18]. Amino acid alignments indicate that 14 amino acid residues extend out from the position equivalent to the start methionine of the Burmese, Mexican and US isolates. In other isolates of HEV, ORF 2 begins 41 bases downstream of ORF1. However, the ORF1 and ORF2 of Ct1 overlap by 1 base and in Jra1 ORF2 begins 9 bases downstream from ORF1. This variation in Ct1 and Jra1 is a result of a single nucleotide substitution (U) at positions 5159 and 5090, respectively. Amino acid identities between the two human US isolates or between the Burmese-like isolates are greater than 98%. In contrast, amino acid identities between isolates from different genotypic groups range between 90.1 and 93.3%. A hydrophobic signal peptide sequence has been identified at the extreme amino terminus of the protein, with potential cleavage that is conserved between isolates [48,49,53]. The immunodominant epitope located at the carboxyl end of the protein (designated 3-2) is also highly conserved between the US, Burmese-like, Mexican, and Chinese isolates, as is the overall hydrophobicity profile.

As a whole the ORF 3 protein is the least conserved ORF product between isolates of HEV. The amino acid identity between the two US isolates is 95.9%. This is slightly more variable than the amino acid identity observed between Burmese-like ORF 3 products (96.8–100%). The US ORF 3 proteins are 84 and 80% identical to the Burmese-like and M1 proteins, respectively. The Burmese-like ORF 3 products are more similar to the M1 ORF 3, with 85.4–87% identity. The Chinese Ct1 isolate is 74–80% identical to the Burmese, Mexican and US isolates. Both of the US isolates and the Japanese isolate, JRA1, have a unique three-nucleotide deletion just downstream of the ORF 3 start codon producing a protein of 122 amino acids vs. the 123 amino acid product identified in the Burmese-like and Mexican isolates. The Chinese isolate has a unique six-nucleotide deletion further downstream. In addition, the Chinese isolate is 27 nucleotides shorter at the $5'$-end of the ORF. This is due to the nucleotide insertion at position 5160, which results in an initiation codon further downstream than occurs in the other isolates of HEV [48]. The Japanese isolate also has an insertion of a single nucleotide at position 5090 [18]. These deletions result in a much shorter protein of 112 amino acids for the Chinese, Ct1 and Japanese, isolates. The hydrophobicity profiles are very similar between the US, Burmese-like, and Mexican isolates indicating two hydrophobic domains at the amino half of the protein [48,49].

Identification of novel isolates

The discovery of a novel strain of HEV from the US, an area not considered endemic for HEV, lead to the identification of a number of isolates from other non-endemic regions. Primers based on the sequence from the original US isolate of HEV were utilized to identify a second patient from the US that was infected with a very similar isolate [7]. Serum from this patient was used as an inoculum in a cynomolgus macaque and shown to be infectious in this animal model for HEV [49]. In addition to the characteristic elevations in ALT, HEV RNA was detected in both the serum and stools from the infected animal during the acute phase of the disease and both an IgM and IgG antibody response to HEV was observed (Fig. 3). These findings indicate that although the US isolate was quite different from the Mexican and Burmese isolates, experimental infection results in a response similar to what was observed for other isolates of HEV.

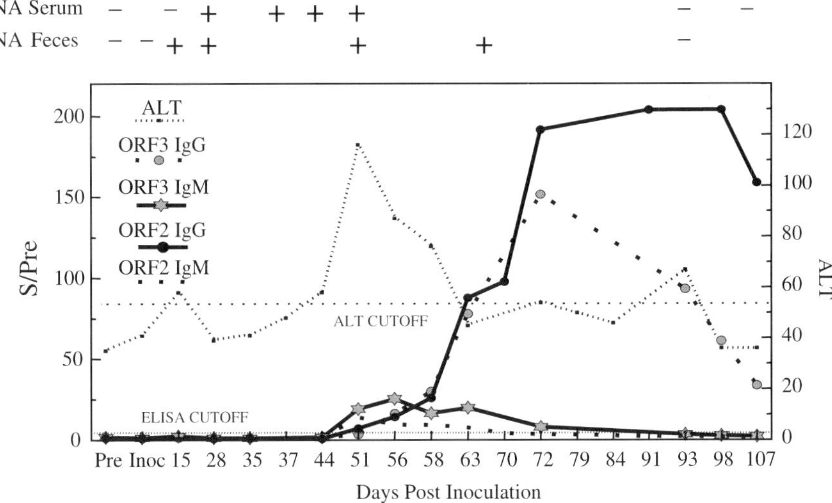

Fig. 3. Serial profile of an experimental HEV infection with strain US2 into a cynomolgus macaque. The value of the ALT level is indicated by the dotted line. The levels of the IgM antibody responses to the ORF 2 and ORF 3 HEV antigens are indicated by the dashed line and the starred-solid line, respectively. The levels of the IgG antibody response to the ORF 2 and ORF 3 HEV antigens are indicated by the dotted-solid line and the dotted-dashed line, respectively. Samples that were tested for HEV RNA by RT-PCR and the results obtained are shown above the graph.

Primers based on the US sequence also were utilized to identify patients from Italy and Greece, both of whom were infected with unique variants of HEV. Both isolates are quite distinct from the original isolates from Burma and Mexico, as well as those from the US [8,9,36]. Again, neither of these patients reported any travel to areas considered endemic for HEV. The methodology in these studies used amplification conditions known as touchdown PCR that can potentially compensate for diversity in target sequence [54]. The annealing temperature is gradually decreased after every cycle of amplification to a minimum of 40°C, enhancing the amplification of sequences of minimal mismatch relative to those with a greater degree of variation. Thus, amplification of variants can occur with a minimal amplification of non-specific background.

A second patient from Greece, who was also classified as a "non-traveler to HEV endemic regions" and who had previously been reported PCR positive for HEV RNA, was confirmed PCR positive with the US primers and also found to be infected with a unique isolate of HEV [7,55]. These results were confirmed utilizing a new set of degenerate PCR primers that incorporated sequence from a number of isolates from Asia, the Mexican isolate and the US isolate, and have been found to be useful in identifying a wide range of HEV isolates [56]. Even though the European isolates are most closely related to the US isolate, they are not subtypes of the US isolates but represent three new groups of HEV. Prior to these findings, there appeared to be a geographical distribution of genotypes. The identification of the two distinct isolates from Greece indicates that there could be significant diversity between isolates from the same region.

Recently, two isolates from patients from Argentina have been identified using the degenerate primers amplifying a portion of the genome from ORF 1 and ORF 2 [11]. As was the case in the patients from Europe and the US, neither patient reported any recent travel to areas classically considered endemic for HEV. In Argentina the seroprevalence rate for HEV in blood donors is similar to the seroprevalence in other industrialized nations, ranging from 1 to 2% [57]. These two isolates are the first isolates of HEV to be identified in South America and are most closely related to each other. The Argentine isolates are also distinct from the most geographically related isolate from Mexico, as well as all other isolates from both endemic and non-endemic regions that have been identified to date.

A recent report from Austria of a patient with acute hepatitis with no apparent travel history to endemic regions indicated a serologic profile suggestive of hepatitis E [58]. Sequencing of PCR products obtained from acute phase serum from this patient showed that this was not an endemic strain of HEV [12]. Interestingly, the isolate is more closely related to the two variants isolated from the Argentine patients than to the other European isolates [59].

The awareness of the potential link between HEV and acute hepatitis cases in non-endemic areas has resulted in the evaluation of many acute hepatitis cases of unknown etiology worldwide. Sequences of HEV were recently identified in serum from a patient from the United Kingdom (UK) who presented with acute hepatitis and who had not traveled outside of the UK for 4 years [60]. The greatest number of cases has recently been identified in patients from Japan. These are from both recent presentations of the disease as well as cases that have been reinvestigated as far back as 1982 [13–18].

Interestingly, although all patients were from Japan, these isolates are distributed over three different genotypes (see "Genotypic distribution").

Evidence of a more extensive diversity between HEV isolates from endemic areas was first suggested from the identification of a short sequence from two patients with acute hepatitis from China [61]. Additional isolates have been recently identified from China and Taiwan, which are also distinct from other Chinese isolates of HEV that are closely related to the original genotype I Burmese isolate [48,62–64]. All of these novel isolates appear to be most closely related to the Chinese genotype IV isolate Ct1.

To date, only one strain of HEV was isolated from the Mexican outbreaks of HEV. However, several isolates from four patients from Nigeria, have been found to be most closely related to the Mexican isolate [65]. As was observed for the Argentine, Austrian and Japanese isolates, genetic relatedness may not be an indicator of geographical distribution.

Genotypic distribution

Pairwise sequence comparisons and phylogenetic analysis have been used to determine the genetic relationships between various isolates of HEV. The preferred method for genetic comparison utilizes full-length sequence. However, limitations due to the amounts of patient sera available for analyses have necessitated genetic comparisons based on less than full-length sequence from different ORFs of the viral genome. The regions of the genome that have been used in several recent analyses, are depicted in Fig. 1 [8,9,11,12,22,59,64,66]. The nucleotide positions are based on the numbering of the original Burmese isolate [43]. The largest region from ORF 1 that was available for comparison of the novel isolates from non-endemic regions is the 371 base segment, spanning nucleotides 80–450 (Fig. 4). Several smaller internal segments have also been used so that published sequence from additional isolates, for which less sequence is available, could be included in order to maximize the number of isolates in the analyses. Figure 5 shows an analysis from a 287 base segment for this region. The region from ORF 2 that has been used for analyses spans a distance of 148 bases between nucleotide positions 6322 and 6469. Smaller internal segments of this region have also been compared to maximize the use of published sequences.

Pairwise comparisons

Pairwise comparisons are utilized to determine the difference, similarity, identity and genetic distance between sequences of different isolates. Viewed as a matrix, such pairwise comparisons have been useful in grouping together isolates with similar values. The nucleotide or amino acid identity is the number of nucleotides or amino acid residues that are identical over a given length of sequence and is typically presented as a percentage of the length of sequence compared. The genetic distance between two isolates also takes into account the number of nucleotide substitutions between two sequences, with the addition of corrections for multiple substitutions at a site. The results are then expressed as the number of substitutions per base.

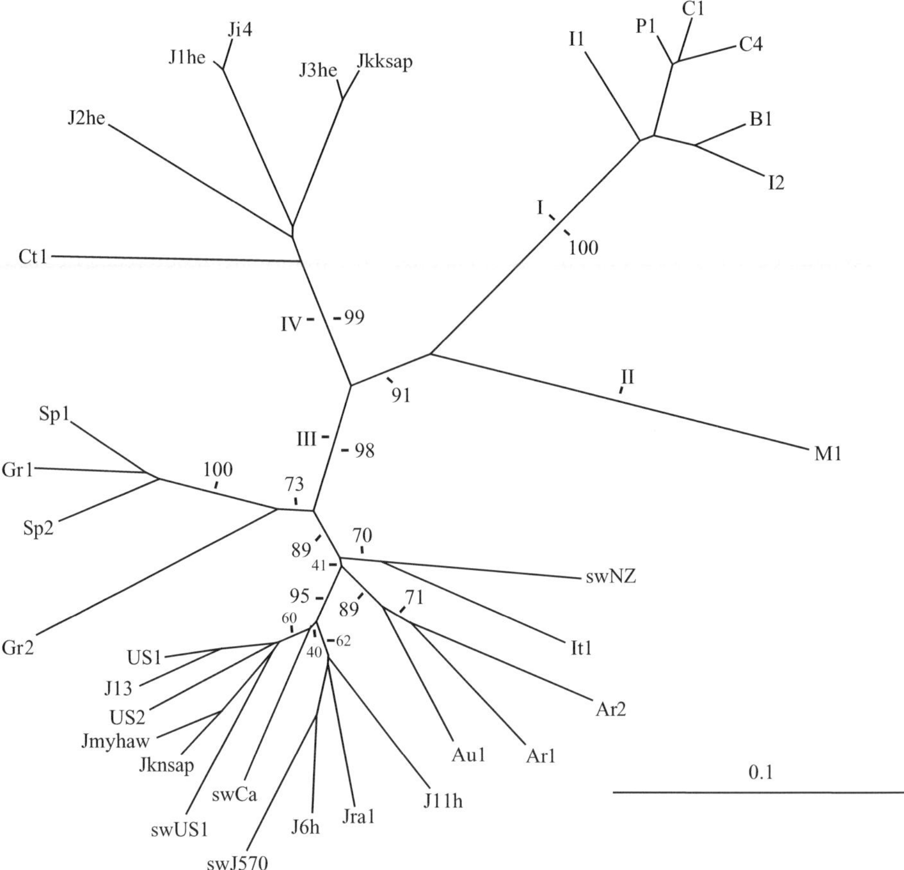

Fig. 4. Unrooted phylogenetic tree depicting the relationship of nucleotide sequences over a 371 base fragment of ORF 1. Branch lengths are proportional to the evolutionary relationship between sequences. The scale representing nucleotide substitutions per position is shown. The internal node numbers indicate the bootstrap values as a percentage of trees obtained from 100 replicates. Human isolates represented are from Burma (B1), China (C1, C4, Ct1), Pakistan, (P1), India (I1, I2), Mexico (M1), United States (US1, US2), Greece (Gr1, Gr2), Spain (Sp1, Sp2), Italy (It1), Argentina (Ar1, Ar2), Austria (Au1), and Japan (Jmyhaw, Jknsap, Jra1, J6h, J11h, JI3, Jkksap, J1he, J2he, J3he). Swine isolates represented are from the United States (swUS1), New Zealand (swNZ1), Canada (swCa), and Japan (swJ570).

The nucleotide identities of full-length sequences and the amino acid identities for the 3 ORFs from a number of HEV isolates representative of different genotypes of HEV are shown in Table 2. The nucleotide and deduced amino acid identities from shorter segments from ORF 1 and ORF 2 for which sequence is available from a number of additional isolates are shown in Tables 3 and 4. The shaded and boxed areas indicate how the different isolates group based on similar nucleotide and amino acid identities between isolates. A discrete heterogeneity again becomes apparent with the longer ORF 1

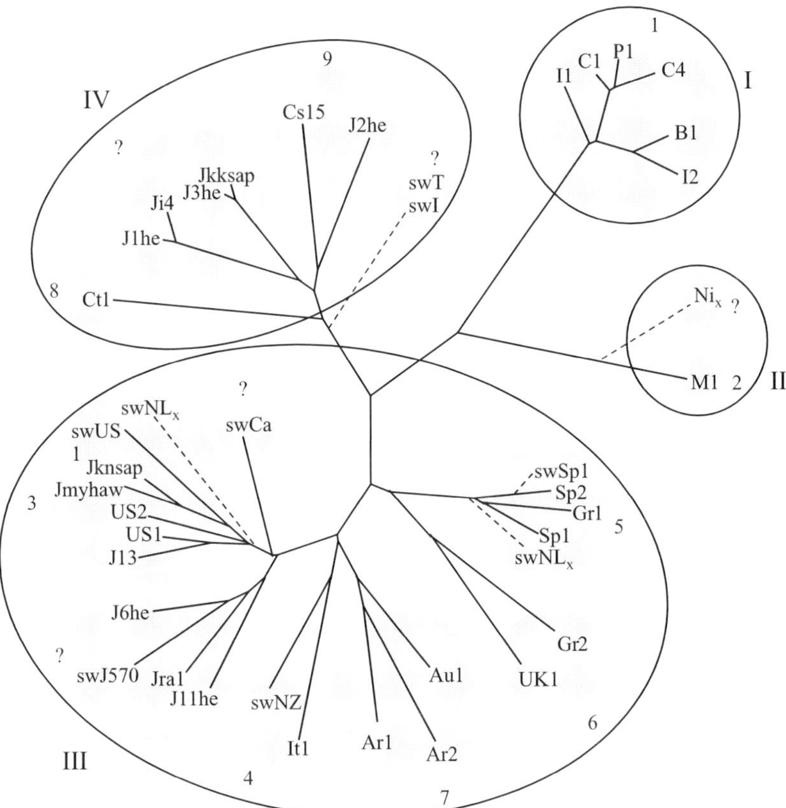

Fig. 5. Unrooted phylogenetic tree depicting the relationship of nucleotide sequences over a 287 base fragment of ORF 1. Human isolates represented are from Burma (B1), China (C1, C4, Ctl, Cs15), Pakistan, (P1), India (I1, I2), Mexico (M1), United States (US1, US2), Greece (Gr1, Gr2), Spain (Sp1, Sp2), Italy (It1), Argentina (Ar1, Ar2), Austria (Au1), Japan (Jmyhaw, Jknsap, Jra1, J6h, J11h, JI3, Jkksap, J1he, J2he, J3he), and United Kingdom (UK1). Swine isolates represented are from the United States (swUS1), New Zealand (swNZ1), Canada (swCa), and Japan (swJ570). The Roman numeral adjacent to the open eclipses and the Arabic number adjacent to the shaded circles indicate potential genotypic designations. The dashed lines indicate the putative position based on similarities of other analyses, of additional human isolates from Nigeria (Nix), swine isolates from Taiwan (swT) and India (swI), swine sewage from Spain (swSp), and additional swine isolates from the Netherlands.

fragment. The isolates can be grouped into 4 major groups (boxed areas in Tables 3 and 4), with isolates from different groups having identities of less than 81%. It has been proposed that phylogenetic groups could be defined as having divergence of greater than 15% [67]. Based on this definition, the Burmese-like isolates form group I; the Mexican isolate represents group II; group III includes the US, European, and South American isolates; and Group IV is represented by the Chinese isolate. An alternative view of the distribution could define a genotype as having an identity of less than 88% with another

210

Table 2

Pairwise identities of the amino acid sequences of ORF1, ORF2, and ORF3 and the nucleotide sequences of the full-length genomes[a,b]

	US1	US2	B1	B2	I2	C1	C2	C3	P1	I1	M1	CT1
	Amino acid identity across the three open reading frames (%)											
US1-ORF1		97.7	85.6	85.4	83.9	85.4	85.7	84.4	85.5	85.7	84.8	88.5
ORF2		98.3	93.3	93.0	93.0	93.5	93.2	92.9	93.2	92.6	91.5	91.9
ORF3		95.9	85.3	84.4	85.3	85.3	83.6	85.3	85.3	85.3	79.5	83.3
US2-ORF1	92.0		86.0	85.9	84.4	85.9	86.1	84.9	86.1	86.2	85.5	88.7
ORF2			93.3	93.0	93.3	93.3	93.3	93.0	93.3	92.7	91.7	93.0
ORF3			85.3	84.4	85.3	85.3	83.6	83.6	85.3	85.3	81.2	79.6
B1-ORF1	73.9	74.0		98.7	96.8	98.4	98.6	97.1	98.5	98.2	86.9	86.5
ORF2				98.9	99.1	99.9	99.2	99.2	99.6	98.9	94.8	92.1
ORF3				98.4	100.0	100.0	98.4	98.4	100.0	98.4	87.0	74.3
C1-ORF1	74.2	74.3	93.5	93.0	92.2		99.5	97.9	99.4	98.2	86.8	86.7
ORF2							99.4	99.1	99.7	99.1	95.0	92.4
ORF3							98.4	98.4	100.0	98.4	87.0	77.8
P1-ORF1	74.1	74.1	93.6	92.8	92.0	98.8	98.3	96.7		98.4	86.9	86.5
ORF2										99.1	95.0	92.1
ORF3										98.4	87.0	76.9
I1-ORF1	74.4	74.4	93.5	93.0	92.0	94.0	97.9	93.5	93.9		87.2	86.7
ORF2											94.7	91.9
ORF3											88.6	76.9
M1-ORF1	74.7	74.7	75.9	75.7	75.9	75.9	75.9	75.7	76.1	75.7		85.0
ORF2												90.1
ORF3												75.0
CT1-ORF1	75.3	76.3	75.5	75.5	74.8	75.3	75.1	75.2	75.2	74.8	74.5	
	Nucleotide identity (%) across the full-length genome											

[a]Erker et al. (1999); Wang et al. (2000).
[b]Wang et al. (2000).

group, only 3% less divergent than the previously proposed definition of phylogenetic group [67]. Based on this definition, the genotypic distribution would yield 8 major groups (shaded areas in Tables 3 and 4): the Burmese-like, Mexican, US, Chinese, Italian, Greek1, Greek2, and the Argentine/Austrian isolates [11,59].

The genetic distances based upon alignments of the 371 base overlapping region from ORF 1, are shown in Table 5. Examination of the distances between the different isolates of HEV demonstrates that there is considerable evolutionary distance between isolates from the proposed four major groups of HEV that were defined by full-length analyses. The distance between any two isolates from different groups, range between 0.2334 and 0.3392. The distances calculated also show the close relationship between isolates within a group. The distances within the group containing the original Burmese isolate range between 0.0136 and 0.0907. The distance between US isolates range between 0.0849 and 0.1091. Similar genetic distances are also observed between the Argentine and Austrian

Table 3

Nucleotide and deduced amino acid identity between isolates of HEV over 371 base (123 amino acid) ORF 1 fragment

Nucleotide Identity

Ar1	88.4	89.8	84.4	81.9	85.4	86.5	85.7	85.2	82.5	76.6	76.6	79.0	78.2	79.2	77.1	78.4	76.8	78.7	77.6	77.6
98.4	**Ar2**	87.9	80.6	81.1	84.4	84.6	84.9	85.4	83.3	76.0	76.6	74.9	75.7	75.7	74.1	77.4	76.8	75.7	76.3	76.0
100.0	98.4	**Au1**	85.2	81.1	86.0	87.1	87.9	87.1	84.6	77.1	77.6	76.8	76.6	77.6	76.6	77.4	78.2	77.1	78.4	78.2
99.1	97.6	99.1	**G1**	84.4	84.1	84.9	81.9	82.5	83.0	78.4	77.9	77.6	78.2	77.9	76.6	77.6	78.2	77.4	79.0	76.6
98.4	96.7	98.4	99.2	**G2**	81.7	83.8	83.8	83.0	81.9	78.2	77.6	78.2	77.9	78.4	77.1	77.9	77.6	78.7	78.7	76.8
97.6	95.9	97.6	96.7	96.7	**It1**	88.1	84.6	86.8	84.9	77.6	77.6	77.1	77.4	77.4	76.3	77.6	77.4	77.6	76.3	78.4
98.4	96.7	98.4	97.6	96.7	97.6	**swNZ1**	86.3	84.1	83.8	78.4	77.4	77.4	77.6	78.6	76.6	79.0	77.6	77.9	78.4	76.8
99.2	97.6	99.2	98.4	97.6	96.7	99.2	**US1**	91.9	90.8	75.5	74.9	75.2	75.2	75.7	75.2	76.6	75.5	76.0	78.7	76.6
100.0	98.4	100.0	99.2	98.4	97.6	98.4	99.2	**US2**	89.9	75.2	75.4	75.2	75.4	76.0	74.9	75.7	75.7	76.3	78.7	77.6
97.6	95.9	97.6	96.7	95.9	95.1	95.9	96.7	97.6	**swUS1**	76.6	76.6	74.7	74.9	74.4	73.6	75.2	74.9	75.2	79.0	75.7
91.1	90.2	91.1	90.2	90.2	92.7	92.7	91.9	91.1	88.6	**B1**	98.7	94.6	94.6	94.9	94.3	94.3	96.0	94.6	79.0	79.0
91.1	90.2	91.1	90.2	90.2	92.7	91.1	90.2	91.1	89.6	98.4	**B2**	93.8	93.8	94.1	93.5	93.5	95.1	93.8	78.4	78.4
89.4	88.6	89.4	88.6	88.6	91.1	91.1	90.2	89.4	87.0	98.4	96.7	**C1**	97.8	98.1	96.8	92.7	91.6	97.3	76.6	79.8
90.2	89.4	90.2	89.4	89.4	91.9	91.9	91.1	90.2	87.8	99.2	97.6	99.2	**C2**	98.7	97.6	93.8	91.6	98.4	77.1	79.5
90.2	89.4	90.2	89.4	89.4	91.9	91.9	91.1	90.2	87.8	99.2	97.6	99.2	100.0	**C3**	97.3	93.5	91.9	98.1	77.6	79.5
89.4	88.6	89.4	88.6	88.6	91.9	91.1	90.2	89.4	87.0	98.4	96.7	99.2	99.2	99.2	**C4**	93.0	91.9	97.0	77.4	78.7
90.2	89.4	90.2	89.4	89.4	91.9	91.9	91.1	90.2	87.8	99.2	97.6	97.6	98.4	98.4	97.6	**I1**	91.4	93.3	77.9	79.2
88.6	87.8	88.6	87.8	87.8	90.2	90.2	89.4	88.6	86.2	97.6	95.9	95.9	96.7	96.7	95.9	96.7	**I2**	91.6	76.8	78.4
90.2	89.4	90.2	89.4	89.4	91.9	91.9	91.1	90.2	87.8	99.2	97.6	99.2	100.0	100.0	99.2	98.4	96.7	**P1**	77.1	78.7
95.1	93.5	95.1	94.3	94.3	95.1	94.3	94.3	95.1	92.7	89.4	90.2	87.8	88.6	88.6	87.8	88.6	87.0	88.6	**Ct1**	78.7
92.7	91.1	92.7	91.9	91.9	94.3	94.3	93.5	92.7	90.2	95.9	95.1	94.3	95.1	95.1	94.3	95.1	93.5	95.1	91.9	**M1**

Amino Acid Identity

Table 4

Nucleotide and deduced amino acid identity between isolates of HEV over 148 base (49 amino acid)[a] ORF 2 fragment

Nucleotide Identity

Ar1	Ar2	Au1	G1	G2	It1	swNZ1	US1	US2	swUS1	B1	B2	C1	C2	C3	C4	I1	I2	P1	Iakl	Cht3	Eg93	Eg94	Ct1	M1
Ar1	91.8	87.8	81.6	82.7	83.7	89.8	80.6	82.7	87.8	80.6	80.6	80.6	80.6	80.6	80.0	82.7	79.6	80.6	82.7	79.6	80.6	79.6	76.5	80.6
100	**Ar2**	88.5	83.8	86.5	87.2	92.3	85.8	85.1	90.5	85.8	85.1	83.8	85.1	85.1	84.5	85.1	85.1	86.5	84.5	81.8	82.4	81.1	79.7	82.4
100	100	**Au1**	83.1	88.5	89.2	88.0	83.1	85.8	87.8	85.1	83.8	83.1	83.8	83.8	83.1	84.5	83.1	82.4	85.1	83.1	81.1	82.4	81.8	79.1
100	100	100	**G1**	87.2	87.8	85.9	84.5	85.1	85.1	84.5	82.4	82.4	83.1	83.1	82.4	83.1	82.4	83.8	84.5	83.8	83.8	83.1	81.8	81.1
100	100	100	100	**G2**	83.1	85.2	82.4	85.1	87.8	85.1	84.5	82.4	83.8	83.8	83.8	83.8	83.1	84.5	83.8	85.1	81.0	83.8	81.1	79.1
100	100	100	100	100	**It1**	88.7	87.8	85.8	85.8	83.8	83.1	82.4	83.1	83.1	82.4	83.8	81.8	82.4	83.8	81.8	82.4	82.4	80.4	79.7
96.9	97.9	97.9	97.9	97.9	97.9	**swNZ1**	86.6	86.6	90.8	85.2	84.5	83.8	84.5	84.5	83.8	85.9	84.5	85.2	85.2	81.0	83.1	81.7	79.6	82.4
96.9	98.0	98.0	98.0	98.0	98.0	95.7	**US1**	93.9	90.5	79.0	78.4	76.4	77.0	77.0	76.4	77.0	78.4	77.7	79.1	78.4	77.0	78.4	79.1	79.7
96.9	98.0	98.0	98.0	98.0	98.0	95.7	100	**US2**	91.2	82.4	80.4	79.7	80.4	80.4	79.3	80.4	81.8	81.1	82.4	83.1	79.0	81.8	81.1	81.8
96.9	98.0	98.0	98.0	98.0	98.0	95.7	100	100	**swUS1**	83.8	84.5	82.4	83.1	83.1	82.4	83.1	83.1	83.8	83.8	84.5	81.8	83.1	79.1	83.8
96.9	98.0	98.0	98.0	98.0	98.0	95.7	95.9	95.9	95.9	**B1**	98.0	94.6	95.3	95.3	94.6	96.6	97.3	93.9	98.6	89.2	89.9	89.9	81.1	82.4
96.9	95.9	95.9	95.9	95.9	95.9	93.6	93.9	93.9	93.9	98.0	**B2**	93.9	94.6	94.6	93.9	95.9	95.3	93.2	96.6	88.5	89.2	89.2	79.7	80.4
96.9	98.0	98.0	98.0	98.0	98.0	95.7	95.9	95.9	95.9	100	98.0	**C1**	98.0	98.0	96.6	96.6	91.9	96.6	95.3	91.2	93.2	93.2	81.1	81.8
96.9	98.0	98.0	98.0	98.0	98.0	95.7	95.9	95.9	95.9	100	98.0	100	**C2**	100	98.6	97.3	92.6	98.6	95.3	93.2	93.2	93.2	80.4	82.4
96.9	98.0	98.0	98.0	98.0	98.0	95.7	95.9	95.9	95.9	100	98.0	100	100	**C3**	98.6	97.3	92.6	98.6	95.3	93.2	93.2	93.2	80.4	82.4
96.9	98.0	98.0	98.0	98.0	98.0	95.7	95.9	95.9	95.9	100	98.0	100	100	100	**C4**	96.6	91.9	97.3	94.6	91.9	93.2	93.2	79.7	81.8
96.9	98.0	98.0	98.0	98.0	98.0	95.7	95.9	95.9	95.9	100	98.0	100	100	100	100	**I1**	93.9	95.9	96.6	91.2	91.9	91.9	80.4	83.8
93.8	95.9	95.9	95.9	95.9	95.9	93.6	93.9	93.9	93.9	98.0	95.9	98.0	98.0	98.0	98.0	98.0	**I2**	91.2	95.9	87.8	88.5	88.5	80.4	83.8
96.9	98.0	98.0	98.0	98.0	98.0	95.7	95.9	95.9	95.9	100	98.0	100	100	100	100	100	98.0	**P1**	93.9	93.2	93.2	93.2	81.1	83.1
96.9	98.0	98.0	98.0	98.0	98.0	95.7	95.9	95.9	95.9	100	98.0	100	100	100	100	100	98.0	100	**Iakl**	89.2	91.2	91.2	82.4	83.1
96.9	98.0	98.0	98.0	98.0	98.0	95.7	95.9	95.9	95.9	100	98.0	100	100	100	100	100	98.0	100	100	**Cht3**	93.2	95.9	80.4	84.5
96.9	98.0	98.0	98.0	98.0	98.0	95.7	95.9	95.9	95.9	100	98.0	100	100	100	100	100	98.0	100	100	100	**Eg93**	97.3	79.7	81.8
96.9	98.0	98.0	98.0	98.0	98.0	95.7	95.9	95.9	95.9	100	98.0	100	100	100	100	100	98.0	100	100	100	100	**Eg94**	78.4	81.8
93.8	93.9	93.9	93.9	93.9	93.9	91.5	91.8	91.8	91.8	95.9	93.9	95.9	95.9	95.9	95.9	95.9	93.9	95.9	95.9	95.9	95.9	95.9	**Ct1**	81.8
96.9	98.0	98.0	98.0	98.0	98.0	95.7	95.9	95.9	95.9	95.9	93.9	95.9	95.9	95.9	95.9	95.9	93.9	95.9	95.9	95.9	95.9	95.9	95.9	**M1**

Amino Acid Identity

[a]Over 98 base (32 amino acid) fragment for Ar1 and 142 base (47 amino acid) fragment for swNZ1.

Table 5

Genetic distances between isolates of HEV over 371 base region of ORF 1

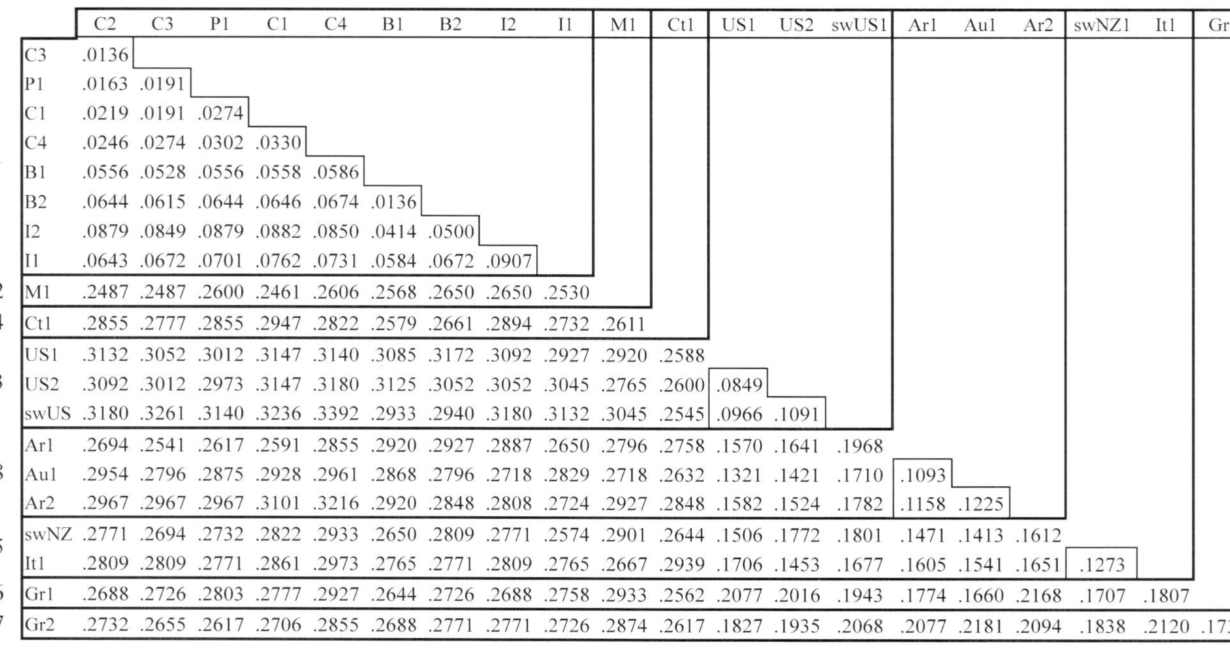

		C2	C3	P1	C1	C4	B1	B2	I2	I1	M1	Ct1	US1	US2	swUS1	Ar1	Au1	Ar2	swNZ1	It1	Gr1
1	C3	.0136																			
	P1	.0163	.0191																		
	C1	.0219	.0191	.0274																	
	C4	.0246	.0274	.0302	.0330																
	B1	.0556	.0528	.0556	.0558	.0586															
	B2	.0644	.0615	.0644	.0646	.0674	.0136														
	I2	.0879	.0849	.0879	.0882	.0850	.0414	.0500													
	I1	.0643	.0672	.0701	.0762	.0731	.0584	.0672	.0907												
2	M1	.2487	.2487	.2600	.2461	.2606	.2568	.2650	.2650	.2530											
4	Ct1	.2855	.2777	.2855	.2947	.2822	.2579	.2661	.2894	.2732	.2611										
3	US1	.3132	.3052	.3012	.3147	.3140	.3085	.3172	.3092	.2927	.2920	.2588									
	US2	.3092	.3012	.2973	.3147	.3180	.3125	.3052	.3052	.3045	.2765	.2600	.0849								
	swUS	.3180	.3261	.3140	.3236	.3392	.2933	.2940	.3180	.3132	.3045	.2545	.0966	.1091							
8	Ar1	.2694	.2541	.2617	.2591	.2855	.2920	.2927	.2887	.2650	.2796	.2758	.1570	.1641	.1968						
	Au1	.2954	.2796	.2875	.2928	.2961	.2868	.2796	.2718	.2829	.2718	.2632	.1321	.1421	.1710	.1093					
	Ar2	.2967	.2967	.2967	.3101	.3216	.2920	.2848	.2808	.2724	.2927	.2848	.1582	.1524	.1782	.1158	.1225				
5	swNZ	.2771	.2694	.2732	.2822	.2933	.2650	.2809	.2771	.2574	.2901	.2644	.1506	.1772	.1801	.1471	.1413	.1612			
	It1	.2809	.2809	.2771	.2861	.2973	.2765	.2771	.2809	.2765	.2667	.2939	.1706	.1453	.1677	.1605	.1541	.1651	.1273		
6	Gr1	.2688	.2726	.2803	.2777	.2927	.2644	.2726	.2688	.2758	.2933	.2562	.2077	.2016	.1943	.1774	.1660	.2168	.1707	.1807	
7	Gr2	.2732	.2655	.2617	.2706	.2855	.2688	.2771	.2771	.2726	.2874	.2617	.1827	.1935	.2068	.2077	.2181	.2094	.1838	.2120	.1739

isolates (0.1093–0.1225); and the New Zealand swine and the Italian isolates (0.1273). As indicated in Fig. 5, the pairwise distances are consistent with the observation based on nucleotide comparisons in grouping isolates into as many as 8 different groups.

Phylogenetic analysis

The genetic relationship demonstrated by the pairwise nucleotide comparisons become apparent upon phylogenetic analyses of multiple sequence alignments of these sequences. Phylogenetic analysis takes into account multiple sequences and demonstrates the relative relationship between different isolates. These analyses have demonstrated the close relationship between the isolates first identified from endemic regions in Asia. The phylogenetic analysis of the full-length sequences initially indicated that at least 3 major groups can be differentiated: group 1, the Burmese-like isolates; group 2, the Mexican isolate; and group 3, the US isolates [49]. A similar distribution of these three groups was demonstrated when using the 371 base region from ORF 1. The use of this region permits the inclusion of a number of additional isolates that have been identified in patients from non-endemic and endemic areas (Fig. 4). At least eight different branches were generated in this analysis. Bootstrap analysis, which gives statistical support for phylogenetic groupings, resulted in significant support for 6 of these branches. This distribution correlated with the groupings defined in the pairwise comparisons using sequence identity and genetic distances. A similar topography has also been reported when sequences from ORF 2 were used to generate the tree [59].

The sequencing of the full-length genome of the Ct1 isolate from China indicated that it is representative of a number of isolates that have been previously identified in patients from Taiwan and a number of different provinces in China for which partial genome sequence was reported. These isolates are different from the Burmese-like Chinese isolates and form a distinct group of isolates. In addition, variants from several Nigerian patients with acute hepatitis have recently been identified. Based on phylogenetic comparisons of the $3'$-end of ORF 2, the Nigerian isolates are on the same branch as the Mexican isolate [65]. Unlike other African strains, which are Burmese-like, these isolates are most closely related to the Mexican isolate of HEV. The ORF 1 and ORF 2 fragments that were isolated are 87 and 80% identical to the Mexican isolate. These results suggest that the Nigerian HEV may be a diverse subgroup of the Mexican isolate or potentially form a distinct group of its own.

Several recent publications have proposed nomenclature for types and subtypes of HEV isolates. Table 6 summarizes some of the proposed groupings of HEV isolates along with some possible designations for the groupings. These analyses also focused on the definition of subtypes within the Burmese-like isolates, as well as a number of isolates closely related to the Cs15 isolate from China [7,8,11,49,64,66,68–71].

Together these data indicate the changing view of the genotypic distribution of HEV. Full-length analysis indicates at least four genotypes of HEV. The analyses of partial sequence data from human isolates, correlate with this distribution into 4 genotypes (I–IV in Fig. 5). However, inclusion of the additional isolates indicate that even a wider range of diversity exist between isolates of HEV, with the division into at least 9 different

Table 6

Potential genotypic designations for isolates of HEV

Isolates	Figure4/5	Schlauder[a]		Schlauder[b]	Erker[c]	Arankalle[d]	Tsarev[e]
B1, B2, I2	I	I	1	1a	1a	1A	I1b
P1, C1-4	I	I	1	1b	1b	1B	I1a
I1	I	I	1	1c	1c		I1c
Cb6, Cb7, Cs13	I	I	1	1c	1c		
Mo12, Mo23	I	I	1	1e	1e	1C	
Uz, Ki, Chad	I	I	1	1d	1d	1C	I2
M	II	II	2	2	2	2	II
Ni[f]	II	II	(12)[g]				
Us1, US2, swUS1	III	III	3	3	3	3	III
It1	III	III	4	5			
SwNZ1	III	III	4				
G1, Sp1, Sp2	III	III	5	6			
swSp1	III						
Gr2	III	III	6	7			
UK1	III						
Ar1, Ar2	III	III	7	8			
Au1	III	III	7				
Jra1, J6h, J11h, swJ5	III						
swCa	III						
swNL	III						
Cs15, Ch3	IV	IV	9	4		4a	
J2he	IV						
Ct1	IV	IV	8				
Cs5, Cb3, Cb4	IV	IV	(10)			4b	
Cs9	IV	IV	(11)			4c	
Ct705, Ct825, Ct845	IV	IV	(10–11)				
J1he, Ji4	IV						
J3he, Jkksap	IV						
swT, swI	IV						

[a]Schlauder and Mushahwar (2001).
[b]Schlauder et al. (2000).
[c]Erker et al. (1999).
[d]Arankalle et al. (1999).
[e]Tsarev et al. (1999).
[f]Nigerian 1, 4, 5, 6, 7 and 9.
[g]Overlapping sequence not available.

groups (1–9 in Fig. 5). The original Burmese isolate and the related strains from Asia and Africa represent group 1. Group 2 is represented by the Mexican isolate. Group 3 consists of the human US isolates as well as a number of isolates from Japan. Group 4 consists of the lone Italian isolate. Group 5 consists of an isolate from Greece and two isolates from Spain, while group 6 contains the second Greek isolate. An isolate from the UK, although most closely related to the Greek 2 isolate, could represent a distinct group. Group 7 consists of the two Argentine isolates and the isolate from Austria. The two distinct isolates from China, Cs15 and Ct1, would represent groups 8 and 9. Related but distinct isolates from China, India, Japan and Taiwan could represent at least three or more additional groups residing off the same branch as the Cs15 and Ct1 isolates. In addition the Nigerian isolates could represent a group distinct from the Mexican isolate.

Alternatively, the distribution could be defined as four major genotypes, with genotypes II, III, and IV, consisting of isolates with a wide degree of diversity. In contrast genotype I exhibits a low level of diversity but does show a genotypic distribution based on the geographical origin of isolates. The possibility of alternative combinations may become apparent as additional isolates and sequence are identified. The ultimate definition of type and subtype awaits a final decision by the International Committee on the Taxonomy of Viruses.

Serology

Commercially available enzyme-linked immunosorbent assays (ELISAs) have utilized recombinant antigens derived from the Burmese and Mexican strains of HEV. The locations of the antigens are indicated in Fig. 1. They include a region from the carboxyl 32–33 amino acids of ORF 3 and are designated antigen 4-2. Antigen 3-2 is composed of the 42 carboxyl amino acids of ORF 2. Two larger recombinant proteins based on the Burmese strain have also shown commercial utility. They represent the full-length ORF 3 protein, designated 8-5, and the carboxyl 327 amino acids of ORF 2, designated SG3.

Analyses of serum specimens collected during the incubation period, acute phase, convalescence phase, and recovery phase of the disease, have shown a classic order of appearance of IgM and IgG. IgG has been found to persist for 3.5 years at high titers and has been detected as long as 12 years post-infection. Seroepidemiologic studies indicate that IgG class antibodies to HEV proteins are detected in 1–2% of over 10,000 blood donors in the US and Western Europe [38]. Results similar to these have also been reported by other investigators [39,40]. This prevalence was higher than expected even though the prevalence was much lower than what was found in populations from areas were HEV is known to circulate. Interestingly, the prevalence in diagnostic sites from industrialized nations was even higher, close to 10% [72]. The IgM prevalence at these sites, an indicator of acute or recent infection, was 1.3%. Many of these cases were found to be associated with travel to areas were HEV was endemic. In contrast, however, travel to endemic regions could not explain the prevalence in normal donors.

It has been suggested that the prevalence observed in industrialized nations may be due to non-specificity associated with these assays. A significant variation in assay performance has been observed and suggested that seroprevalence data in non-endemic countries may be unreliable and should be interpreted with caution [73]. It has also been suggested that diagnostic evaluations based on these assays may also be in question [6,7]. In the case of the patient from the US with no travel history to endemic areas, IgG, but not IgM class antibodies, were detected with ELISAs based on sequence from the Burmese and Mexican strains of HEV [6]. Since the diagnosis of acute hepatitis is associated with IgM class antibodies to HEV, the patient would not have been diagnosed with acute HEV based on serology with ELISAs derived from the Burmese and Mexican sequences.

However, IgM class antibodies to HEV were identified by utilizing the peptides based on the US isolate sequence, permitting the serologic diagnosis of acute HEV infection [7]. These strain-specific ELISAs were also used to determine the relative usefulness of these synthetic peptides on sera infected with different isolates of HEV. In addition to the 4-2 peptide from ORF 3 an extended version of the ORF 2 peptide, 3-2e, was also used (Fig. 1). As shown in Table 7, an overall enhanced reactivity of sera from individuals infected with the so-called "non-endemic" strains was observed with a peptide ELISA based on the sequence from the US isolate relative to the peptide ELISAs based on the sequences from the Burmese and Mexican isolates [59]. Within the 4-2 epitope located at the carboxyl terminus of ORF 3, only 1 of 33 positions differed between the two US isolates and 2 of 33 positions differed between 9 Burmese-like isolates. However, there is a significant degree of amino acid sequence variability between the US and Burmese-like isolates (10 of 33), US and Mexican (12 of 33) and the Burmese-like and Mexican (8 of 33). Most of the variability is localized within the amino half of the epitope. This apparent geographically distributed sequence variation may affect the ability to identify HEV infection or exposure with current immunoassays and that the presence of acute HEV infection in various populations is being underestimated, due to the lack of appropriate reagents for detection of strain-specific IgM class antibodies. The Chinese isolate T1 is also reported to show a similar variability in these epitopes. The Chinese and Burmese isolates differ by 3 residues over 4-2 and by 8 residues over 3-2. The Chinese and Mexican isolates differ by 4 residues over 4-2 and by 9 residues over 3-2. The Chinese and US2 isolates differ by 6 residues over 4-2 and by 4 residues over 3-2 [48]. Interestingly, patients infected with similar variants from China showed no reactivity to the Burmese and Mexican-based epitopes as soon as 3 months after acute infection. In contrast, patients infected with the Burmese-like strain of HEV were IgG positive with these ELISAs, suggesting the potential of distinct serotypes of HEV [64].

Table 7

IgG/IgM in HEV infected patients tested with peptide ELISAs based on different isolates

| Patient | | IgG/IgM | | | | | |
| | | ORF3 (4-2 peptide sequence) | | | ORF2 (3-2 peptide sequence) | | |
Country	Strain	Burma	Mexico	US	Burma	Mexico	US
Egypt	B1-like	+/+	+/+	+/+	+/+	+/+	+/+
Norway	B1-like	+/+	+/+	+/+	+/+	+/+	+/+
Japan	B1-like	+/+	nd	+/+	+/+	nd	+/+
United States	US1	+/−	+/−	+/+	−/−	+/−	+/+
United States	US2	−/+	−/+	+/+	+/+	+/+	+/+
Italy	It1	nd	−/−	+/+	−/−	−/+	−/+
Greece	Gr1	−/−	−/−	−/−	−/+	−/+	+/+
Greece	Gr2	−/+	−/+	−/+	−/+	−/−	+/+
Argentina	Ar1	nd	−/−	+/−	nd	−/−	−/−
Argentina	Ar2	nd	−/−	−/−	nd	−/+	+/+
Austria	Au1	nd	−/+	−/+	nd	+/−	+/+

218

Swine HEV

HEV is transmitted primarily by an oral–fecal route, frequently by fecal contamination of the drinking water supply. The possibility of zoonotic infections from pigs to humans was first made following the experimental infection of pigs with a human strain of HEV [74]. In addition, antibodies to HEV were detected in sera from a number of domestic farm animals in Thailand, with HEV RNA being detected in pigs [19]. The description of sequence related to HEV isolated from swineherds in the US [20] provided support for these studies, which suggested that zoonotic infections from pigs to human or humans to pigs may occur. The US swine isolate is more closely related to the human HEV strains isolated from patients in the US than to the Burmese and Mexican isolates [7]. Most recently, swine isolates have been isolated from swine from Japan, Taiwan, India, Canada, New Zealand and the Netherlands. These findings clearly indicate that human and swine HEV strains co-exist in many different geographic regions.

Recombinant fusion proteins based on the US genotype of HEV from the US2 isolate have been expressed in *E. coli*. The proteins represent the carboxyl half of ORF 2 (327 amino acids) and the full-length ORF 3 (123 amino acids) and were used to determine the prevalence in swineherds from the US. As shown in Table 8, total exposure rates for these antigens range from 28 to 93% in herds from different parts of the US [75]. A similar prevalence was observed in swineherds from the US Midwest region from which the US swine strain was isolated [20].

HEV isolates have also been identified in swine from Taiwan and have been shown to exhibit a close homology to human HEV isolates from those regions [20,21,23]. An isolate from sewage from a slaughterhouse in Spain, which mainly processed swine, was most closely related to a human isolate from Spain [10]. An isolate from a Japanese swine was closely related to a human isolate from that country [24]. Novel isolates have also been identified in swine from Canada [28], the Netherlands [25] and New Zealand [22]. However, to date no human isolates have been identified from these countries. Phylogenetic analysis indicates that the New Zealand swine isolate is most closely related to the Italian isolate of HEV and while several isolates from the Netherlands are related to the Spanish isolates and US isolates (Figs. 4 and 5). In addition, Indian pigs were found not to be infected with the endemic genotype I that circulates in India in humans, but with genotype IV, which is most similar to newer strains isolated in China and Japan [26]. The significance of this is unclear, however, the finding of isolates from South America that

Table 8

HEV US2 exposure in swine herds from different states in the United States

	Iowa	New Jersey	Iowa	Oregon	Total
Exposure[a]	18/64	13/14	18/64	11/36	83/164
Percent (%)	28	93	28	31	51

[a]Sum of reactivity of ELISAs using antigens from ORF 2 and ORF 3.

are also closely related to other European strains, and the close relationship between the US and some Japanese isolates, suggest that some strains may have a broad distribution across many continents in both humans and swine.

Summary

The rapid increase in the identification of novel human isolates of HEV and the identification of HEV in swine has given rise to a number of new questions concerning this virus. Previously considered a disease associated with underdeveloped areas and presumably due to the fecal contamination of drinking water, early sequencing data from different HEV isolates indicated that the virus exhibited little genetic diversity. The majority of strains were very closely related to the original isolate from Burma [43]. The only genetically diverse isolate was from an outbreak in Mexico [44]. However, the identification of isolates from patients with acute hepatitis from regions not considered endemic for HEV, has raised the question as to the source of the infection [5–18]. The finding of the virus in swine and other domestic animals indicates the potential of zoonosis [19–29,76]. Also of interest is the high level of diversity that is becoming apparent with the identification of these novel isolates.

New isolates of the virus have been identified in a number of patients from regions considered non-endemic for HEV [5–18]. In addition, HEV-related sequences have been identified in swine from countries both non-endemic and endemic for HEV [19–29].

Phylogenetic analyses using full-length sequences give evidence for at least 4 genotypes of HEV. The phylogenetic analyses using the overlapping region from ORF 1 correlate with the distribution of the full-length genomes and clearly demonstrates that with the addition of the novel HEV variants from Europe, Asia, North America and South America, HEV can be separated into at least 9 major groups if not more. In addition, pairwise comparisons of the genetic distances and nucleotide identities resulted in distributions that were consistent with the phylogenetic groupings.

The genetic diversity observed between the isolates from Europe, Japan and the US may be indicative of the mode of transmission or the reservoir for these isolates. It is interesting that in comparison, the isolates within group I have originated from geographically diverse regions. Although there is some geographic relatedness between many of these strains, the genetic diversity is nowhere near that which is observed between the isolates identified from non-endemic regions. There is also a similar diversity observed between the isolates from China and Taiwan in (group 8,9/group IV). All of these isolates were also reported to be isolated from sporadic cases of hepatitis E.

As an increase in surveillance of HEV-related acute hepatitis infection occurs, especially in areas not considered endemic for HEV, additional patients may be identified with ongoing infections, which will add to the database of HEV sequences. The identification of additional variants will certainly occur and either complicate the issue even more or hopefully, clarify the genetic relationships between these viruses.

220

References

1. Khuroo MS. Am J Med 1980; 68: 818–824.
2. Bradley DW. Br Med Bull 1990; 46: 442–461.
3. Krawczynski K. Hepatology 1993; 17: 932–941.
4. Mushahwar IK, Schlauder GG, Dawson GJ. In: Viral Hepatitis and Liver Disease, Proceedings of the IX Triennial International Symposium on Viral Hepatitis and Liver Disease, Rome (Rizetto M, Purcell R, Gerin J, Verme G, editors). Turin: Edizoni Minerva Medica, S.P.A.; 1997; pp. 317–320.
5. Schlauder GG, Dawson GJ, Kwo PY, Gabriel GS, Mushahwar IK. IX Triennial Int Symp Viral Hep Liver Dis 1996; A103: 78.
6. Kwo PY, Schlauder GG, Carpenter HA, Murphy PJ, Rosenblatt JE, Dawson GJ, Mast EE, Krawczynski K, Balan V. Mayo Clin Proc 1997; 72: 1133–1136.
7. Schlauder GG, Dawson GJ, Erker JC, Kwo PY, Knigge MF, Smalley DL, Desai SM, Mushahwar IK. J Gen Virol 1998; 79: 447–456.
8. Schlauder GG, Desai SM, Zanetti AR, Tassopoulos NC, Mushahwar IK. J Med Virol 1999; 57: 243–251.
9. Zanetti AR, Schlauder GG, Ramano L, Tanzi E, Fabris P, Dawson GJ, Mushahwar IK. J Med Virol 1999; 57: 356–360.
10. Pina S, Buti M, Cotrina M, Piella J, Girones R. J Hepatol 2000; 33: 826–833.
11. Schlauder GG, Frider B, Sookoian S, Castano GC, Mushahwar IK. J Infect Dis 2000; 182: 294–297.
12. Worm HC, Schlauder GG, Wurzer H, Mushahwar IK. J Gen Virol 2000; 81: 2885–2890.
13. Takahashi K, Iwata K, Watanabe N, Hatahara T, Ohta Y, Baba K, Mishiro S. Virology 2001; 287: 9–12.
14. Aikawa T, Kojima M, Takahashi M, Nishizawa T, Okamoto H. J Infect Dis 2002; 186: 1535–1537.
15. Mizuo H, Suzuki K, Takikawa Y, Sugai Y, Tokita H, Akahane Y, Itoh K, Gotanda Y, Takahashi M, Nishizaw T, Okamoto H. J Clin Microbiol 2002; 40: 3209–3218.
16. Suzuki K, Aikawa T, Okamoto H. N Engl J Med 2002; 347: 1456.
17. Takahashi K, Kang J-H, Ohnishi S, Hino K, Mishiro S. J Infect Dis 2002; 185: 1342–1345.
18. Takahashi M, Nishizawa T, Yoshikawa A, Sato S, Isoda N, Ido K, Sugano K, Okamoto H. J Gen Virol 2002; 83: 1931–1940.
19. Clayson ET, Snitbhan R, Ngarmpochana M, Vaughn DW, Shrestha MP. In: Enterically Transmitted Hepatitis Viruses (Buisson Y, Coursaget P, Kane M, editors). Joue-les-Tours: La Simarre; 1996; pp. 329–335.
20. Meng XJ, Purcell RH, Halbur PG, Lehman JR, Webb DM, Tsareva TS, Haynes JS, Thacker BJ, Emerson SU. Proc Natl Acad Sci USA 1997; 94: 9860–9865.
21. Hsieh S, Meng X, Wu Y, Liu S, Tam A, Lin D, Liaw Y. J Clin Microbiol 1999; 37: 3828–3834.
22. Garkavenko O, Anderson D, Schroeder B, Croxson C. Antiviral Ther 2000; 5: E10.
23. Wu JC, Chen CM, Chiang TY, Sheen IJ, Chen JY, Tsai WH, Huang YH, Lee SD. J Med Virol 2000; 60: 166–171.
24. Okamoto H, Takahashi M, Nishizaw T, Fukai K, Muamatsu U, Yoshikawa A. Biochem Biophys Res Commun 2001; 289: 929–936.
25. van der Poel WHM, Verschoor F, van der Heide R, Herrera MI, Vivo A, Kooreman M, de Roda Husman AM. Emerg Infect Dis 2001; 7: 970–976.

26. Arankalle VA, Chobe LP, Joshi MV, Chadha MS, Kundu B, Walimbe AM. J Hepatol 2002; 36: 417–425.
27. Huang FF, Haqshenas G, Guenette DK, Halbur PG, Schommer SK, Pierson FW, Toth TE, Meng XJ. J Clin Microbiol 2002; 40: 1326–1332.
28. Pei Y, Yoo D. J Clin Microbiol 2002; 40: 4021–4029.
29. Wu JC, Chen CM, Chiang TY, Tsai WH, Jeng WJ, Sheen IJ, Lin CC, Meng XJ. J Med Virol 2002; 66: 488–492.
30. Tsarev SA, Emerson SU, Reyes GR, Tsareva TS, Legters LJ, Malik IA, Iqbal M, Purcell RH. Proc Natl Acad Sci USA 1992; 89: 559–563.
31. Chauhan A, Jameel S, Dilawari JB, Chawla YK, Kaur U, Ganguly NK. Lancet 1993; 341: 149–150.
32. Bader TK, Krawczynski K, Polish L, Favorov M. N Engl J Med 1991; 325: 1659.
33. Dawson GJ, Mushahwar IK, Chau KH, Gitnick GL. Lancet 1992; 340: 326.
34. Smalley DL, Brewer SC, Dawson GJ, Kyrk C, Waters B. South Med J 1996; 89: 993–996.
35. Zanetti AR, Dawson GJ. The Hepatitis Group. J Med Virol 1994; 42: 318–320.
36. Tassopoulos NC, Krawczynski K, Hatzakis A, Katsoulidou A, Delladetsima I, Koutelou MG, Trichopoulos D. J Med Virol 1994; 42: 124–128.
37. Zaijer HL, Kok M, Lelie PN, Timmerman RJ, Chau K, VanderPol HJ. Lancet 1993; 341: 286.
38. Dawson GJ, Chau CH, Cabal CM, Yarbough PO, Reyes GR, Mushahwar IK. J Virol Methods 1992; 38: 175–186.
39. Quiroga JA, Cotonat T, Castillo I, Carreno V. J Med Virol 1996; 50: 16–19.
40. Mast EE, Kuramoto IK, Favorov MO, Sehocing VR, Burkholder BT, Shapirs CN, Holland PV. J Infect Dis 1997; 176: 34–40.
41. Bradley DW, Maynard JE. Semin Liver Dis 1986; 6: 56–66.
42. Reyes GR, Purdy MA, Kim JP, Luk KC, Young LM, Fry KE, Bradley DW. Science 1990; 247: 1335–1339.
43. Tam AW, Smith MM, Guerra ME, Huang CC, Bradley DW, Fry KE, Reyes GR. Virology 1991; 185: 120–131.
44. Huang CC, Nguyen D, Fernandez J, Yun KY, Fry KE, Bradley DW, Tam AW, Reyes GR. Virology 1992; 191: 550–558.
45. Zafrullah M, Ozdener MH, Panda SK, Jameel S. J Virol 1997; 71: 9045–9053.
46. Pringle CR. Arch Virol 1998; 143: 1449–1459.
47. Yin S, Tsarev SA, Purcell RH, Emerson SU. J Med Virol 1993; 41: 230–234.
48. Wang Y, Zhang H, Ling R, Li H, Harrison TJ. J Gen Virol 2000; 81: 1675–1686.
49. Erker JC, Desai SM, Schlauder GG, Dawson GJ, Mushahwar IK. Virology 1999; 80: 681–690.
50. Gorbalenya AE, Koonin EV, Lai MM-C. FEBS 1991; 288: 201–205.
51. Fry KE, Tam AW, Smith MM, Kim JP, Luk K-C, Young LM, Piatak M, Feldman RA, Yun KY, Purdy MA, McCaustland KA, Bradley DW, Reyes GR. Virus Genes 1992; 6: 173–185.
52. Koonin EV, Gorbalenya AE, Purdy MA. Proc Natl Acad Sci USA 1992; 9: 8259–8263.
53. Nielsen H, Engelbrecht J, Brunak S, Heijne Gv. Protein Engng 1997; 10: 1–6.
54. Roux KH. Biotechniques 1995; 16: 812–814.
55. Psichogiou MA, Tassopoulus N, Papatheodoridis GV, Tzala E, Klarmann R, Witteler H, Schlauder GG, Troonen H, Hatzakis A. J Hepatol 1995; 23: 668–673.
56. Erker JC, Desai SM, Mushahwar IK. J Virol Methods 1999; 81: 109–113.
57. Rey JA, Findor JA, Daruich JR, Velazco CC, Igartua EB, Schmee E, Kohan AI. J Travel Med 1997; 4: 100–101.
58. Worm HC, Wurzer H, Frosner G. N. Engl J Med 1998; 339: 1554–1555.

59. Schlauder GG, Knigge MF, Gutierrez R, Sookoian S, Castano G, Worm H, Frider B, Mushahwar IK. Antiviral Ther 2000; 5: E10.

60. Wang Y, Levine DF, Bendall RP, Teo C-G, Harrison TJ. J Med Virol 2001; 65: 706–709.

61. Huang R, Nakazono N, Ishii K, Kawamata O, Kawaguchi R, Tsukada Y. J Med Virol 1995; 47: 303–308.

62. Hsieh SY, Yang PY, Ho YP, Chu CM, Liaw YF. J Med Virol 1998; 55: 300–304.

63. Wu JC, Sheen IJ, Chiang TY, Sheng WY, Wang YJ, Chan CY, Lee SD. Hepatology 1998; 27: 1415–1420.

64. Wang Y, Ling R, Erker JC, Zhang H, Li H, Desai SM, Mushahwar IK, Harrison TJ. J Gen Virol 1999; 80: 169–177.

65. Buisson Y, Grandadam M, Nicand E, Cheval P, vanCuyck-Gandre H, Innis B, Rehel P, Coursaget P, Teyssou R, Tsarev S. J Gen Virol 2000; 81: 903–909.

66. Schlauder GG, Mushahwar IK. J Med Virol 2001; 65: 282–292.

67. Lemon SM, Robertson BH. Semin Virol 1993; 4: 285–295.

68. Chatterjee RS, Tsarev S, Pillot J, Coursaget P, Emerson SU, Purcell RH. J Med Virol 1997; 53: 139–144.

69. VanCuyck-Grande HH, Zhang HY, Tsarev SA, Clements NJ, Cohen SJ, Caudill JD, Buisson Y, Coursaget P, Warren RL, Longer CF. J Med Virol 1997; 53: 340–347.

70. Arankalle VA, Paranjape S, Emerson SU, Purcell RH, Walimbe AM. J Gen Virol 1999; 80: 1691–1700.

71. Tsarev SA, Binn LN, Gomatos PJ, Arthur RR, Monier MK, Cuyck-Gandre Hv, Longer CF, Innis BL. J Med Virol 1999; 57: 68–74.

72. Dawson GJ, Schlauder GG, Mushahwar IK. In: Enterically Transmitted Hepatitis Viruses (Buisson Y, Coursaget P, Kane M, editors). Joue-les-Tours: La Simarre; 1996; pp. 167–174.

73. Mast EE, Alter MJ, Holland PV, Purcell RH. Hepatology 1998; 27: 857–861.

74. Balayan MS, Usmanov RK, Zamayatina NA, Djumalieva DI, Karas FR. J Med Virol 1990; 32: 58–59.

75. Schlauder GG, Gutierrez RA, Kyrk CR, Erker JC, Surowy TK, Dille BJ, Mushahwar IK. FASEB J 1999; 13(A15): #LB67.

76. Kabrane-Lazizi Y, Fine JB, Elm J, Glass GE, Higa H, Diwan A, Gibbs Jr CJ, Meng XJ, Emerson SU, Purcell RH. Am J Trop Med Hyg 1999; 61: 331–335.

Viral Hepatitis
I.K. Mushahwar (editor)
© 2004 Elsevier B.V. All rights reserved.

GB virus C

Thomas P. Leary and Isa K. Mushahwar
Core R&D Infectious Diseases, Abbott Diagnostics Division, Abbott Laboratories, Abbott Park, IL, USA

Introduction

The GB agent was first described in 1967 [1]. Serum obtained from a surgeon experiencing acute hepatitis was inoculated into tamarins that subsequently developed acute biochemical hepatitis. The agent was filterable and could be serially passed, though isolation of the causative agent was not successful despite repeated attempts over several decades. In 1995, the causative agent was finally isolated utilizing molecular technologies that did not exist at the time of the initial description [2]. The isolated agent was not a single virus, but two related viruses, termed GB virus A (GBV-A) and GB virus B (GBV-B). Further molecular studies designed to determine the prevalence of these viruses in the human population resulted in the isolation of a third related virus, GB virus C (GBV-C) [3,4]. This virus was later described as Hepatitis G virus [5]. Determination of the genome-length sequences of the GB viruses revealed that each were genetically related to one another, and to Hepatitis C virus (HCV) [4,6]. Currently, GBV-A and GBV-B have only been detected in non-human primates, while GBV-C has only been detected naturally in humans. Assay systems designed to detect the presence of GBV-C have been utilized to determine that the virus is found at very low levels in normal blood donors, with elevated rates of infection occurring in those at risk for exposure to parentally transmitted viruses [7].

Genomic organization

GBV-C is a flavivirus-like enveloped RNA virus measuring $50-100$ nm in diameter and bands in sucrose at a peak density of $1.08-1.13$ g/cm^3 [8]. The particle is associated with a significant amount of lipoprotein and also contains varying levels of carbohydrate moieties. The genome is greater than 9300 nucleotides in length (Fig. 1), is single-stranded and of positive polarity, and contains a long open reading frame (ORF) of 2843 amino acids [4]. The ORF encodes a large polyprotein that is post-translationally cleaved into the individual viral proteins by a combination of either host or viral proteases. Additionally, non-coding sequences are found both upstream and downstream of the ORFs. Based on this criteria, as well as similarities with other viruses, GBV-C has been classified within a distinct genus of the *Flaviviridae* [4,6].

224

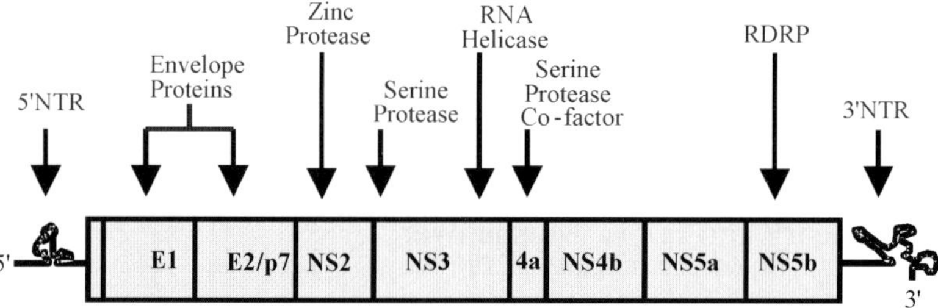

Fig. 1. Genomic organization of GB virus C. Dark lines designate non-coding regions. Coding regions are shaded with the specific viral proteins specified. RDRP: RNA-dependent RNA polymerase; NTR: Non-translated region.

Non-translated virus sequences

GBV-C possesses long, non-translated sequences both upstream and downstream the ORF, 533 and 312 nucleotides, respectively. These sequences are thought to be of importance in genome replication and packaging, as well as in the regulation of gene expression. Indicative of their importance, they are highly conserved between distinct isolates of the virus [9], and to some degree between the three GB viruses and other members of the *Flaviviridae* [6]. Within the 5′NTR, multiple inverted repeat sequences are present. These sequences suggest a highly ordered structure and are consistent with the presence of an internal ribosome entry site (IRES) that has been attributed to these sequences [10,11]. In this manner, translation of the large ORF occurs in a cap-independent fashion. This is as opposed to the cap-dependent translation that is utilized by eucaryotic cells and many other viruses. In addition to the IRES, the 5′NTR of HCV has also been demonstrated to be a translational inhibitor [12]. While a similar characteristic has not been formally demonstrated for GBV-C, it can likely be inferred based on the similar folding pattern that occurs between the viruses [10,11].

Sequences within the 3′NTR are thought to be important in viral packaging and replication. Like HCV, it has been speculated that substantial structure exists in these regions much like that found in the 5′NTR [13,14]. Because these viruses do not utilize a DNA intermediate during genome replication, by default, an antigenomic intermediate is necessary for the RNA-dependent RNA-polymerase to produce genomic length RNA molecules for polyprotein synthesis and the production of progeny virus. The structure that has been described within the 3′NTR is thought to confer a docking site for the viral replicase encoded by the NS5B gene. Indicative of the importance of these sequences to virus replication, the 3′ sequences have been shown to be indispensable in the construction of an infectious molecular clone of GBV-B [14]. An additional point of interest is that GBV-B contains a poly-U tract just downstream of the translational stop codon, identical to that found in some isolates of HCV [6]. This is as opposed to host mRNAs that contain a similar poly-A tract. While the significance of this finding is unclear, it is possible that some unknown advantage is conferred upon the genome in terms of stability, replication or enhanced protein synthesis.

Structural proteins

Typical of the *Flaviviridae*, the structural proteins are located in the N-terminal one-third of the polyprotein and are processed into the individual virion components by host-encoded proteases [4]. Hydropathy plots of the GB viruses within the structural region, as compared to other members of the *Flaviviridae*, demonstrate a significant degree of structural similarity between distinct viruses [6]. Despite this observed similarity, it is also clear from these analyses that GBV-C, as well as GBV-A, is distinct from most other members of the *Flaviviridae*. In this case, no apparent core protein is encoded by GBV-C. While a potential eucaryotic signal cleavage site is present just upstream of the putative E1 protein, only a very short peptide is synthesized in this region. Apparently, this short peptide is translated to allow translocation of the E1 protein into the host cell endoplasmic reticulum. While extensive studies have been performed to elucidate a core protein within this region of the viral genome, its presence has not been convincingly demonstrated. While it is likely that a core-like protein is present within the intact virion, the source of that protein is unclear. It is possible that the protein is encoded elsewhere on the viral genome, but it is more likely that a host cell protein is performing this function.

GBV-C encodes at least two distinct envelope proteins (E1 and E2), and evidence suggests further post-translational cleavage of E2. Both proteins contain sugar moieties that substantially increase their molecular mass, and a number of cysteine residues present suggest that they possess extensive tertiary structure [15]. The E1 protein of GBV-C is highly conserved between different isolates of the virus [16], and a conserved Asn-Cys-Cys motif is absolutely conserved between each of the GB viruses at the E1 N-terminus [6,16]. These levels of conservation between isolates suggest a level of functional significance, or perhaps the lack of significant host immune selective pressures. Net hydrophobicity analyses of the GBV-C E1 protein suggest that it may be buried on the virion surface, protected from the host environment. Regardless of the reason, the conservation observed in E1 is in direct contrast to that of HCV where the level of diversity is as great as that found elsewhere in the polyprotein [16].

The GBV-C E2 protein is similar in that multiple cysteine residues are conserved among distinct isolates of the virus, and the molecular mass is dramatically enhanced by the addition of sugar moieties [15]. Variability within the E2 protein is significantly greater than at any other region of the polyprotein, though the observed variability in this region is not nearly as extensive as that found within the E2 region of various HCV isolates [16]. This could reflect a lack of host immune selective pressures or a lack of tolerance for substitutions in this protein. In HCV, the host immune response appears to be a driver of variability within E2 [17,18]. The lack of variability within the E2 protein is interesting because patients infected with GBV-C can remain viremic for extended periods of time in the absence of a detectable anti-E2 response. However, the virus is eliminated upon development of such a response [19,20]. This would support the hypothesis that GBV-C has the inability to escape immune clearance due to a lack of tolerance for mutations in E2.

Nonstructural proteins

The nonstructural proteins of GBV-C are encoded on the $3'$ two-thirds of the genome, as is the case with other members of the *Flaviviridae* [4]. They are cleaved from the polyprotein into the individual viral proteins by way of viral proteases encoded within the nonstructural genes. Additionally, a helicase and a replicase are encoded within the nonstructural proteins of GBV-C. In all, the nonstructural proteins are involved in genome replication and the production of structural proteins for nascent virions.

The first of the individual proteins encoded within the nonstructural genes is NS2. This protein is a zinc-dependent metalloproteinase that is responsible for the cleavage of NS2 from NS3 in the nascent polyprotein. Although little identity exists between GBV-C and HCV within NS2, the histidine and cysteine residues essential to the function of the HCV NS2 protease [21,22] are spacially conserved in the GB viruses [4,6,23]. Additionally, a point mutation introduced at either position in the GBV-C protease abrogates cleavage of NS2 from NS3 in a recombinant system [24]. This would indicate that despite the observed amino acid sequence differences, each virus processes the NS2/NS3 cleavage in an identical fashion.

The remainder of the GBV-C polyprotein is processed into the individual viral proteins utilizing a serine protease found at the N-terminus of NS3. As is the case with HCV and GBV-B [25,26], the GBV-C NS3 serine protease in all likelihood functions in conjunction with a co-factor provided by the NS4A protein. As in NS2, little amino acid sequence conservation is found between the GB viruses and HCV in this region. Regardless, the amino acid residues that constitute the active site of the HCV serine protease are conserved and appropriately spaced in the GB viruses [4,6,23]. The functional sites of serine proteases are composed of histidine, aspartic acid and serine residues. In HCV, the aspartic acid residue is located 24 amino acids downstream of the histidine, while the serine residue is located 58 amino acids downstream of the aspartic acid. As in the HCV serine protease [21], alteration of the serine moiety eliminates processing of NS3 cleavage sites in GBV-C [24]. In total, these studies would suggest that GBV-C process the downstream polyprotein cleavages by way of a serine protease located within the NS3 gene, as is the case with HCV.

In GBV-C, the carboxy-terminus of NS3 contains an RNA helicase that is located in this same region of the polyprotein for all members of the *Flaviviridae*. These helicases are members of the supergroup II RNA helicases based on general amino acid sequence conservation, and also signature sequence motifs found in nucleic acid helicases (DECHXXD) and NTPases (GXGKS). Detailed studies demonstrate that the GBV-C helicase is similar to HCV in that it proceeds in a $3'-5'$ direction, possesses NTPase activity, and is dependent on ATP and divalent metal ions [27].

The final protein encoded within GBV-C polyprotein is the NS5B protein. Though little characterization of this protein has been performed at this time, other members of the *Flaviviridae* encode an RNA-dependent RNA polymerase utilized in genome replication at this position of the genome. In all likelihood, this is also the case with GBV-C as sequence motifs conserved by members of the supergroup II replicases are found in GBV-C. Additionally, a GDD signature sequence, thought to be the enzymatic active site, is present in this sequence. While detailed studies on the GBV-C replicase have not been reported, the extensive sequence conservation would suggest the functional significance of the protein.

GBV-C Diversity

Despite the relatively low level of overall amino acid sequence identity between GBV-C and other members of the *Flaviviridae*, significant regional identity occurs. Identity approaching 50% is found within the NS3 helicase and the NS5B replicase genes [4,28], and additional levels of lesser, but significant identity occur within E2, the NS3 protease, NS4A and NS5A. Polyprotein comparisons between GBV-A and GBV-C, viruses that are more closely related, demonstrate much greater localized identity. Although localized identity can provide insights into the functional role of the individual viral proteins, phylogenetic analysis is necessary to determine the exact evolutionary relationship of these proteins to one another, as well as the relatedness of the viruses as a whole. As shown in Fig. 2, phylogenetic anaylsis of the NS3 helicase region unequivocally demonstrates the relationship of GBV-C to other members of the *Flaviviridae*. The GB viruses and HCV group on a distinct segment of the tree, while the classic flaviviruses and the pestiviruses group to other segments of the tree. Finally, several plant viruses that group to the *Flaviviridae* segment to a distinct branch.

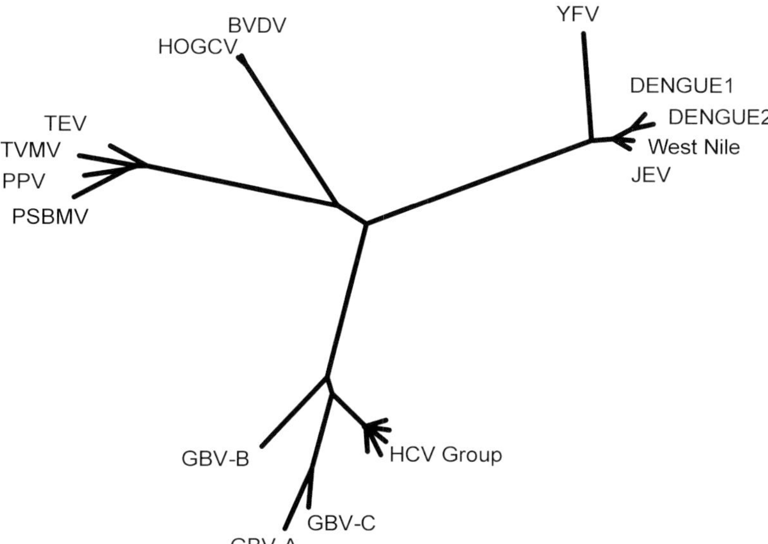

Fig. 2. Phylogenic tree depicting the *Flaviviridae* NS3 protease domain. See Ref. [6] for methodology. JEV: japanese encephalitis virus; WNV: west nile virus; YFV: yellow fever virus; BVDV: bovine viral diarrhea virus; HOGCV: hog cholera virus; TEV: tobacco etch virus; TVMV: tobacco vein mottling virus; PPV: plum pox virus; PSBMV: pea sees-borne mosaic virus.

Phylogenetic analysis within the $5'$ non-coding region of GBV-C has been used to delineate specific genotypes of the virus that correlate with the geographic region in which the particular isolate was recovered [29]. To this point, five GBV-C genotypes have been described. Genotype 1 isolates are found predominantly in West Africa, while genotype 2 (subtypes 2a, 2b and 2c) isolates are found in Europe and North America. The majority of isolates from Asia are of type 3. Finally, newer isolates have been identified as type 4 and type 5, respectively [30,31]. It is of note that the GBV-C genotypes are identified within the non-coding region of the virus. In fact, the coding regions of the virus do not provide dependable analyses. This is distinct from HCV genotyping in which reliable analysis can only be performed within the coding sequences of the virus, routinely NB5b.

GBV-C Detection and prevalence

To accurately assess the prevalence of GBV-C in the population, both serological- and probe-based assays are required. This results from the fact that detection in either system is for the most part mutually exclusive of the other [19]. Probe-based assays for the detection of GBV-C are markers of active infection. In these assays, nucleic acids are extracted from a small volume of plasma or serum, followed by reverse transcription and PCR. A number of distinct assays have been characterized that amplify sequences from both coding and non-coding regions of the genome. The first of these assays was designed within conserved regions of the NS3 gene [28]. Degenerate primers and altered cycling conditions were utilized in an attempt to overcome detection problems associated with genetic variation in this region of the virus genome. The more current PCR assays for the detection of GBV-C amplify sequences within the $5'$NTR [29]. Sequences within the $5'$NTR of GBV-C are much more conserved than in coding regions of the virus. In fact, several regions are absolutely conserved between most known virus isolates. Such conservation provides excellent sites for primer design, resulting in much improved reliability. An additional advantage of this assay is that amplified products can be readily sequenced to provide the genotype identity.

The lone serological assay available for the detection of GBV-C utilizes the second envelope protein for the detection of anti-E2 antibodies [19,20]. This assay has been utilized to demonstrate that GBV-C seroprevalence is between 3 and 8%, and that individuals with increased exposure to parenterally transmitted viruses have much higher seroprevalence rates. Of interest is the fact that individuals possessing anti-E2 antibodies are almost never RNA-positive utilizing probe-based assays. Thus, the appearance of anti-viral antibodies successfully results in viral clearance. Therefore, it is important that both probe and serological assays be utilized when GBV-C prevalence is assessed. In such an analysis, the probe-based assays identify those currently infected, while the serological assay identifies those who were previously infected, but have subsequently eliminated the infection.

GBV-C Epidemiology

GBV-C RNA has been found in blood and blood products [32], human saliva [33] and semen [34]. Thus, the predominant routes of transmission appear to be mostly parenteral, but also other routes of transmission such as sexual [35], intraspousal [36], vertical/perinatal [37,38] and nosocomial [39] have been reported.

Prevalence of GBV-C RNA

Current epidemiological studies [40] show the presence of GBV-C RNA obtained from a variety of sources. These include sera from aplastic anemia patients [41], fulminant hepatitis cases [42], hemophiliacs [43], HIV [44], Hepatitis B virus (HBV) and HCV-[45] infected individuals, intravenous drug abusers [46], multiply transfused individuals [47], non-A-E-infected patients [7], patients on maintenance hemodialysis [48], thalassemic patients [49], kidney, liver and bone marrow transplant (BMT) patients and donors [50–52]. Examination of serum specimens collected from healthy volunteer blood donors from different parts of the world confirmed the presence of GBV-C RNA in 2–4% of the specimens. The infection is thus spread globally in healthy and diseased populations.

As mentioned above, exposure to GBV-C infection is determined using a combination of tests for RNA and antibody presence. This allows for the estimation of both the exposure and the active infection rate. These two rates were determined in several populations [53]. It was found that most individuals exposed to GBV-C were either GBV-C RNA-positive/anti-E2-negative or anti-E2-positive/GBV-C RNA-negative. Exposure, therefore, was calculated as the sum of GBV-C RNA and anti-E2-positive individuals. Exposure (Table 1) to GBV-C in commercial plasmaphoresis donors, intravenous drug abusers (IVDA), acute non-A-E patients, chronic non-A-E patients, acute HCV-infected individuals and volunteer blood donors were 40.5, 89.2, 43.1, 41.4, 73.2 and 6.0%, respectively [53]. The exposure rate among all groups tested was found to

Table 1

GBV-C exposure and viremic rates in the United States

Groups tested	Total	GBV-C RNA + [*]		Anti-E2		Exposure rate (%)	Exposure viremic rate
		No.	%-	No.	%-		
Volunteer blood donors	199	3	1.5	9	4.5	6.0	4.0
Commercial plasma donors	711	93	13.1	195	27.4	40.5	3.1
Acute non-A-E	51	4	7.8	18	35.3	43.1	5.5
Chronic non-A-E	41	3	7.3	14	34.1	41.4	5.7
Acute HCV	41	8	19.5	22	53.7	73.2	2.8
IVDAs	102	15	14.7	76	74.5	89.2	6.1

[*]Active viremic rate.

230

be 2.8–6.1 times the rate of viremia, and that the rates of both exposure and active viremia are considerably higher among IVDAs and acute HCV infection cases. Furthermore, serial bleeds tested for GBV-C RNA indicated that some patients remained viremic for more than three years and failed to produce detectable antibodies to GBV-C E2. Antibodies to E2 appear to be long-lived with a fairly constant titer. In a similar study [20], anti-E2 was performed on six individuals who acquired GBV-C infection through transfusion from GBV-C RNA-positive donors. Of the six blood recipients, four (67%) became anti-E2-positive and cleared the viremia. In four of the patients, appearance of anti-E2 corresponded to a loss of detectable GBV-C RNA. Three of four individuals were dually reactive for GBV-C RNA and anti-E2 for one or more of the serial specimens tested. In two patients, GBV-C RNA was detectable for the entire study period (265–280 days) with no detectable anti-E2. The six GBV-C RNA-positive donors were all negative for anti-E2 antibodies. These data suggest that anti-E2 is associated with the clearance of GBV-C. Analysis of the serial serum specimens from the six recipients of GBV-C RNA-positive blood donors together with other serial serum specimens from GBV-C RNA-positive individuals collected during the incubation period, acute phase, convalescence, and recovery, and/or chronic disease state have shown two serologic profiles: one is of acute infection followed by recovery; the other is of acute infection progressing to chronicity [54].

Protective immunity

Evidence for the protective immunity of anti-E2 has resulted from several studies. These included studies in kidney transplant recipients [55]. In this study, anti-E2-positive kidney recipients did not acquire GBV-C infection after transplantation from GBV-C RNA-positive kidney donors, while anti-E2-negative kidney recipients did acquire GBV-C infection following transplantation from GBV-C RNA-positive donors. Similar results have been reported in orthotopic liver transplant (OLT) patients [56,57]. In intravenous drug users, it has been reported [58] that no new GBV-C infections were seen among intravenous drug users with anti-E2 despite ongoing intravenous drug use and needle sharing.

GBV-C Tissue tropism

It has been established that GBV-C [55,59–62], contrary to previous reports [63–65], is indeed a hepatotropic, as well as a lymphotropic virus [59,60], when specific, appropriate and specialized techniques such as fluorescence *in situ* hybridization and immunohisto-chemical staining. Other methods using PCR techniques on strand-specific detection of viral genomes, including those using conventional RT-PCR [66–68], tagged RT-PCR [69] and rTth methods [69–71] are not reliable. These techniques, as described by Kobayashi et al. [61], have limitations. Poor specificity for the detection of the negative strand RNA in conventional RT-PCR results from false-priming or self-priming of the positive strand [68,72]. With the rTth method, an incorrect strand is detected [61] by a lapse of strand specificity at levels 10,000-fold higher than are required to detect the correct strand [71]. Therefore, it is necessary to utilize techniques such as immunohistochemical straining and *in situ* hybridization rather than PCR when studying the tissue tropism of viruses. Utilizing these techniques, it was shown that only the livers of two immunopromised GBV-C RNA-positive patients were the site of replication of GBV-C [55], having more than 8×10^5 copies of GBV-C RNA per 2×10^4 copies of glycerol phosphate dehydrogenase. The GBV-C-positive livers showed significant fluorescence localized within individual hepatocytes. These results were confirmed by Seipp et al. [62], who investigated tissue samples from several explanted livers from GBV-C-infected patients. These tissues were examined using nested RT-PCR and fluorescence *in situ* hybridization on liver cryosections. Five of 6 liver specimens were found to be positive for GBV-C plus-strand RNA, and viral minus-strand RNA was detected in 4 of 6 liver specimens. In two specimens, GBV-C infection was identified by *in situ* hybridization. Furthermore, virus infection appeared to be restricted to hepatocytes and the detection of minus-strand RNA showed viral replication in a few highly infected liver cells. *In vitro* infection of HepG2 or HuH7 cells confirmed these findings by the release of virions into the culture supernatant. The above studies [62] do establish that GBV-C is a hepatotropic virus, infecting human cells of hepatic origin both *in vivo* and *in vitro*. Besides its replication in human hepatocytes and human cells of hepatic origin, GBV-C has also been shown to replicate in human peripheral blood mononuclear cells [59,60] using similar appropriate methodologies.

Pathogenesis, persistence, treatment and disease

Most GBV-C infections appear to be asymptomatic, transient, and self-limiting, with slight or no elevation of alanine aminotransferase (ALT) levels. Most of these subclinical cases resolve after the loss of serum GBV-C RNA with a concomitant appearance of antibody to the E2 protein of GBV-C [56–58,73]. These infections are rarely identified and very difficult to evaluate when studied in multitransfused patients and/or in patients with HBV or HCV infections.

Persistence

GBV-C has been found to induce persistent infection in about 5–10% of infected individuals. Eight hemodialysis patients with GBV-C infection were followed retrospectively for 7–16 years [74]. In two of the patients, GBV-C was found at the beginning of hemodialysis. One patient had a history of transfusion, and GBV-C RNA persisted over a period of 16 years. The other patient cleared GBV-C RNA after 10 years. In five patients, GBV-C RNA was detected 3–20 weeks after blood transfusion and persisted for up to 13 years. A recent report [52] on the persistence and clinical outcome of GBV-C infection in pediatric BMT recipients and children treated for hematological malignancy showed the persistence of GBV-C viremia over a follow-up period of 3–10 years. Thus, the possibility of more severe long-term consequences of GBV-C infection should not be precluded. Elucidating the viral mechanism that led to the establishment and maintenance of the persistent state is very crucial for the understanding

232

of the pathogenesis of GBV-C [75]. In the study above [52], it is postulated that since these patients with hematological malignancy and BMT recipients suffer from severe immunosuppression; and immunosuppression itself may have a significant role on the clearance of GBV-C viremia and the production of a GBV-C anti-E2 antibody response.

Treatment

Several studies [76–80] indicate that GBV-C is sensitive to IFN while the patient is receiving treatment. However, most cases relapse on withdrawal of treatment. Lau et al. [81] reported that IFN-alpha treatment caused a marked, usually transient, reduction in serum GBV-C RNA, and that ribavirin had, at most, a modest anti-viral effect. On the other hand, Marrone [82] showed that GBV-C RNA levels were suppressed but not eradicated by IFN-alpha and were unaffected by ribavirin treatment. However, the spontaneous loss of GBV-C RNA occurred over time in a proportion of patients with GBV-C infection. From these studies, it is believed that a larger study of GBV-C RNA-positive population is required to more clearly define the key variables for the establishment of an effective anti-viral therapy for GBV-C-infected individuals. These variables should include such parameters as duration, level (dose) of anti-viral treatment, viral load and genotype, and possibly the effect of dual infection with either hepatitis B or C.

Fulminant hepatitis

The role of GBV-C in the etiology of fulminant hepatitis has not been fully established [73]. Yoshiba et al. [83] documented the presence of GBV-C RNA in 3 of 6 patients with fulminant hepatitis without evidence of infection with other known hepatitis viruses. Since that report, doubts concerning the association of GBV-C with acute liver failure [84] have arisen. Specifically, of whether GBV-C was an "innocent bystander" transmitted by transfusions given to the 3 patients prior to the onset of fulminant hepatitis. Additional studies [85], however, showed that few of the 63 fulminant hepatitis patients studied had received blood transfusions prior to the onset of fulminant hepatitis. In a similar study [86], GBV-C RNA was detected in 3 (20%) of 15 patients with HBV infection and 3 (12%) of 25 patients without markers of non-A-E infection. Overall, GBV-C RNA was detected in 6 of 44 (14%) patients with fulminant hepatitis, a frequency significantly higher ($P < 0.001$) than that in three of 326 (0.9%) aged matched blood donors. Of the six patients with GBV-C RNA, only three had a history of transfusion and all of these patients were coinfected with HBV. These results, according to Tameda et al., [86] indicate a role of GBV-C in inducing fulminant hepatitis either by itself or in concert with other hepatitis viruses. Other recent studies [42,87–89] either concur or refute the involvement of GBV-C fulminant hepatitis in non-A-E hepatitis patients.

Acute and chronic GBV-C infection and organ transplantation

A study implicating GBV-C in a significant number of acute and chronic cases of non-A-E hepatitis has been reported [90]. In 6 chronic hepatitis patients with GBV-C infection, the histology of the liver samples revealed chronic active hepatitis in one patient and chronic persistent hepatitis in five others. All these patients had elevated ALT levels between 89 and 478 U/l. Among the 11 acute hepatitis cases with GBV-C infection, the ALT levels varied between 615 and 2477 U/l. However, this study has not been confirmed by other investigations.

In another study, GBV-C RNA was investigated in 67 non-A-E patients with liver disease [91]. These investigators found that the spectrum of liver disease associated with GBV-C infection in these patients was wide, with a variety of histological liver lesions ranging from steatosis to fibrosis and cirrhosis. Of interest, nonspecific inflammatory bile duct lesions were found in 50% of the patients with GBV-C infection, and that infection was significantly more often associated with elevated cholestatic enzymes, namely, gamma-glutamyl transpeptidase and alkaline phosphatase. Other similar studies [92] have shown that GBV-C infection could affect the clinical course and outcome after OLT by the development of severe cholestasis due to bile duct damage and bile duct loss. Extensive studies on the association between recurrent or *de novo* GBV-C infection and severe post-transplant cholestatis and ductopenia need to be extended to further confirm that patients with chronic cholestatis are GBV-C-positive [73].

A comparative study of the acquisition of GBV-C, HCV and HBV in OLT patients found that HBV-infected and uninfected OLT patients were more susceptible to GBV-C infection than HCV-infected patients [93]. These findings demonstrate possible viral interference between GBV-C and HCV infection. The effects of GBV-C in other organ transplantation such as kidney [50,93–95] and BMT [52], shed light on the modes of GBV-C transmission among end-stage renal disease and BMT patients. In one study, GBV-C RNA-positive/HCV-positive renal transplant patients experienced acute rejection more frequently than GBV-C RNA-negative/HCV-positive renal transplant patients [93]. However, further studies are required to determine the risk of transmission of GBV-C by organ transplantation and its role in post-transplantation liver disease.

HCV Coinfection

The clinicopathological characteristics of patients with chronic hepatitis C-coinfected with GBV-C were studied in 214 patients with chronic HCV, 18 of which were also positive for GBV-C RNA [96]. It was found that the histological features of GBV-C RNA patients tended to show slightly stronger intralobular inflammatory cell infiltration, portal lymphocytic reaction, and significantly stronger damage of bile duct, perivenular and pericellular fibrosis. Even in healthy blood donors (without HCV coinfection) with isolated GBV-C viremia, steatosis and mild portal inflammatory lesions were found in liver biopsies [97].

Numerous studies have shown that in HCV-coinfected individuals, GBV-C does not affect HCV replication, HCV RNA concentration, and liver disease. On the other hand,

Manolakopoulos et al. [98] showed the influence of GBV-C viremia on the clinical, virological, and histological features of early hepatitis C-related hepatic disease. It was found that although coinfection with GBV-C did not alter the biochemical and virological profiles of patients with HCV hepatitis, there was an association between GBV-C and HCV viremia and portal and periportal inflammation. It was noted that the duration of HCV/GBV-C coinfection may be an important factor in progression of liver disease. It was also observed that inflammation with necrosis in the portal and periportal tracts was significantly higher in patients with combined viremia compared to those with HCV infection alone. These findings suggest that GBV-C in patients with HCV infection might accelerate liver injury. In concurrence, a trend has been noted towards more severe fibrosis in patients with dual infection. Diamantis et al. [99] reported that virus load and ALT levels did not differ significantly in patients coinfected with HCV and GBV-C. However, mild fibrosis correlated with GBV-C coinfection. More recently it was shown that chronic GBV-C infection of HCV patients worsens the histological features of liver disease, including bile duct damage, perivenular fibrosis, pericellular fibrosis and IR [100]. Similarly, comparing histological features between GBV-C-positive and GBV-C-negative chronic hepatitis C showed that the histological feature of GBV-C RNA-positive patients tends to show slightly stronger intralobular inflammatory cell infiltration, portal lymphocytic reaction, and significantly stronger damage of bile duct, perivenular, and pericellular fibrosis [98]. In total, GBV-C coinfection was suggested to worsen hepatic histological features, while chronic GBV-C coinfection did not seem to modify the serological features of chronic hepatitis.

Finally, a study on the incidence of transfusion-associated GBV-C infection and its relation to liver disease in multitransfused patients has been reported [101]. The conclusion was that most GBV-C infections were not associated with hepatitis and that GBV-C did not worsen the course of concurrent HCV infection; however, this result was based only on the evaluation of ALT levels. This result is not surprising since multitransfused patients are not ideal to study virus pathogenicity, especially without data regarding the number of units these patients received and if any of the units were positive for the protective antibody, anti-GBV-C E2 [56–58,73]. Similarly, another study on community-acquired hepatitis concluded that persistent infection with GBV-C was common, but it did not lead to chronic disease and did not affect the clinical course in patients with hepatitis A, B, or C, based on ALT levels [102]. However, the authors note that they only evaluated patients with clinically apparent disease and, therefore, conclusions regarding GBV-C infection alone could not be made.

Hepatocellular carcinoma (HCC)

Several studies [103–107] concluded that GBV-C is unlikely to be a major etiologic agent in HCC. However, in one study, the authors report that the presence of GBV-C RNA in serum was associated with a statistically significant (5.4-fold) risk of HCC among non-Asians in Los Angeles County, California [108]. The excess risk for HCC in GBV-C-infected individuals was independent of the effects of both hepatitis B and hepatitis C infections. It was concluded that chronic infection with GBV-C may play a role in the development of HCC, and if the observed statistical association is a casual one, then infection with GBV-C may account for approximately 8% of HCC cases. Regardless, more studies are required to establish a link between HCC and GBV-C infection.

HIV Coinfection

The prevalence of GBV-C markers, namely, GBV-C RNA and E2 anti-GBV-C in human immunodeficiency virus 1-infected patients was found to be high in many studies (Table 2), ranging from 14.4 to 37.7% for GBV-C RNA and from 14.5 to 56.9% for anti-GBV-C E2 [109–119]. Of interest is the study of Heringlake et al. [110] that GBV-C infection is a favorable prognostic factor in HIV-infected patients where CD4 cell counts were significantly higher in GBV-C viremic patients. Furthermore, Kaplan–Meier analysis demonstrated significantly better cumulative survival in GBV-C RNA-positive HIV-infected patients, suggesting that GBV-C might be a favorable prognostic factor in HIV disease. This observation was later substantiated from studies that demonstrate GBV-C RNA is associated with a slower progression of HIV disease in coinfected patients, due to a higher CD4 + lymphocyte count [113]. Similarly, Yeo et al. [118] reported that the risk of AIDS was 40% lower for GBV-C-positive patients with hemophilia. Compared with GBV-C RNA-negative patients, the GBV-C RNA-positive patients had higher CD4 + lymphocyte counts despite similar age and viral loads. These investigators speculated that GBV-C could indirectly affect AIDS risk through induction of various chemokines and other soluble factors, or by altered expression of chemokine receptors, which are essential co-receptors for HIV infection.

Table 2

Prevalence of GBV-C markers in HIV type 1 infected individuals

Study	No. of patients	GBV-C RNA		Anti-GBV-C E2		Reference
		No.	Percent	No.	Percent	
1	58	12	20.7	20	30.4	[109]
2	197	33	16.8	112	56.9	[110]
3	79	18	22.8	NT*		[119]
4	80					[111]
5	138	30	21.7	20	14.5	[112]
6	95	23	24.2	18	17.0	[113]
7	56	16	28.6	17	30.4	[114]
8	379	143	37.7	NT*		[115]
9	41	11	26.8	NT*		[116]
10	160	23	14.4	NT*		[117]
11	131	19	14.5	38	29.0	[118]

*NT: not tested.

Future studies

At present, attempts to grow virus in tissue culture or to infect convenient and available laboratory animals have hampered the study of GBV-C. For these reasons, the development of vaccine or immune serum globulin (IgG) has been difficult to achieve. However, since high titers of GBV-C E2 antibodies have been shown to be protective in humans [55–58], the possibility of using human IgG for such a purpose is intriguing. Other possibilities, including the use of recombinant antigens, such as GBV-C E2 glycoprotein as a synthetic subunit vaccine, may need to be explored. However, based on the most recent studies regarding HIV coinfection, the elimination of GBV-C from a patient may not be a desirable outcome. Future studies will need to address these issues.

References

1. Deinhardt F, Holmes AW, Capps RB, Popper H. J Exp Med 1967; 125: 673–687.
2. Simons JN, Pilot-Matias TJ, Leary TP, Dawson GJ, Desai SM, Schlauder GG, Muerhoff AS, Erker JC, Buijk SL, Chalmers ML, van Sant CL, Mushahwar IK. Proc Natl Acad Sci USA 1995; 92: 3401–3405.
3. Simons JN, Leary TP, Dawson GJ, Pilot-Matias TJ, Muerhoff AS, Schlauder GG, Desai SM, Mushahwar IK. Nature Med 1995; 1: 564–569.
4. Leary TP, Muerhoff AS, Simons JN, Pilot-Matias TJ, Erker JC, Chalmers MC, Schlauder GG, Dawson GJ, Desai SM, Mushahwar IK. J Med Virol 1996; 48: 60–67.
5. Linnen J, Wages J, Zhang-Keck Z-Y, Fry KE, Krawczynski KZ, Alter H, Koonin E, Gallagher M, Alter M, Hadziyannis S, Karayiannis P, Fung K, Nakatsuji Y, Shih JW-K, Young L Jr., Hoover C, Fernandez J, Chen S, Zou J-C, Morris T, Hyams KC, Ismay S, Lifson JD, Hess G, Foung SKH, Thomas H, Bradley D, Margolis H, Kim JP. Science 1996; 271: 505–508.
6. Muerhoff AS, Leary TP, Simons JN, Pilot-Matias TJ, Erker JC, Chalmers MC, Schlauder GG, Dawson GJ, Desai SM, Mushahwar IK. J Virol 1995; 69: 5621–5630.
7. Dawson GJ, Schlauder GG, Pilot-Matias TJ, Thiele D, Leary TP, Murphy P, Rosenblatt JE, Simons JN, Martinson FE, Gutierrez RA, Lentino JR, Pachucki C, Muerhoff AS, Widell A, Tegtmeier G, Desai S, Mushahwar IK. J Med Virol 1996; 50: 97–103.
8. Melvin SL, Dawson GJ, Carrick RJ, Schlauder GG, Heynen CA, Mushahwar IK. J Virol Methods 1998; 71: 147–157.
9. Muerhoff AS, Simons JN, Erker JC, Desai SM, Mushahwar IK. J Virol Methods 1996; 62: 55–62.
10. Simons JN, Desai SM, Schultz DE, Lemon SM, Mushahwar IK. J Virol 1996; 70: 6129–6135.
11. Rijnbrand R, Abell G, Lemon SM. J Virol 2000; 74: 773–783.
12. Yoo BJ, Spaete RR, Geballe AP, Selby M, Houghton M, Han JH. Virology 1992; 191: 889–899.
13. Sbardellati A, Scarselli E, Tomei L, Kekule AS, Traboni C. J Virol 1999; 73: 10546–10550.
14. Bukh J, Apgar CL, Yanagi M. Virology 1999; 262: 470–478.
15. Surowy TK, Leary TP, Carrick RJ, Knigge MF, Pilot-Matias TJ, Heynen C, Gutierrez RA, Desai SM, Dawson GJ, Mushahwar IK. J Gen Virol 1997; 78: 1851–1859.
16. Erker JC, Simons JN, Muerhoff AS, Leary TP, Chalmers ML, Desai SM, Mushahwar IK. J Gen Virol 1996; 77: 2713–2720.

17. Weiner AJ, Brauer MJ, Rosenblatt J, Richman KH, Tung J, Crawford K, Bonino F, Saracco G, Choo QL, Houghton M, Han JH. Virol 1991 1991; 180: 842–848.

18. Hijikata M, Kato N, Ootsuyama Y, Nakagawa M, Ohkoshi S, Shimotohno K. Biochem Biophys Res Commun 1991; 175: 220–228.

19. Pilot-Matias TJ, Carrick RJ, Coleman PF, Leary TP, Surowy TK, Simons JN, Muerhoff AS, Buijk SL, Chalmers ML, Dawson GJ, Desai SM, Mushahwar IK. Virology 1996; 225: 282–292.

20. Dille BJ, Surowy TK, Gutierrez RA, Coleman PF, Knigge MF, Carrick RJ, Aach RD, Hollinger FB, Stevens CE, Barbosa LH, Nemo GJ, Mosley JW, Dawson GJ, Mushahwar IK. J Infect Dis 1997; 175: 458–461.

21. Grakoui A, McCourt DW, Wychowski C, Feinstone SM, Rice CM. Proc Natl Acad Sci USA 1993; 90: 10583–10587.

22. Grakoui A, McCourt DW, Wychowski C, Feinstone SM, Rice CM. J Virol 1993; 67: 2832–2843.

23. Leary TP, Desai SM, Erker JC, Mushahwar IK. J Gen Virol 1997; 78: 2307–2313.

24. Belyaev A, Chong S, Novikov A, Kongpachith A, Masiarz FR, Lim M, Kim JP. J Virol 1998; 72: 868–872.

25. Butkiewicz N, Yao N, Zhong W, Wright-Minogue J, Ingravallo P, Zhang R, Durkin J, Standring DN, Baroudy BM, Sangar DV, Lemon SM, Lau JY, Hong Z. J Virol 2000; 74: 4291–4301.

26. Sbardellati A, Scarselli E, Amati V, Falcinelli S, Kekule AS, Traboni C. J Gen Virol 2000; 81: 2183–2188.

27. Gwack Y, Yoo H, Song I, Choe J, Han JH. J Virol 1999; 73: 2909–2915.

28. Leary TP, Muerhoff AS, Simons JN, Pilot-Matias TJ, Erker JC, Chalmers ML, Schlauder GG, Dawson GJ, Desai SM, Mushahwar IK. J Virol Methods 1996; 56: 119–121.

29. Muerhoff AS, Simons JN, Leary TP, Erker JC, Chalmers ML, Pilot-Matias TJ, Dawson GJ, Desai SM, Mushahwar IK. J Hepatol 1996; 25: 379–384.

30. Tucker TJ, Smuts H, Eickhaus P, Robson SC, Kirsch RE. J Med Virol 1999; 59: 52–59.

31. Tucker TJ, Smuts HE. J Med Virol 2000; 62: 82–83.

32. Simons JN, Mushahwar IK. Hepatol Clin 1996; 4: 1–16.

33. Seemayer CA, Viazov S, Philipp T, Roggendorf M. Infection 1998; 26: 39–41.

34. Semprini AE, Persico T, Thiers V, Oneta M, Tuveri R, Serafini P, Boschini A, Giuntelli S, Pardi G, Brechot C. J Infect Dis 1998; 177: 848–854.

35. Scallan MF, Clutterbuck D, Jarvis LM, Scott GR, Simmonds P. J Med Virol 1998; 55: 203–208.

36. Kao JH, Liu CJ, Chen PJ, Chen W, Hsiang SC, Lai MY, Chen DS. J Med Virol 1997; 53: 348–353.

37. Hino K, Moriya T, Ohno N, Takahashi K, Hoshino H, Ishiyama N, Katayama K, Yoshizawa H, Mishiro S. Arch Virol 1998; 143: 65–72.

38. Zanetti AR, Tanzi E, Romano L, Principi N, Zuin G, Minola E, Zapparoli B, Palmieri M, Marini A, Ghisotti D, Friedman P, Hunt J, Laffler T. J Med Virol 1998; 54: 107–112.

39. Kao JH, Huang CH, Chen V, Tsai TJ, Lee SH, Hung KY, Chen DS. J Infect Dis 1999; 180: 191–194.

40. Leary TP, Mushahwar IK. In: Viral Hepatitis: Diagnosis, Therapy and Prevention (Spector S, editor). Totoway, NJ: Humana Press, Inc.; 1999; pp. 175–191.

41. Moriyama K, Okamura T, Nakano S. Br J Haematol 1997; 96: 864–867.

42. Inoue K, Yoshiba M, Sekiyama K, Kohara M. J Hepatol 1999; 30: 801–806.
43. Nishiyama Y, Nakashima H, Hino K, Mori K, Ishikawa M, Okamoto H. Transfusion 1996; 36: 669.
44. Nerurkar VR, Chua PK, Hoffmann PR, Dashwood WM, Shikuma CM, Yanagihara R. J Med Virol 1998; 56: 123–127.
45. Berg T, Naumann U, Fukumoto T, Bechstein WO, Neuhaus P, Lobeck H, Hohne M, Schreier E, Hopf U. Transplantation 1996; 62: 711–714.
46. Viazov S, Riffelmann M, Sarr S, Ballauff A, Meisel H, Roggendorf M. J Hepatol 1997; 27: 85–90.
47. De Filippi F, Colombo M, Rumi MG, Tradati F, Prati D, Zanella A, Mannucci PM. Blood 1997; 90: 4634–4637.
48. de Lamballerie X, Charrel RN, Dussol B. N Engl J Med 1996; 334: 1549.
49. Prati D, Zanella A, Bosoni P, Rebulla P, Farma E, De Mattei C, Capelli C, Mozzi F, Gallisai D, Magnano C, Melevendi C, Sirchia G. Blood 1998; 91: 774–777.
50. Murthy BV, Muerhoff AS, Desai SM, Natov SN, Bouthot BA, Ruthazer R, Schmid CH, Levey AS, Mushahwar IK, Pereira BJ. Kidney Hit 1998; 53: 1769–1774.
51. Kallinowski B, Buhrmann C, Seipp S, Goeser T, Stremmel W, Otto G, Theilmann L. Liver Transpl Surg 1998; 4: 28–33.
52. Maki YO, Sumazaki R, Tsuchida M, Koike K, Fukushima T, Matsui A. Blood 1999; 92: 721–727.
53. Gutierrez RA, Dawson GJ, Mushahwar IK. J Virol Methods 1997; 69: 1–6.
54. Mushahwar IK, Simons JN, Dawson GJ. In: Viral Hepat (Zuckerman AJ, Thomas HC, editors). London: Churchill Livingston; 1998; pp. 469–476.
55. Mushahwar IK, Erker JC, Muerhoff AS, Desai SM. Hepatol Clin 1998; 6: 23–27.
56. Tillmann HL, Heringlake S, Trautwein C, Meissner D, Nashan B, Schlitt HJ, Kratochvil J, Hunt J, Qiu X, Lou SC, Pichlmayr R, Manns MP. Hepatology 1998; 28: 379–384.
57. Hassoba HM, Pessoa MG, Terrault NA, Lewis NJ, Hayden M, Hunt JC, Qiu X, Lou SC, Wright TL. J Med Virol 1998; 56: 253–258.
58. Thomas DL, Vlahov D, Alter HJ, Hunt JC, Marshall R, Astemborski J, Nelson KE. J Infect Dis 1998; 177: 539–542.
59. Madejon A, Fogeda M, Bartolome J, Pardo M, Gonzalez C, Cotant T, Carreno V. Gastroenterology 1997; 113: 573–578.
60. Saito S, Tanaka K, Kondo M, Morita K, Kitamura T, Kiba T, Numata K, Sekihara H. Biochem Biophys Res Commun 1997; 237: 288–291.
61. Kobayashi M, Tanaka E, Nakayama J, Furuwatari C, Katsuyama T, Kawasaki S, Kiyosawa K. J Med Virol 1999; 57: 114–121.
62. Seipp S, Scheidel M, Hofmann WJ, Tox U, Theilmann L, Goeser T, Kallinowski B. J Hepatol 1999; 30: 570–579.
63. Laskus T, Radkowski M, Wang LF, Vargas H, Rakela J. J Virol 1997; 71: 7804–7806.
64. Pessoa MG, Terrault NA, Detmer J, Kolberg J, Collins M, Hassoba HM, Wright TL. Hepatology 1998; 27: 877–880.
65. Laras A, Zacharakis G, Hadziyannis SJ. J Hepatol 1999; 30: 383–388.
66. Saleh MG, Tibbs CJ, Koskinas J, Pereira LM, Bomford AB, Portmann BC, McFarlane IG, Williams R. Hepatology 1994; 20: 1399–1404.
67. Chang TT, Young KC, Yang YJ, Lei HY, Wu HL. Hepatology 1996; 23: 977–981.
68. Kao JH, Chen PJ, Lai MY, Wang TH, Chen DS. J Med Virol 1997; 52: 270–274.
69. Lanford RE, Sureau C, Jacob JR, White R, Fuers TR. Virology 1994; 205: 2321–2324.

70. Young KK, Resnick RM, Myers TW. J Clin Microbiol 1993; 31: 882–886.

71. Lanford RE, Chavez D, Chisari FV, Sureau C. J Virol 1995; 69: 8079–8083.

72. McGuinness PH, Bishop GA, McCaughn GW, Trowbridge R, Gowans EJ. Lancet 1994; 343: 551–552.

73. Mushahwar IK, Zuckerman AJ. J Med Virol 1998; 56: 1–3.

74. Masuko K, Mitsui T, Iwano K, Yamazaki C, Okuda K, Meguro T, Murayama N, Inoue T, Tsuda F, Okamoto H, Miyakawa Y, Mayumi M. N Engl J Med 1996; 334: 1485–1490.

75. Kuhn J. Intervirology 1996; 39: 139–150.

76. Orito E, Mizokami M, Yasuda K, Sugihara K, Nakamura M, Mukaide M, Ohba KI, Nakano T, Kato T, Kondo Y, Kumada T, Ueda R, Iino S. J Hepatol 1997; 27: 603–612.

77. Enomoto M, Nishiguchi S, Fukuda K, Kuroki T, Tanaka M, Otani S, Ogami M, Monna T. Hepatology 1998; 27: 1388–1393.

78. Kao JH, Lai MY, Chen W, Chen PJ, Chen DS. J Gastroenterol Hepatol 1998; 13: 1249–1253.

79. Jarvis LM, Bell H, Simmonds P, Hawkins A, Hellum K, Harthug S, Maeland A, Ritland S, Myrvang B, von der Lippe B, Raknerud N, Skaug K. Scand J Gastroenterol 1998; 33: 195–200.

80. Pawlotsky JM, Roudot-Thoraval F, Muerhoff AS, Pellerin M, Germanidis G, Desai SM, Bastie A, Darthuy F, Remire J, Zafrani ES, Soussy CJ, Mushahwar IK, Dhumeaux D. J Med Virol 1998; 54: 26–37.

81. Lau JY, Qian K, Detmer J, Collins ML, Orito E, Kolberg JA, Urdea MS, Mizokami M, Davis GL. J Infect Dis 1997; 76: 421–426.

82. Marrone A, Shih JW, Nakatsuji Y, Alter HJ, Lau D, Vergalla J, Hoofnagle JH. Am J Gastroenterol 1997; 92: 1992–1996.

83. Yoshiba M, Okamoto H, Mishiro S. Lancet 1995; 346: 1131–1132.

84. Alter HJ. N Engl J Med 1996; 334: 1536–1537.

85. Yoshiba M, Inoue K, Sekiyama K. N Engl J Med 1996; 335: 1392–1393.

86. Tameda Y, Kosaka Y, Tagawa S, Takase K, Sawada N, Nakao H, Tsuda F, Tanaka T, Okamoto H, Miyakawa Y, Mayumi M. J Hepatol 1996; 25: 842–847.

87. Kanda T, Yokosuka O, Ehata T, Maru Y, Imazeki F, Saisho H, Shiratori Y, Omata M. Hepatology 1997; 25: 1261–1265.

88. Wu JC, Chiang TY, Huang YH, Huo TI, Hwang SJ, Huang IS, Sheng WY, Lee SD. J Med Virol 1998; 56: 118–122.

89. Liu CJ, Kao JH, Lai MX, Chen PJ, Chu JS, Chen W, Chen DS. J Gastroenterol Hepatol 1999; 14: 352–357.

90. Fiordalisi G, Zanella I, Mantero G, Bettinardi A, Stellini R, Paraninfo G, Cadeo G, Primi D. J Infect Dis 1996; 174: 181–183.

91. Colombatto P, Randone A, Civitico C, Gorin JM, Dolci L, Medaina N, Calleri G, Oliveri F, Baldi M, Tappero G, Volpes R, David E, Verme G, Smedile A, Bonino F, Brunetto MR. J Viral Hepat 1997; 4(Suppl 1): 55–60.

92. Ross RS, Viazov S, Kruppenbacher JP, Elsner S, Sarr S, Lange R, Eigler FW, Roggendorf M. Liver 1997; 17: 238–243.

93. Rostaing L, Izopet J, Arnaud C, Cisterne JM, Alric L, Rumeau JL, Duffaut M, Durand D. Transplantation 1999; 67: 556–560.

94. Murthy BVR, Muerhoff AS, Desai SM, Yamaguchi J, Mushahwar IK, Schmid CH, Levey AS, Pereira BJG. J Am Soc Nephrol 1997; 8: 1164–1173.

95. Murthy BV, Muerhoff AS, Desai SM, Lund J, Schmid CH, Levey AS, Mushahwar IK, Pereira BJ. Transplantation 1997; 63: 346–351.

96. Moriyama M, Shibahara N, Matsumura H, Kaneko M, Shioda A, Komine F. Gastroenterology 1998; 114: L0439.

97. Bjorkman P, Sundstrom G, Veress B, Widell A. Vox Sang 2000; 78: 143–148.

98. Manolakopoulos S, Morris A, Davies S, Brown D, Hajat S, Dusheiko G. J Hepatol 1998; 28: 173–178.

99. Diamantis ID, Kouroumalis E, Koulentaki M, Fasler-Kan E, Schmid PA, Hirsch HH, Buhler H, Gyr K, Battegay M. Eur J Clin Microbiol Infect Ibis 1997; 16: 916–919.

100. Moriyama M, Matsumura H, Shimizu T, Shioda A, Kaneko M, Saito H, Miyazawa K, Tanaka N, Sugitani M, Komiyama K, Arakawa Y. Liver 2000; 20: 397–404.

101. Alter HJ, Nakatsuji Y, Melpolder J, Wages J, Wesley R, Shih JW-K, Kim JP. N Engl J Med 1997; 336: 747–754.

102. Alter MJ, Gallagher M, Morris TT, Moyer LA, Meeks EL, Krawczynksi K, Kim JP, Margolis HS. N Engl J Med 1997; 336: 741–746.

103. Kitamoto M, Moriya T, Sasaki F, Katayama K, Yoshizawa H, Asahara T, Nakanishi T. Nippon Rinsho 1997; 55: 583–586.

104. Kubo S, Nishiguchi S, Kuroki T, Hirohashi K, Tanaka H, Tsukamoto T, Shuto T, Kinoshita H. J Hepatol 1997; 27: 91–95.

105. Brechot C, Jaffredo F, Lagorce D, Gerken G, Meyer zum Buschenfelde K, Papakonstontinou A, Hadziyannis S, Romeo R, Colombo M, Rodes J, Bruix J, Williams R, Naoumov N. J Hepatol 1998; 29: 173–183.

106. Lightfoot K, Skelton M, Kew MC, Yu MC, Kedda MA, Coppin A, Hodkinson J. Hepatology 1998; 26: 740–742.

107. Nishiyama Y, Hino K, Sawada N, Tsuda F. J Gastroenterol 1999; 34: 494–497.

108. Yuan JM, Govindarajan S, Ross RK, Yu MC. Cancer 1999; 86: 936–943.

109. de Martino M, Azzari C, Resti M, Moriondo M, Rossi ME, Galli L, Vierucci A. J Infect Dis 1998; 178: 862–865.

110. Heringlake S, Ockenga J, Tillman HL, Trautwein C, Solomon N, Schmidt RE, Manns MP. J Infect Dis 1998; 177: 1723–1726.

111. Bourlet T, Guglielminotti C, Evrard M, Berthelot P, Grattard F, Fresard A, Lucht FR, Pozzetto B. J Med Virol 1999; 58: 373–377.

112. Goubau P, Liu HF, Goderniaux E, Burtonboy G. J Med Virol 1999; 57: 367–369.

113. Lefrere JJ, Roudot-Thoraval F, Morand-Joubert L, Petit JC, Lerable J, Thauvin M, Mariotti M. J Infect Dis 1999; 179: 783–789.

114. Palomba E, Bairo A, Tovo PA. Acta Paediatr 1999; 88: 1392–1395.

115. Rey D, Fraize S, Vidinic J, Meyer P, Fritsch S, Labouret N, Schmitt C, Lang JM, Stoll-Keller F. J Med Virol 1999; 57: 75–79.

116. Toyoda H, Fukuda Y, Hayakawa T, Takamatsu J, Saito H. J Acquir Immune Defic Syndr Hum Retrovirol 1998; 17: 209–213.

117. Puig-Basagoiti F, Cabana M, Guilera M, Gimenez-Barcons M, Sirera G, Tural C, Clotet B, Sanchez-Tapias JM, Rodes J, Saiz JC, Martinez MA. J Acquir Immune Defic Syndr 2000; 23: 89–94.

118. Yeo AE, Matsumoto A, Hisada M, Shih JW, Alter HJ, Goedert JJ. Multicenter Hemophilia Cohort Study. Ann Intern Med 2000; 132: 959–963.

119. Zehender G, De Maddalena C, Bosisio AB, Gianotto M, Santambrogio S, Moroni M, Galli M. J Hum Virol 1998; 1: 96–100.

Viral Hepatitis
I.K. Mushahwar (editor)
© 2004 Elsevier B.V. All rights reserved.

Torque Teno Virus (TTV): molecular virology and clinical implications

Hiroaki Okamoto,* Tsutomu Nishizawa and Masaharu Takahashi
Division of Virology, Department of Infection and Immunity, Jichi Medical School, Minamikawachi-Machi, Tochigi-Ken, Japan

Abbreviations: aa, amino acids; ALT, alanine aminotransferase; α-GST, α-glutathione-*S*-transferase; BFDV, psittacine beak and feather disease virus; CAV, chicken anemia virus; FHF, fulminant hepatic failure; HAV, hepatitis A virus; HBV, hepatitis B virus; HCV, hepatitis C virus; HCC, hepatocellular carcinoma; HIV, human immunodeficiency virus type 1; ICTV, International Committee on Taxonomy of Viruses; kb, kilobases; nt, nucleotides; ORF, open reading frame; PCR, polymerase chain reaction; PCV1, porcine circovirus type 1; PCV2, porcine circovirus type 2; PI, postinoculation; Rep proteins, replication-associated proteins; SENV, SEN virus; TTV, Torque Teno virus; TTMV, Torque Teno mini virus; UTR, untranslated region.

Introduction

In 1997, a novel DNA virus unrelated to the known human viruses was isolated from the serum of a patient with post-transfusion acute hepatitis of unknown etiology, and it was provisionally named TT virus (TTV) after the initials (T.T.) of the index patient [1,2]. Due to the lack of a suitable cell culture system to support the growth of TTV, the biology of TTV is unknown. However, it has been elucidated that TTV is a unique unenveloped virus with a circular single-stranded DNA genome of 3.6–3.9 kilobases (kb) [3–5], and is characterized by ubiquitous distribution and wide genetic variability [6–12]. TTV infection is characterized by persistent viremia and the presence of replicating virus in a wide range of tissues and organs including the liver, bone marrow, lung tissues, lymph nodes, spleen and pancreas [13–15], as well as the presence of activated peripheral blood mononuclear cells [16,17], indicating that TTV infection may be associated with hepatic and extrahepatic diseases according to the TTV genotype. This review summarizes our current knowledge on the molecular virology of TTV, and provides an overview of the clinical significance of TTV infections in humans.

**Corresponding author.*

242

Nomenclature of TTV and related viruses

The genome of TTV consists of a single-stranded DNA with a circular structure. Virus families with a single-stranded circular DNA genome are well known [18]; they include *Microviridae* and *Inoviridae* among bacterial viruses, *Geminiviridae* and *Nanoviridae* among plant viruses, and *Circoviridae* among animal viruses. The *Circoviridae* family includes at least seven animal viruses that infect several domesticated animals and bird species: chicken anemia virus (CAV) in the *Gyrovirus* genus; and porcine circovirus type 1 (PCV1), porcine circovirus type 2 (PCV2), psittacine beak and feather disease virus (BFDV), pigeon circovirus, goose circovirus, and canary circovirus in the *Circovirus* genus [19–21].

TTV is the first human virus with a single-stranded circular DNA genome to be identified, and its structure is partially similar to that of CAV. Namely, TTV and CAV possess negative sense genomes, in contrast to the ambisense genomes possessed by other circoviruses such as PCV1, PCV2 and BFDV [22–24]. The putative capsid protein of TTV possesses amino acid motifs that are characteristic of Rep proteins for rolling circle replication [3,25], similar to that of CAV. Therefore, TTV was originally classified as a member of the *Circoviridae* family. However, TTV differs from CAV in that it uses at least three spliced transcripts, whereas a single unspliced polycistronic transcript has been detected for CAV [18,26,27]. After the discovery of the original TTV isolate, many TTV variants with marked genetic variability were identified and segregated into at least 39 genotypes or 5 major genetic groups [12]. In 2000, a small virus that was distantly related to TTV and provisionally named as TTV-like mini virus, was discovered by polymerase chain reaction (PCR) using TTV-specific primers that partially matched homologous sequences in TTV-like mini virus [28].

A rule of the International Committee on Taxonomy of Viruses (ICTV) states, "No person's name shall be used when devising names for new taxa." Recently, taxonomic nomenclatures for TTV and TTV-like mini virus have been proposed by the ICTV Circoviridae Study Group, and they are designated as Torque Teno virus (TTV) and Torque Teno mini virus (TTMV), respectively. This name is derived from "Torque," which means *the necklace*, and "Tenuis," which means *thin*, and relates to the circular, single-stranded nature of its DNA genome, thus maintaining the widely used term, TTV. TTV and TTMV are classified into a novel floating genus, the *Anellovirus* genus, which has not yet been assigned to a family. The name of the genus, *Anellovirus*, is derived from "anello," which means *the ring*, and relates to the circular nature of the DNA genome [29].

TTV and TTMV are not restricted to humans as hosts, and species-specific TTV or TTMV strains have been detected in non-human primates (chimpanzees, macaques, tamarins, douroucoulis), tupaias, cats and dogs, and in farm animals [25,30–34]. Analysis of the complete sequences of TTVs from animals revealed high heterogeneity with regard to the size of the viral genome (2.0–3.9 kb), along with high genetic divergence when compared with human isolates [25,32,33]. However, the presumed genomic organization and transcriptional profiles correspond to those found in human isolates. The precise criteria demarcating species in the *Anellovirus* genus are currently being developed by the ICTV Circoviridae Study Group.

Virological characteristics

TTV and TTMV are each an unenveloped, small, spherical particle with a diameter of 30–32 nm and <30 nm, respectively [28,35]. The buoyant density of virions in cesium chloride is estimated to be 1.31–1.34 g/cm^3 for TTV in serum and 1.27–1.28 g/cm^3 for TTMV in serum. Both TTV and TTMV virions contain circular single-stranded DNA of negative polarity. TTV particles in the circulation are bound to IgG, forming immune complexes [35]. Therefore, TTV-associated particles recovered from the sera of infected humans are observed as aggregates of various sizes on electron microscopy. In contrast, TTV particles in feces exist as free virions banding at 1.33–1.35 g/cm^3. TTMV has not yet been visualized.

Genomic organization, mRNAs and proteins of TTV and TTMV

TTV and TTMV—each have a circular, single (minus)-stranded DNA of 3.6–3.9 and 2.8–2.9 kb, respectively [9,12,28,36–44]. The proposed genomic organization of the prototype TTV and TTMV isolates [TA278 (Accession No. AB017610) and CBD231 (AB026930), respectively] are illustrated in Fig. 1. The coding region including the four open reading frames (ORFs: ORF1, ORF2, ORF3 and ORF4) occupies 71% of the TA278 genome and 82% of the CBD231 genome. Two main ORFs, ORF1 and ORF2, may be directly deduced from the nucleotide sequence of each prototype. These two ORFs partially overlap and their estimated sizes slightly differ between the two isolates. As TTV and TTMV share a similar genomic organization, two additional coding regions, ORF3 and ORF4, are presumed to be created by the two shorter spliced mRNAs in each prototype (see below). The TTV and TTMV genomes, each contain a region of about 80–160 nucleotides (nt) with high GC content (approximately 90%), which have a high degree of similarity among the extremely divergent TTV variants (see below).

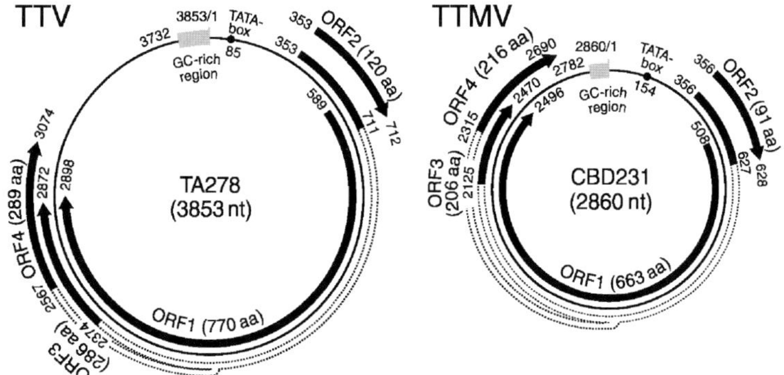

Fig. 1. Genomic organization of the prototype TTV and TTMV isolates. The closed arrows represent ORFs (ORF1–ORF4). The open boxes located between an upstream closed box and downstream closed arrow in ORF3 and ORF4 encoding joint proteins, represent areas corresponding to introns in the two shorter mRNAs. The shaded box indicates the GC-rich region, which forms characteristic stem-and-loop structures.

This sequence forms a secondary structure composed of stems and loops, suggesting the origin of DNA replication [11].

Knowledge about the genomic expression and replication of TTV and TTMV is limited mainly due to the lack of a cell culture propagation system for TTV and TTMV. However, double-stranded, replicative intermediate forms of TTV DNA have been detected not only in liver tissues but also in other tissues and organs including the bone marrow, lymph nodes, thyroid gland, lung, spleen, pancreas, kidney and muscle of infected individuals [13–15], as circoviruses generate a circular, double-stranded replicative intermediate during the replication and transcription of viral DNA [18]. TTV-specific mRNAs have also been detected in various tissues and organs of infected humans [14,15]. It has been demonstrated *in vitro* and *in vivo* that at least three spliced mRNAs of distinct sizes (3.0, 1.2, and 1.0 kb) with common 5′- and 3′-termini, are transcribed from the negative strand of the putative circular double-stranded replicative form TTV DNA [14,26]. These three mRNAs have in common a short splicing of 91 nt at the 5′-terminal region (Fig. 2). The splicing in the 3.0-kb mRNA does not affect ORF1 that is predicted in frame 1, since the splicing site is located upstream of the first ATG codon (nt 589). In contrast, ORF2 is deduced to start at the third ATG codon (nt 353) in frame 2 and encode 120 amino acids (aa) because the first and second ATG codons (nt 236 and 257) are located in the spliced region. The 1.2- and 1.0-kb mRNAs possess another splicing to join nt 711 with nt 2374 or nt 2567, leading to the creation of two novel ORFs (ORF3 encoding 286 aa or ORF4 encoding 289 aa, respectively) [14,26]. All three TTV mRNAs are transcribed from a common, internal promoter, i.e. a TATA-box (ATATAA), in the TTV genome and are linked with poly(A) stretches, immediately downstream of the poly(A) signal (AATAAA). The genomic organization and transcription profile are preserved among all TTV isolates of various genotypes [12]. The transcription profile of TTMV is not known, but the fact that TTV and TTMV share similar genomic organizations is highly suggestive that at least three mRNAs are expressed in TTMV.

The proteins of TTV and TTMV have not been molecularly characterized. The expression of three ORF2-related proteins remains to be demonstrated, and their function and virological significance need to be clarified in future studies. ORF1 is composed of

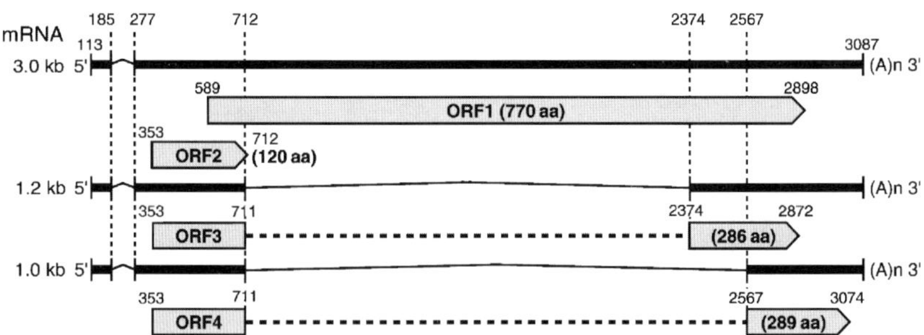

Fig. 2. Configuration of TTV mRNAs of three different sizes (3.0, 1.2, and 1.0 kb) with one or two splicings, and the predicted ORFs (ORF1–ORF4). The closed bars indicate exons, bent lines indicate introns, and shaded arrows represent coding regions.

770 aa in TTV (the TA278 isolate) and 663 aa in TTMV (the CBD231 isolate) (Fig. 1), and has been deduced to encode the putative capsid protein and Rep protein of TTV and TTMV. The N-terminal sequence of the putative ORF1 protein contains multiple positively charged amino acid residues (enriched with Arg residues), which are considered to mediate binding with viral DNA and transportation to the nucleus of the infected cell [11]. ORF2 protein is composed of 120 aa in TTV and 91 aa in TTMV, and the ORF2 proteins of TTV and TTMV share a common motif, $WX_7HX_3CXCX_5H$, as that in CAV [40]. The ORF2 proteins of TTV and TTMV contain an amino acid sequence that is similar to that of protein-tyrosine phosphatase alpha protein, as that of CAV [45]. Although the function of the TTV ORF3 protein has yet to be elucidated, it has been shown in an *in vitro* expression system using COS-1 cells that ORF3 protein is phosphorylated at serine residues in its C-terminal portion [46].

Genomic heterogeneity of TTV

Comparison of the complete nucleotide sequences of TTV variants, as well as the corresponding amino acid sequences of ORF1 and ORF2, reveals very high genetic variability for a DNA virus [9,11,12,47]. At least 39 genotypes including the prototype genotype (genotype 1 for the index case T.T.), with over 30% nucleotide diversity, have been identified [12]. Based on sequence similarity, eight genotypes (A–H) of the SEN virus (SENV), which was first reported as a new hepatitis virus in the *New York Times* on July 20, 1999, were found to be distantly related to the prototype genotype 1 TTV [43]. Among these eight SENV genotypes, SENV-A shows marked similarity with TTV genotype 9; SENV-B with genotype 10; SENV-D with genotype 12; SENV-E with genotype 14; SENV-C with genotype 15; and SENV-H with genotype 16 [48]. A total of 39 TTV genotypes can be classified into five main genetic groups (Groups 1–5) based on phylogenetic analysis (Fig. 3), and the five genetic groups show at least 50% nucleotide sequence divergence from each other. Group 1 includes genotype 1 of the prototype TTV and five additional genotypes (2–6). Group 2 consists of TTV isolates of genotypes 7, 8, 17, 22, 23 and 28. Group 3 consists of the largest number of genotypes (9–16, 18–20, and 24–27) and includes SANBAN (AB025946), which is the Japanese word meaning "the third," and the TUS01 (AB017613), TJN01 (AB028668), and TYM9 (AB050448) isolates as well as 8 genotypes of SENV [43]. Group 4 comprises the YONBAN isolates, whose name is derived from the Japanese word meaning "the fourth" [44]. Although this group initially consisted of only one genotype (genotype 21), eight genotypes (tentatively designated as genotypes 29–36 in this review article) identified in a later study [14] were added to this group. Furthermore, TTV variants of genotypes 37, 38 and 39 have recently been identified and segregated into the fifth genetic group (Group 5). It is likely that the number of TTV genotypes in each group and/or the number of groups will increase with the accumulation of sequence data of TTV isolates in future studies. TTMV shows genetic heterogeneity comparable to or even greater than that of TTV, although only a limited number of TTMV isolates have been isolated to date [28,36,44].

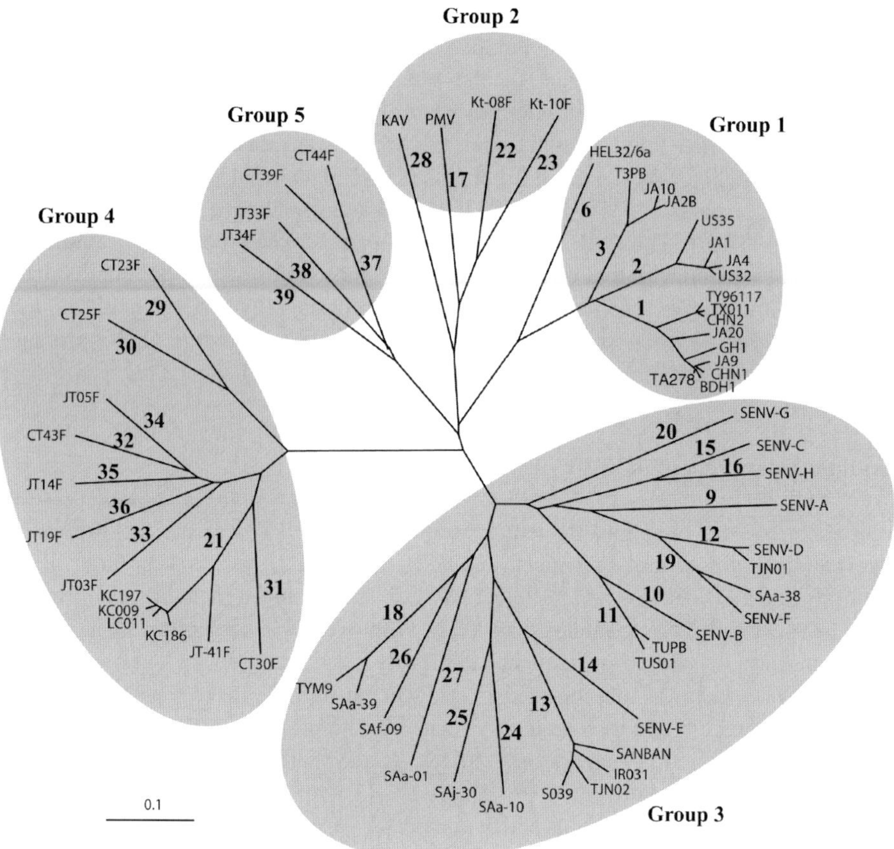

Fig. 3. Phylogenetic tree constructed based on the nearly entire nucleotide sequences of 60 TTV isolates by the neighbor-joining method. The TTV variants are classifiable into 5 main genetic groups, with over 50% nucleotide diversity. The entire or nearly entire nucleotide sequence has not yet been determined for TTV isolates of genotypes 4 and 5 in Group 1, and those of genotypes 7 and 8 in Group 2.

Detection of TTV DNA

Detection of TTV is mainly carried out by PCR. The reported prevalence of TTV viremia among healthy individuals and patients with or without liver disease vary widely in the examined countries in the world [10]. This seems to be largely attributable to the PCR method used. The extent to which the TTV sequence has been preserved varies by the genomic region; the coding region is remarkably variable, whereas the untranslated region (UTR) is well preserved. In particular, the 169–174 nt sequence downstream of the TATA-box in the UTR is highly conserved among TTV isolates [12]. Hence, the genomic areas and genotypes that are selected for designing primers for PCR amplification of TTV DNA, considerably influence the rate of detection of TTV DNA by PCR [9,52–54]. Two distinct PCR methods are widely used [11]. PCR for amplification of the N22 region within ORF1 (called N22 PCR), with primers derived from TTVs of genotypes 1 and 2, can detect mainly group 1 TTVs (genotypes 1–6). Positivity for TTV by N22 PCR in the general population has been reported to differ by geographic region, being 1–11% in the United States, 8–26% in Japan, and 86% in Gambia (detailed review in Ref. [10]). In contrast, PCR amplification of the highly conserved area of the UTR (called UTR PCR) can detect essentially all known TTV strains. On UTR PCR, TTV was detected in over 95% of the general population of the examined countries in the world. Since the highly conserved area of the UTR of TTMV is homologous to that of TTV and since TTMV infection is also common in humans, UTR primers for TTV should be designed so as not to co-amplify TTMV DNA, for the specific detection of TTV DNA [25,49,50].

Multiple genotypes of TTV may be found within an infected individual, often with different genotype combinations predominating in different tissues [15]. If only certain TTV genotypes or group of TTV genotypes is pathogenic and others are not, as is well known for human papillomaviruses and adenoviruses [51], epidemiological studies using universal or general TTV PCR will dilute the result. Therefore, the development of genotype-dependent or, at least, group-dependent PCR with less broad specificity is needed for epidemiological studies on TTV infection. Besides N22 primers for Group 1 TTVs, primers for the specific detection of Group 4 or Group 5 TTVs have been described [12].

Mode of TTV transmission and replication sites

TTV can infect the host either transiently or persistently [1]. TTV is most often detected in serum/plasma, and is shed into the feces via secretion from the liver into the bile [55, 56]: the infectivity of TTV in serum and feces has been demonstrated in chimpanzee transmission studies [3,57]. In addition, TTV has been detected in saliva, on throat swab, and in other body fluids including semen, tears and breast milk, suggesting other modes of transmission such as air-borne (spread by saliva droplets), mouth-to-mouth, or sexual transmission of the virus [58–62]. Hence, extensive spread of TTV in the general population is suspected to occur through multiple pathways. It is uncertain whether mother-to-infant transmission of TTV occurs, but horizontal infection is more likely to occur than vertical infection in mother-to-child transmission of TTV. TTV may be acquired in early childhood [12,63–66] and is prevalent among adults, who are frequently co-infected with TTVs of distinct genotypes/genetic groups [9,12,15].

TTV replicates not only in the liver but also in other tissues and organs such as the bone marrow, lung tissues, lymph nodes, spleen and pancreas [13–15]. As an indication of its great replicative capacity *in vivo*, a study on the kinetics of clearance of TTV viremia during interferon-α therapy suggested that $>10^{10}$ TTV virions are produced everyday, with 90% of the virions in the plasma cleared and replenished during a 24-h period [67].

248

Chimpanzee transmission study of TTV

Mushahwar et al. [3] first reported a transmission study of TTV to chimpanzees. They demonstrated that human TTV can be transmitted to chimpanzees, although they did not find biochemical or histological evidence of hepatitis in the infected chimpanzees. However, our data from a transmission study of TTV to chimpanzees [57] were not consistent with those of the previous study. In our study, a serum sample obtained from an 11-month-old infant with acute hepatitis of unknown etiology who was transiently infected with genotype 1 TTV (10^5 copies/ml), was inoculated intravenously into a naive chimpanzee (No. 228). As illustrated in Fig. 4, TTV DNA was transiently detected in chimpanzee 228 at 5–15 weeks postinoculation (PI), with the titer peaking at 12–13 weeks PI. This viremia was accompanied by an abrupt elevation of the serum α-glutathione-S-transferase (α-GST) level, which has been reported to be a sensitive marker of hepatocellular damage [68]. Mild elevation of the serum alanine aminotransferase (ALT) level and histological changes in biopsied liver samples (ballooning degeneration of hepatocytes) were observed in association with the reduction in TTV DNA titer and the appearance of IgM-class and IgG-class anti-TTV (genotype 1) antibodies. Complete

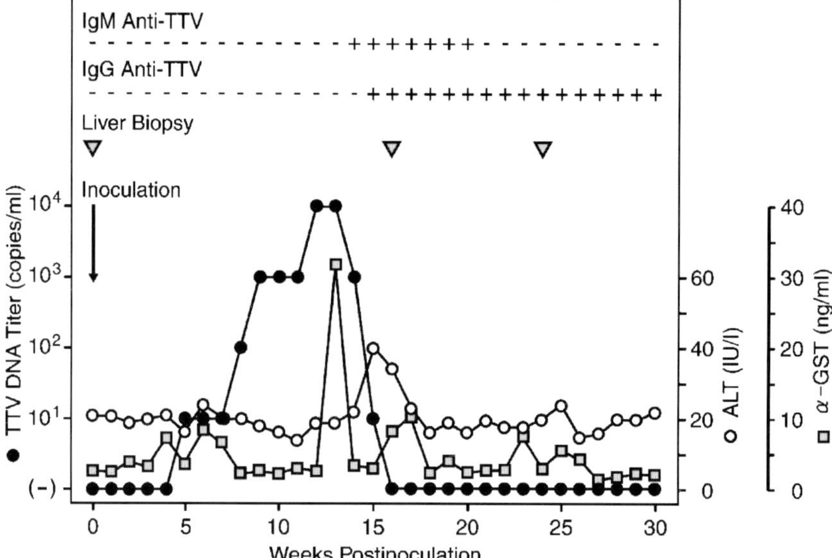

Fig. 4. Course of the laboratory parameters over 30 weeks in a chimpanzee inoculated with TTV of genotype 1. Serum (0.5 ml) containing TTV of genotype 1 (10^5 copies/ml) from an 11-month-old infant with acute hepatitis of unknown etiology, was intravenously inoculated into a naive chimpanzee (No. 228). Serum samples were obtained weekly and tested for TTV DNA by genotype 1-specific PCR. Liver tissues were biopsied at three time points, and histological changes were only observed at the second time point. The serum samples were tested for the presence of IgM-class or IgG-class anti-TTV antibodies with specificity for genotype 1 was determined. The serum level of α-GST was measured by enzyme immunoassay using a commercial kit (Biotrin International Ltd., Dublin, Ireland). (Modified from Ref. [57].)

conservation of the TTV sequence between the source and recipient was confirmed, indicating that TTV infection was derived from the inoculum. These results suggest that TTV has hepatitis-inducing capacity.

Clinical implications of TTV infection

The role of TTV in hepatic diseases has not yet been defined and remains controversial [69–77]. However, clinical studies on TTV-infected patients in relation to its genotype and viral load are warranted. TTV of genotype 1 was detected in three of five patients with post-transfusion acute hepatitis of unknown etiology, and the presence of TTV genotype 1 was closely associated with the serum ALT level [1]: this observation was corroborated by other investigators [78]. It has been suggested that TTV detectable by N22 PCR may play a role in the pathogenesis of non-A, -B or -C fulminant hepatic failure (FHF), since the TTV-positive rate was significantly higher among the group with non-A, -B or -C FHF (6/7 or 86%) than among the group with hepatitis A virus (HAV)-, hepatitis B virus (HBV)- or hepatitis C virus (HCV)-associated FHF (6/17 or 35%; $P = 0.0247$) or with non-A, -B or -C acute hepatitis (4/17 or 24%; $P = 0.005$) [79]. It has also been suggested that infection of certain TTV genotypes detectable by N22 PCR or genotype-1 PCR may interfere with the improvement of liver function following the start of abstinence, in patients with alcoholic liver disease [80], and that it influences the necrosis and inflammation of hepatocytes and liver fibrosis in patients with nonalcoholic fatty liver disease including nonalcoholic steatohepatitis [81]. Tajiri et al. [82] presented three infants diagnosed with idiopathic neonatal hepatitis and intrahepatic fatty degeneration and whose livers were infected with TTV detectable by N22 PCR. Several other studies in Japan suggest that TTV of genotype 1 may be more pathogenic than other genotypes of TTV in children with liver disease of unknown etiology [83,84]. On the other hand, infection with TTV genotype 12 or 16, which was described as SENV-D and SENV-H, respectively, was found to be much more prevalent among patients with transfusion-associated non-A to -E hepatitis than among transfused patients without hepatitis in the United States (92 vs. 24%; $P < 0.001$) [85]. Foschini et al. [86] reported an Italian case of TTV (genotype 13)-related acute recurrent hepatitis, with clinico-pathological findings reinforcing the suggestion that TTV can be responsible for a mild form of liver disease.

Tokita et al. [87] reported that a high TTV viral load was independently associated with the complication of hepatocellular carcinoma (HCC) and that it may have prognostic significance in patients with HCV-related chronic liver disease. There are two possible explanations for the findings in this report. One explanation is that high TTV viremia has an adverse effect on the progression of chronic liver disease in concert with concurrent HCV infection and may be associated with the development of HCC. Zein et al. [75] reported that TTV infection was more prevalent among patients with advanced HCV-associated liver disease (decompensated cirrhosis and HCC) than among those with stable disease (chronic hepatitis and compensated cirrhosis). Moriyama et al. [77] reported that the score of irregular regeneration of hepatocytes among TTV-infected cirrhotic patients with chronic hepatitis C was higher than that among patients who were not infected with TTV. These findings indicate that TTV plays a role in the development of cirrhosis and subsequent complications. However, another explanation is possible. A correlation between high TTV titer and a low CD4 T-cell count among patients infected with human immunodeficiency virus type 1 (HIV), and the possible prognostic significance of TTV viral load in immunocompromised patients have been reported [88,89]. A possible relationship between the prevalence of elevated TTV viral load and the level of immunocompetence of the populations studied among TTV-infected patients on maintenance hemodialysis or with diabetes mellitus has also been suggested [90]. Therefore, it is likely that an impaired immune system or suppression of the immune system is involved in elevated TTV viremia in HCC patients.

Reflecting the evidence of TTV replication in a wide range of tissues in addition to the liver, the implications of TTV infection in extrahepatic diseases have also been suggested. Of note, Yokoyama et al. [91] presented the pathological changes of renal epithelial cells of mice transgenic for the TTV ORF1 gene, providing new insights in the investigation of TTV pathogenicity. The influence of TTV replication on thrombocytopenia and aplastic anemia has been reported [92–94]. Although the precipitating factors of idiopathic pulmonary fibrosis have not been elucidated, Bando et al. [95] first reported the influence of TTV infection on the disease activity and prognosis of idiopathic pulmonary fibrosis.

Recently, Biagini et al. [96] performed a follow-up study of the natural history of a newborn with TTV infection in parallel with TTV DNA detection, and reported the association of TTV primary infection with clinical symptoms of benign "viral" rhinitis in the newborn. Maggi et al. [97] examined nasal and blood specimens from children with a clinical diagnosis of acute respiratory disease for the presence, load, and genotype of TTV, and found evidence that the average TTV load was considerably higher among patients with bronchopneumonia than among those with milder acute respiratory disease. This raises interesting questions about the pathophysiological significance of TTV in the respiratory tract of infected humans.

Summary and conclusions

TTV is a unique, unenveloped human virus with a genome of circular, single (minus)-stranded DNA of a total length of 3.6–3.9 kb, and it has a characteristic genomic organization and transcription profile that are not observed in known members of the *Circoviridae* family. It has been proposed by the ICTV Circoviridae Study Group that TTV is to be named as Torque Teno virus classifiable into a novel genus *Anellovirus*, unassigned to any family. At least 39 genotypes of TTV have been identified, and they are classified into five distantly-related groups (1–5). The natural history and pathogenic potential of TTV are currently under intensive investigation. Although the precise relationship between TTV infection and viral hepatitis remains to be established, it is likely that certain genotypes of TTV or some strains of TTV cause hepatitis in some clinical and epidemiological settings. As TTV replicates not only in the liver but also in other tissues such as the bone marrow and lung tissue, TTV infection may also be associated with extrahepatic disorders as suggested in several recent studies [91–97]. Therefore, detailed cross-sectional and longitudinal studies are required to address the exact role of TTV in children and adults with cryptogenic liver disease or other diseases that are currently of unknown etiology, with regard to the genotypes and viral load of TTV.

References

1. Nishizawa T, Okamoto H, Konishi K, Yoshizawa H, Miyakawa Y, Mayumi M. Biochem Biophys Res Commun 1997; 241: 92–97.
2. Okamoto H, Nishizawa T, Kato N, Ukita M, Ikeda H, Iizuka H, Miyakawa M, Mamyumi M. Hepatol Res 1998; 10: 1–16.
3. Mushahwar IK, Erker JC, Muerhoff AS, Leary TP, Simons JN, Birkenmeyer LG, Chalmers ML, Pilot-Matias TJ, Desai SM. Proc Natl Acad Sci USA 1999; 96: 3177–3182.
4. Miyata H, Tsunoda H, Kazi A, Yamada A, Khan MA, Murakami J, Kamahora T, Shiraki K, Hino S. J Virol 1999; 73: 3582–3586.
5. Okamoto H, Nishizawa T, Ukita M, Takahashi M, Fukuda M, Iizuka H, Miyakawa M, Mayumi M. Virology 1999; 259: 437–448.
6. Prescott LE, Simmonds P, International Collaborators. N Engl J Med 1998; 339: 776–777.
7. Abe K, Inami T, Asano K, Miyoshi C, Masaki N, Hayashi S, Ishikawa K, Takabe Y, Win KM, El-Zayadi AR, Han KH, Zhang DY. J Clin Microbiol 1999; 37: 2703–2705.
8. Okamoto H, Nishizawa T, Ukita M. Intervirology 1999; 42: 196–204.
9. Okamoto H, Takahashi M, Nishizawa T, Ukita M, Fukuda M, Tsuda F, Miyakawa Y, Mayumi M. Virology 1999; 259: 428–436.
10. Bendinelli M, Pistello M, Maggi F, Fornal C, Freer G, Vatteroni ML. Clin Microbiol Rev 2001; 14: 98–113.
11. Okamoto H, Mayumi M. J Gastroenterol 2001; 36: 519–529.
12. Peng YH, Nishizawa T, Takahashi M, Ishikawa T, Yoshikawa A, Okamoto H. Arch Virol 2002; 147: 21–41.
13. Okamoto H, Ukita M, Nishizawa T, Kishimoto J, Hoshi Y, Mizuo H, Tanaka T, Miyakawa Y, Mayumi M. J Virol 2000; 74: 5161–5167.
14. Okamoto H, Takahashi M, Nishizawa T, Tawara A, Sugai Y, Sai T, Tanaka T, Tsuda F. Biochem Biophys Res Commun 2000; 270: 657–662.
15. Okamoto H, Nishizawa T, Takahashi M, Asabe S, Tsuda F, Yoshikawa A. Virology 2001; 288: 358–368.
16. Maggi F, Fornai C, Zaccaro L, Morrica A, Vatteroni ML, Isola P, Marchi S, Ricchiuti A, Pistello M, Bendinelli M. J Med Virol 2001; 64: 190–194.
17. Mariscal LF, Lopez-Alcorocho JM, Rodriguez-Inigo E, Ortiz-Movilla N, de Lucas S, Bartolome J, Carreno V. Virology 2002; 301: 121–129.
18. Todd D, McNulty MS, Mankertz A, Lukert PD, Randles JW, Dale JL. Family *Circoviridae*. In: Virus Taxonomy: Classification and Nomenclature of Viruses (van Regenmortel MHV, Fauquet CM, Bishop DHL, Carstens EB, Estes MK, Lemon SM, Maniloff J, Mayo MA, McGeoch DJ, Pringle CR, Wickner RB, editors), Seventh Report of the International Committee on Taxonomy of Viruses. San Diego: Academic Press; 2000; pp. 299–303.
19. Mankertz A, Hattermann K, Ehlers B, Soike D. Arch Virol 2000; 145: 2469–2479.
20. Todd D, Weston JH, Soike D, Smyth JA. Virology 2001; 286: 354–362.
21. Phenix KV, Weston JH, Ypelaar I, Lavazza A, Smyth JA, Todd D, Wilcox GE, Raidal SR. J Gen Virol 2001; 82: 2805–2809.
22. Pringle CR. Arch Virol 1999; 144: 2065–2070.

23. Noteborn MHM, Koch G. Avian Pathol 1995; 24: 11–31.
24. Niagro FD, Forstoefel AN, Lawther RP, Kamalanathan L, Ritchie BW, Latimer KS, Lukert PD. Arch Virol 1998; 143: 1723–1744.
25. Okamoto H, Nishizawa T, Tawara A, Peng Y, Takahashi M, Kishimoto J, Tanaka T, Miyakawa Y, Mayumi M. Virology 2000; 277: 368–378.
26. Kamahora T, Hino S, Miyata H. J Virol 2000; 74: 9980–9986.
27. Okamoto H, Nishizawa T, Tawara A, Takahashi M, Kishimoto J, Sai T, Sugai Y. Biochem Biophys Res Commun 2000; 279: 700–707.
28. Takahashi K, Iwasa Y, Hijikata M, Mishiro S. Arch Virol 2000; 145: 979–993.
29. Todd D, Bendinelli M, Biagini P, Hino S, Mankertz A, Mishiro S, Niel C, Okamoto H, Raidal S, Ritchie BW, Teo GC. Eighth Report of the International Committee on Taxonomy of Viruses. Genus Anellovirus; in press.
30. Okamoto H, Fukuda M, Tawara A, Nishizawa T, Itoh Y, Hayasaka I, Tsuda F, Tanaka T, Miyakawa Y, Mayumi M. J Virol 2000; 74: 1132–1139.
31. Inami T, Obara T, Moriyama M, Arakawa Y, Abe K. Virology 2000; 277: 330–335.
32. Okamoto H, Nishizawa T, Takahashi M, Tawara A, Peng Y, Kishimoto J, Wang Y. J Gen Virol 2001; 82: 2041–2050.
33. Okamoto H, Takahahshi M, Nishizawa T, Tawara A, Fukai K, Muramatsu U, Naito Y, Yoshikawa A. J Gen Virol 2002; 83: 1291–1297.
34. Thom K, Morrison C, Lewis JCM, Simmonds P. Virology 2003; 306: 324–333.
35. Itoh Y, Takahashi M, Fukuda M, Shibayama T, Ishikawa T, Tsuda F, Tanaka T, Nishizawa T, Okamoto H. Biochem Biophys Res Commun 2000; 279: 718–724.
36. Biagini P, Gallian P, Attoui H, Touinssi M, Cantaloube J, de Micco P, de Lamballerie X. J Gen Virol 2001; 82: 379–383.
37. Muljono DH, Nishizawa T, Tsuda F, Takahashi M, Okamoto H. Arch Virol 2001; 146: 1249–1266.
38. Ukita M, Okamoto H, Nishizawa T, Tawara A, Takahashi M, Iizuka H, Miyakawa Y, Mayumi M. Arch Virol 2000; 145: 1543–1559.
39. Erker JC, Leary TP, Desai SM, Chalmers ML, Mushahwar IK. J Gen Virol 1999; 80: 1743–1750.
40. Hijikata M, Takahashi K, Mishiro S. Virology 1999; 260: 17–22.
41. Biagini P, Attoui H, Gallian P, Touinssi M, Cantaloube JF, de Micco P, de Lamballerie X. Biochem Biophys Res Commun 2000; 271: 837–841.
42. Hallett RL, Clewley JP, Bobet F, McKiernan PJ, Teo CG. J Gen Virol 2000; 81: 2273–2279.
43. Tanaka Y, Primi D, Wang RYH, Umemura T, Yeo AET, Mizokami M, Alter HJ, Shih JWK. J Infect Dis 2001; 183: 359–367.
44. Takahashi K, Hijikata M, Samokhvalov EI, Mishiro S. Intervirology 2000; 43: 119–123.
45. Peters MA, Jackson DC, Crabb BS, Browning GF. J Biol Chem 2002; 277: 39566–39573.
46. Asabe S, Nishizawa T, Iwanari H, Okamoto H. Biochem Biophys Res Commun 2001; 286: 298–304.
47. Khudyakov YE, Cong ME, Nichols B, Reed D, Dou XG, Viazov SO, Chang J, Fried NW, Williams I, Bower W, Lambert S, Purdy M, Roggendorf M, Fields HA. J Virol 2000; 74: 2990–3000.
48. Mushahwar IK. J Med Virol 2000; 62: 399–404.
49. Tokita H, Murai S, Kamitsukasa H, Yagura M, Harada H, Takahashi M, Okamoto H. J Med Virol 2002; 67: 501–509.
50. Moen EM, Sleboda J, Grinde B. J Virol Methods 2002; 104: 59–67.

51. zur Hausen H. Lancet 2001; 357: 381–384.

52. Takahashi K, Hoshino H, Ohta Y, Yoshida N, Mishiro S. Hepatol Res 1998; 12: 233–239.

53. Irving WL, Ball JK, Berridge S, Curran R, Grabowska AM, Jameson CL, Neal KR, Ryder SD, Thomson BJ. J Infect Dis 1999; 180: 27–34.

54. Itoh K, Takahashi M, Ukita M, Nishizawa T, Okamoto H. J Infect Dis 1999; 180: 1750–1751.

55. Okamoto H, Akahane Y, Ukita M, Fukuda M, Tsuda F, Miyakawa Y, Mayumi M. J Med Virol 1998; 56: 128–132.

56. Ukita M, Okamoto H, Kato N, Miyakawa Y, Mayumi M. J Infect Dis 1999; 179: 1245–1248.

57. Tawara A, Akahane Y, Takahashi M, Nishizawa T, Ishikawa T, Okamoto H. Biochem Biophys Res Commun 2000; 278: 470–476.

58. Ross RS, Viazov S, Runde V, Schaefer UW, Roggendorf M. J Clin Virol 1999; 13: 181–184.

59. Deng X, Terunuma H, Handema R, Sakamoto M, Kitamura T, Ito M, Akahane Y. J Med Virol 2000; 62: 531–537.

60. Ishikawa T, Hamano Y, Okamoto H. Infection 1999; 27: 298.

61. Inami T, Konomi N, Arakawa Y, Abe K. J Clin Microbiol 2000; 38: 2407–2408.

62. Matsubara H, Michitaka K, Horiike N, Yano M, Akbar SM, Torisu M, Onji M. Intervirology 2000; 18: 122–131.

63. Davidson F, MacDonald D, Mokili JLK, Prescott LE, Graham S, Simmonds P. J Infect Dis 1999; 179: 1070–1076.

64. Saback FL, Gomes SA, de Paula VS, da Silva RRS, Lewis-Ximenez LL, Niel C. J Med Virol 1999; 59: 318–322.

65. Gerner P, Oettinger R, Gerner W, Falbrede J, Wirth S. Pediatr Infect Dis J 2000; 19: 1074–1077.

66. Ohto H, Ujiie N, Takeuchi C, Sato A, Hayashi A, Ishiko H, Nishizawa T, Okamoto H, The Vertical Transmission of Hepatitis Viruses Collaborative Study Group. Transfusion 2002; 42: 892–898.

67. Maggi F, Pistello M, Vatteroni M, Presciuttini S, Marchi S, Isola P, Fornai C, Fagnani S, Andreoli E, Antonelli G, Bendinelli M. J Virol 2001; 75: 11999–12004.

68. Vaubourdolle M, Chazouilleres O, Briaud I, Legendre C, Serfaty L, Poupon R, Giboudeau J. Clin Chem 1995; 41: 1716–1719.

69. Charlton M, Adjei P, Poterucha J, Zein N, Moore B, Therneau T, Krom R, Wiesner R. Hepatology 1998; 28: 839–842.

70. Tanaka H, Okamoto H, Luengrojanakul P, Chainuvati T, Tsuda F, Tanaka T, Miyakawa Y, Mayumi M. J Med Virol 1998; 56: 234–238.

71. Matsumoto A, Yeo AE, Shih JW, Tanaka E, Kiyosawa K, Alter HJ. Hepatology 1999; 30: 283–288.

72. Gimenez-Barcons M, Forns X, Ampurdanes S, Guilera M, Soler M, Soguero C, Sanchez-Fueyo A, Mas A, Bruix J, Sanchez-Tapias JM, Rodes J, Saiz JC. J Hepatol 1999; 30: 1028–1034.

73. Ikeda H, Takasu M, Inoue K, Okamoto H, Miyakawa Y, Mayumi M. J Hepatol 1999; 30: 205–212.

74. Tagger A, Donato F, Ribero ML, Binelli G, Gelatti U, Portera G, Albertini A, Fasola M, Chiesa R, Nardi G. Hepatology 1999; 30: 294–299.

75. Zein NN, Arslan M, Li H, Charlton MR, Gross JB Jr., Poterucha JJ, Therneau TM, Kolbert CP, Persing DH. J Gastroenterol 1999; 94: 3020–3027.

76. Trimoulet P, De Ledinghen V, Ekouevi D, Bernard PH, Merel P, Chene G, Couzigou P, Fleury H. Am J Gastroenterol 2000; 95: 1765–1769.

77. Moriyama M, Matsumura H, Shimizu T, Shioda A, Kaneko M, Miyazawa K, Miyata H, Tanaka N, Uchida T, Arakawa Y. J Med Virol 2001; 64: 74–81.
78. Tanaka Y, Hayashi J, Ariyama I, Furusyo N, Etoh Y, Kashiwagi S. Dig Dis Sci 2000; 45: 2214–2220.
79. Shibata M, Morizane T, Baba T, Inoue K, Sekiyama K, Yoshiba M, Mitamura K. Am J Gastroenterol 2000; 95: 3602–3606.
80. Tokita H, Murai S, Kamitsukasa H, Yagura M, Harada H, Tawara A, Takahashi M, Okamoto H. Hepatol Res 2001; 19: 180–193.
81. Tokita H, Murai S, Kamitsukasa H, Yagura M, Harada H, Hebisawa A, Takahahsi M, Okamoto H. Hepatol Res 2001; 19: 197–211.
82. Tajiri H, Tanaka T, Sawada A, Etani Y, Kozaiwa K, Mushiake S, Mishiro S. Intervirology 2001; 44: 364–369.
83. Okamura A, Yoshioka M, Kikuta H, Kubota M, Ma X, Hayashi A, Ishiko H, Kobayashi K. J Med Virol 2000; 62: 104–108.
84. Sugiyama K, Goto K, Ando T, Mizutani F, Terabe K, Yokoyama T, Wada Y. Tohoku J Exp Med 2000; 191: 233–239.
85. Umemura T, Yeo AE, Sottini A, Moratto D, Tanaka Y, Wang RY, Shih JW, Donahue P, Primi D, Alter HJ. Hepatology 2001; 33: 1303–1311.
86. Foschini MP, Morandi L, Macchia S, DalMonte PR, Pession A. Virchows Arch 2001; 439: 752–755.
87. Tokita H, Murai S, Kamitsukasa H, Yagura M, Harada H, Takahashi M, Okamoto H. J Med Virol 2002; 67: 501–509.
88. Christensen JK, Eugen-Olsen J, Sorensen M, Ullum H, Gjedde SB, Pedersen BK, Nielsen JO, Krogsgaard K. J Infect Dis 2000; 181: 1769–1796.
89. Shibayama T, Masuda G, Ajisawa A, Takahashi M, Nishizawa T, Tsuda F, Okamoto H. AIDS 2001; 15: 563–570.
90. Touinssi M, Gallian P, Biagini P, Attoui H, Vialettes B, Berland Y, Tamalet C, Dhiver C, Ravaux I, de Micco P, de Lamballerie X. J Clin Virol 2001; 21: 135–141.
91. Yokoyama H, Yasuda J, Okamoto H, Iwakura Y. J Gen Virol 2002; 83: 141–150.
92. Tokita H, Murai S, Kamitsukasa H, Yagura M, Harada H, Takahashi M, Okamoto H. Hepatol Res 2002; 23: 105–114.
93. Kikuchi K, Miyakawa H, Abe K, Kako M, Katayama K, Fukushi S, Mishiro S. J Med Virol 2000; 61: 165–170.
94. Safadi R, Or R, Ilan Y, Naparstek E, Nagler A, Klein A, Ketzinel-Gilaad M, Ergunay K, Danon D, Shouval D, Galun E. Bone Marrow Transplant 2001; 27: 183–190.
95. Bando M, Ohno S, Oshikawa M, Takahashi M, Okamoto H, Sugiyama Y. Respir Med 2001; 95: 935–942.
96. Biagini P, Charrel RN, de Micco P, de Lamballerie X. Clin Infect Dis 2003; 36: 128.
97. Maggi F, Pifferi M, Fornai C, Andreoli E, Tempestini E, Vatteroni M, Presciuttini S, Marchi S, Pietrobelli A, Boner A, Pistello M, Bendinelli M. J Virol 2003; 77: 2418–2425.

List of contributors

Larry G. Biekenmeyer
 Core Infectious Disease R&D, Abbott Diagnostics Division
 Abbott Laboratories
 Department 9NB, Building AP20,lOO Abott Park Road, Abbott Park
 Illinois
 IL 60064-6015
 USA

Paul Coleman
 Core Infectious Disease R&D, Abbott Diagnostics Division
 Abbott Laboratories
 Abbott Park
 Illinois
 IL 60064-6015
 USA

George J. Dawson
 Core Infectious Disease R&D, Abbott Diagnostics Division
 Abbott Laboratories
 Abbott Park
 Illinois
 IL 60064-6015
 USA

Thomas P. Leary
 Core Infectious Disease R&D, Abbott Diagnostics Division
 Abbott Laboratories
 Abbott Park
 Illinois
 IL 60064-6015
 USA

Daniel Levanchy
 Biosafety Programme, World Health Organization
 Communicable Disease Surveillance and Response
 Av. Appia 20
 CH-1211 Geneva 27
 Switzerland

A. Scott Muerhoff
Core Infectious Disease R&D, Abbott Diagnostics Division
Abbott Laboratories
Abbott Park
Illinois
IL 60064-6015
USA

Isa K. Mushahwar
Viral Discovery Group
Abbott Laboratories
Abbott Park
Illinois
IL 60064-6015
USA

Tsutomu Nishizawa
Division of Virology, Department of Infection and Immunity
Jichi Medical School
3311-1 Yakushiji, Minamikawachi-Machi
Tochigi-Ken 329-0498
Japan

Hiroaki Okamoto
Division of Virology, Department of Infection and Immunity
Jichi Medical School
3311-1 Yakushiji, Minamikawachi-Machi
Tochigi-Ken 329-0498
Japan

Nicoletta Previsani
Biosafety Programme, World Health Organization
Communicable Disease Surveillance and Response
Av. Appia 20
CH-1211 Geneva 27
Switzerland

Mario Rizzetto
Azienda Ospedaliera S. Giovanni Battista di Torino
Torino
Italy

George G. Schlauder
Core Infectious Disease R&D, Abbott Diagnostics Division
Abbott Laboratories

Abbott Park
Illinois
IL 60064-6015
USA

Günter Siegl
Institut fur Klinishce Mikrobiologie und Immunologie
St Gallon
Switzerland

Masaharu Takahashi
Division of Virology, Department of Infection and Immunity
Jichi Medical School
3311-1 Yakushiji, Minamikawachi-Machi
Tochigi-Ken 329-0498
Japan

Arie J. Zuckerman
W.H.O. Centre - Viral Diseases
Royal Free and University College Medical School, University College London
Rowland Hill Street
London
NW3 2PF
UK

Viral Hepatitis
I.K. Mushahwar (editor)
© 2004 Elsevier B.V. All rights reserved.

Index

260